Methods in Molecular Biology

For further volumes:
http://www.springer.com/series/7651

Plant Pattern Recognition Receptors

Methods and Protocols

Edited by

Libo Shan

Department of Plant Pathology and Microbiology, Institute for Plant Genomics and Biotechnology, Texas A&M University, College Station, TX, USA

Ping He

Department of Biochemistry and Biophysics, Institute for Plant Genomics and Biotechnology, Texas A&M University, College Station, TX, USA

Humana Press

Editors
Libo Shan
Department of Plant Pathology and Microbiology
Institute for Plant Genomics and Biotechnology
Texas A&M University
College Station, TX, USA

Ping He
Department of Biochemistry and Biophysics
Institute for Plant Genomics and Biotechnology
Texas A&M University
College Station, TX, USA

ISSN 1064-3745 ISSN 1940-6029 (electronic)
Methods in Molecular Biology
ISBN 978-1-4939-6858-9 ISBN 978-1-4939-6859-6 (eBook)
DOI 10.1007/978-1-4939-6859-6

Library of Congress Control Number: 2017933458

Printed on acid-free paper

This Humana Press imprint is published by Springer Nature
The registered company is Springer Science+Business Media LLC
The registered company address is: 233 Spring Street, New York, NY 10013, U.S.A.

Preface

Sessile plants are under a constant risk of infections by different microorganisms in their natural habitats. The first line of immune response is activated via sensing of the conserved signatures from different microbial species, which are termed as pathogen- or microbe-associated molecular patterns (PAMPs or MAMPs), by cell surface-resident pattern recognition receptors (PRRs). MAMPs were originally named as microbial elicitors which have long been observed to trigger various cellular responses in plants. In recent years, remarkable progresses have been made on the research of their corresponding receptors, signaling mechanism, and involvement in disease resistance. Plant PRRs are often members of receptor-like kinases (RLKs) and receptor-like proteins (RLPs), which mediate PAMP- or MAMP-triggered immunity (PTI or MTI) contributing to host resistance against a broad spectrum of microbial infections.

This book volume will cover a collection of step-by-step protocols on techniques ranging from MAMP isolations from diverse microorganisms, PRR identifications from different plant species, MAMP-PRR binding, and a series of signaling responses and events revealed by various biochemical, cellular, genetic, and bioinformatic tools.

College Station, TX, USA

Libo Shan
Ping He

Contents

Contributors

CAROLINA AGUILERA-GALVEZ • *Plant Breeding, Wageningen University & Research, Wageningen, The Netherlands*

ADI AVNI • *Department of Molecular Biology and Ecology of Plants, Tel-Aviv University, Tel-Aviv, Israel*

LAURA BACETE • *Centro de Biotecnología y Genómica de Plantas, Universidad Politécnica de Madrid (UPM)—Instituto Nacional de Investigación y Tecnología Agraria y Alimentaria (INIA), Campus de Montegancedo UPM, Pozuelo de Alarcón (Madrid), Spain*

MANUEL BENEDETTI • *Dipartimento di Biologia e Biotecnologie "Charles Darwin", Sapienza Università di Roma, Rome, Italy*

UTE BERTSCHE • *Department of Infection Biology, Interfaculty Institute for Microbiology and Infection Medicine Tübingen (IMIT), University of Tübingen, Tübingen, Germany*

GERARD BIJSTERBOSCH • *Plant Breeding, Wageningen University & Research, Wageningen, The Netherlands*

YOGENDRA BORDIYA • *Department of Biology, Texas State University, San Marcos, TX, USA*

MARIE BOUDSOCQ • *Institute of Plant Sciences Paris-Saclay (IPS2), CNRS, INRA, Université Paris-Sud, Université d'Evry Val d'Essonne, Université Paris-Diderot, Sorbonne Paris-Cité, Université Paris-Saclay, Orsay, France*

SIMON BRESSENDORFF • *Department of Biology, University of Copenhagen, Copenhagen, Denmark*

OTÁVIO J.B. BRUSTOLINI • *Department of Biochemistry and Molecular Biology, National Institute of Science and Technology in Plant-Pest Interactions, Bioagro, Universidade Federal de Viçosa, Viçosa, MG, Brazil*

TERESA CESERANI • *Department of Molecular, Cellular & Developmental Biology, Yale University, New Haven, CT, USA*

SIXUE CHEN • *Department of Biology, University of Florida, Gainesville, FL, USA; Genetics Institute, University of Florida, Gainesville, FL, USA; Plant Molecular and Cellular Biology Program, University of Florida, Gainesville, FL, USA; Interdisciplinary Center for Biotechnology Research, University of Florida, Gainesville, FL, USA*

HOO SUN CHUNG • *Department of Molecular Biology and Center for Computational and Integrative Biology, Massachusetts General Hospital, Boston, MA, USA; Department of Genetics, Harvard Medical School, Boston, MA, USA*

CHRISTOPHER R. CLARKE • *Department of Plant Pathology, Physiology and Weed Science, Virginia Tech, Blacksburg, VA, USA*

NICOLE K. CLAY • *Department of Molecular, Cellular & Developmental Biology, Yale University, New Haven, CT, USA*

BRAD DAY • *Department of Plant, Soil and Microbial Sciences, Michigan State University, Lansing, MI, USA; Graduate Program in Cell and Molecular Biology, Michigan State University, East Lansing, MI, USA; Graduate Program in Genetics, Michigan State University, East Lansing, MI, USA*

EMMANOUIL DOMAZAKIS • *Plant Breeding, Wageningen University & Research, Wageningen, The Netherlands*

SIMONE FERRARI • *Dipartimento di Biologia e Biotecnologie "Charles Darwin", Sapienza Università di Roma, Rome, Italy*
ELIZABETH P.B. FONTES • *Department of Biochemistry and Molecular Biology, National Institute of Science and Technology in Plant-Pest Interactions, Bioagro, Universidade Federal de Viçosa, Viçosa, MG, Brazil*
GIULIA FURLAN • *Leibniz Institute of Plant Biochemistry, Halle (Saale), Germany*
ANDREA A. GUST • *Department of Plant Biochemistry, ZMBP, University of Tübingen, Tübingen, Germany*
MICHAEL G. HAHN • *Complex Carbohydrate Research Center (CCRC), University of Georgia, Athens, GA, USA*
ANTJE HEESE • *Division of Biochemistry, Interdisciplinary Plant Group (IPG), University of Missouri, Columbia, MO, USA*
YINGNAN HOU • *Department of Plant Pathology and Microbiology, University of California, Riverside, CA, USA; Center for Plant Cell Biology, University of California, Riverside, CA, USA*
JIAN HUA • *Plant Biology Section, School of Integrative Plant Science, Cornell University, Ithaca, NY, USA*
TAKAAKIRA INOKUCHI • *Graduate School of Biological Sciences, Nara Institute of Science and Technology, Ikoma, Japan*
LIN JIN • *Department of Horticulture and Crop Science, The Ohio State University, Columbus, OH, USA*
HONG-GU KANG • *Department of Biology, Texas State University, San Marcos, TX, USA*
FUMIAKI KATAGIRI • *Department of Plant and Microbial Biology, Microbial and Plant Genomics Institute, University of Minnesota, St. Paul, MN, USA*
TSUTOMU KAWASAKI • *Graduate School of Agriculture, Kindai University, Nakamachi, Nara, Japan*
CHANDRA KENCHAPPA • *Deptartment of Biology, University of Copenhagen, Copenhagen, Denmark*
STACEY A. LAWRENCE • *Department of Molecular, Cellular & Developmental Biology, Yale University, New Haven, CT, USA*
MEIRAV LEIBMAN-MARKUS • *Department of Molecular Biology and Ecology of Plants, Tel-Aviv University, Tel-Aviv, Israel*
MICHELLE E. LESLIE • *Division of Biochemistry, Interdisciplinary Plant Group (IPG), University of Missouri, Columbia, MO, USA; Elemental Enzymes Inc., St. Louis, MO, USA*
XIAO LIN • *Plant Breeding, Wageningen University & Research, Wageningen, The Netherlands*
SIMON R. LLOYD • *Department of Crop Genetics, John Innes Centre, Norwich Research Park, Norwich, UK*
LEVI LOWDER • *Department of Biology, University of Maryland, College Park, Greenville, NC, USA*
YI-JU LU • *Department of Plant, Soil and Microbial Sciences, Michigan State University, East Lansing, MI, USA*
WENBO MA • *Department of Plant Pathology and Microbiology, University of California, Riverside, CA, USA; Center for Plant Cell Biology, University of California, Riverside, CA, USA*

ALBERTO P. MACHO • *CAS Center for Excellence in Molecular Plant Sciences, Shanghai Center for Plant Stress Biology, Shanghai Institutes of Biological Sciences, Chinese Academy of Sciences, Shanghai, China*
DAVID M. MACKEY • *Department of Horticulture and Crop Science, The Ohio State University, Columbus, OH, USA; Department of Molecular Genetics, The Ohio State University, Columbus, OH, USA*
AIMEE MALZAHN • *Department of Plant Science and Landscape Architecture, University of Maryland, College Park, MD, USA*
BENEDETTA MATTEI • *Dipartimento di Biologia e Biotecnologie "Charles Darwin", Sapienza Università di Roma, Rome, Italy*
HUGO MÉLIDA • *Centro de Biotecnología y Genómica de Plantas, Universidad Politécnica de Madrid (UPM)—Instituto Nacional de Investigación y Tecnología Agraria y Alimentaria (INIA), Campus de Montegancedo UPM, Pozuelo de Alarcón (Madrid), Spain*
MAELI MELOTTO • *Department of Plant Sciences, University of California, Davis, CA, USA*
EVA MIEDES • *Centro de Biotecnología y Genómica de Plantas, Universidad Politécnica de Madrid (UPM)—Instituto Nacional de Investigación y Tecnología Agraria y Alimentaria (INIA), Campus de Montegancedo UPM, Pozuelo de Alarcón (Madrid), Spain*
ANTONIO MOLINA • *Centro de Biotecnología y Genómica de Plantas, Universidad Politécnica de Madrid (UPM)—Instituto Nacional de Investigación y Tecnología Agraria y Alimentaria (INIA), Campus de Montegancedo UPM, Pozuelo de Alarcón (Madrid), Spain*
JEANINE MONTANO • *Department of Plant Sciences, University of California, Davis, CA, USA*
JOHN MUNDY • *Department Of Biology, University of Copenhagen, Copenhagen, Denmark*
THORSTEN NÜRNBERGER • *Center of Plant Molecular Biology (ZMBP), Eberhard-Karls-University Tübingen, Tübingen, Germany*
JAKOB OLSEN • *Department of Biology, University of Copenhagen, Copenhagen, Denmark*
SIVAKUMAR PATTATHIL • *Complex Carbohydrate Research Center (CCRC), University of Georgia, Athens, GA, USA*
MORTEN PETERSEN • *Department of Biology, University of Copenhagen, Copenhagen, Denmark*
DANIELA PONTIGGIA • *Dipartimento di Biologia e Biotecnologie "Charles Darwin", Sapienza Università di Roma, Rome, Italy*
YIPING QI • *Department of Plant Science and Landscape Architecture, University of Maryland, College Park, MD, USA*
MAGNUS WOHLFAHRT RASMUSSEN • *Department of Biology, University of Copenhagen, Copenhagen, Denmark*
ANDRÉ REINHARD • *Center of Plant Molecular Biology (ZMBP), Eberhard-Karls-University Tübingen, Tübingen, Germany*
CHRISTOPHER J. RIDOUT • *Department of Crop Genetics, John Innes Centre, Norwich Research Park, Norwich, UK*
TINA ROMEIS • *Dahlem Centre of Plant Sciences, Plant Biochemistry, Freie Universität Berlin, Berlin, Germany*
YUSUKE SAIJO • *Graduate School of Biological Sciences, Nara Institute of Science and Technology, Ikoma, Japan; Japan Science and Technology Agency (JST), Precursory Research for Embryonic Science and Technology (PRESTO), Kawaguchi, Japan*

TETSU SAKAMOTO • *Department of Biochemistry and Immunology, Universidade Federal de Minas Gerais, Belo Horizonte, MG, Brazil*
GIANNI SALVI • *Dipartimento di Biologia e Biotecnologie "Charles Darwin", Sapienza Università di Roma, Rome, Italy*
YUYING SANG • *CAS Center for Excellence in Molecular Plant Sciences, Shanghai Center for Plant Stress Biology, Shanghai Institutes of Biological Sciences, Chinese Academy of Sciences, Shanghai, China*
DANIEL VALENTIN SAVATIN • *Dipartimento di Biologia e Biotecnologie "Charles Darwin", Sapienza Università di Roma, Rome, Italy*
JACQUELINE D. SCHNEIDER • *Department of Biology, University of Florida, Gainesville, FL, USA; Department of Chemical Engineering, University of Florida, Gainesville, FL, USA*
HENK-JAN SCHOONBEEK • *Department of Crop Genetics, John Innes Centre, Norwich Research Park, Norwich, UK*
SILVIA SCHUSTER • *Department of Molecular Biology and Ecology of Plants, Tel-Aviv University, Tel-Aviv, Israel*
HEIKE SEYBOLD • *Dahlem Centre of Plant Sciences, Plant Biochemistry, Freie Universität Berlin, Berlin, Germany*
JEN SHEEN • *Department of Molecular Biology and Center for Computational and Integrative Biology, Massachusetts General Hospital, Boston, MA, USA; Department of Genetics, Harvard Medical School, Boston, MA, USA*
RAQUEL AZEVEDO DA SILVA • *Department of Biology, University of Copenhagen, Copenhagen, Denmark*
JOSÉ CLEYDSON F. SILVA • *Department of Biochemistry and Molecular Biology, National Institute of Science and Technology in Plant-Pest Interactions, Bioagro, Universidade Federal de Viçosa, Viçosa, MG, Brazil; Department of Informatics, Universidade Federal de Viçosa, Viçosa, MG, Brazil*
SABRINA STANIMIROVIC • *Department Of Biology, University of Copenhagen, Copenhagen, Denmark*
QI SUN • *Biocomputing Service Unit, Institute of Biotechnology, Cornell University, Ithaca, NY, USA*
MARCO TRUJILLO • *Leibniz Institute of Plant Biochemistry, Halle (Saale), Germany*
BORIS A. VINATZER • *Department of Plant Pathology, Physiology and Weed Science, Virginia Tech, Blacksburg, VA, USA*
VIVIANNE G.A.A. VLEESHOUWERS • *Plant Breeding, Wageningen University & Research, Wageningen, The Netherlands*
SHUAI WANG • *Plant Biology Section, School of Integrative Plant Science, Cornell University, Ithaca, NY, USA*
PIETER J. WOLTERS • *Plant Breeding, Wageningen University & Research, Wageningen, The Netherlands*
DORET WOUTERS • *Plant Breeding, Wageningen University & Research, Wageningen, The Netherlands*
SHUCHI WU • *Department of Horticulture, Virginia Tech, Blacksburg, VA, USA*
KOJI YAMAGUCHI • *Graduate School of Agriculture, Kindai University, Nakamachi, Nara, Japan*
TONG ZHANG • *Department of Biology, University of Florida, Gainesville, FL, USA; Genetics Institute, University of Florida, Gainesville, FL, USA;*
BINGYU ZHAO • *Department of Horticulture, Virginia Tech, Blacksburg, VA, USA*
NING ZHU • *Department of Biology, University of Florida, Gainesville, FL, USA*

Chapter 1

Peptidoglycan Isolation and Binding Studies with LysM-Type Pattern Recognition Receptors

Ute Bertsche and Andrea A. Gust

Abstract

In the last decade, more and more plant receptors for complex carbohydrate structures have been described. However, studies on receptor binding to glycan ligands are often hampered due to the technical challenge to obtain pure preparations of homogeneous carbohydrate ligands such as bacterial peptidoglycan (PGN) in amounts suitable for studying protein–glycan interactions. Also, most approaches rely on the availability of defined soluble ligands, which in the case of glycans can rarely be synthesized but have to be purified from the respective microorganism. In this chapter, we describe the purification of complex PGN from sources such as gram-positive bacteria, from which PGN isolation is facilitated due to its larger content in their cell wall. Insoluble PGN can subsequently be used in simple carbohydrate pull-down assays to test for interaction with plant proteins. In this respect, lysin motif (LysM)-domain containing proteins are of particular interest. All plant receptors described to date to be involved in the perception of *N*-Acetylglucosamine-containing ligands (such as PGN or chitin) have been shown to belong to this protein class. Thus, this chapter will also include the production of recombinant LysM proteins to analyze their PGN interaction.

Key words Peptidoglycan, Chitin, LysM, Carbohydrate affinity assay, Protein–glycan interaction

1 Introduction

Peptidoglycan (PGN) or murein is a unique structure only present in bacteria. It confers rigidity to bacterial cell walls, protects bacteria from their environment, defines bacterial shape, and functions as anchor for proteins and other polymers [1, 2]. Structurally, the PGN backbone consists of a glycan heteromer of alternating *N*-acetylglucosamine (GlcNAc) and *N*-acetylmuramic acid (MurNAc) moieties. To obtain a robust net-like structure, glycan chains are crosslinked via short peptide bridges, the composition of which might vary depending on the bacterial species [1–4]. Determination of PGN composition was initially described for *Escherichia coli* (*E. coli*) and *Staphylococcus aureus* (*S. aureus*) and was based on high-performance liquid chromatography (HPLC) analysis of muropeptides obtained by PGN digest [5, 6]. Recently, a faster method for the PGN isolation and analysis of muropeptide

Libo Shan and Ping He (eds.), *Plant Pattern Recognition Receptors: Methods and Protocols*, Methods in Molecular Biology, vol. 1578, DOI 10.1007/978-1-4939-6859-6_1, © Springer Science+Business Media LLC 2017

structures was established using HPLC or even ultra-performance liquid chromatography (UPLC) that can be combined with mass spectrometry (UPLC-MS) [7].

Its unique composition, such as the presence of the amino sugar MurNAc, a high d-amino acid content in the peptide bridges and unusual amide bonds, as well as its restricted occurrence in bacteria makes PGN a perfect signal for eukaryotic hosts to monitor the presence of potentially harmful bacteria [8, 9]. The immunogenic activity of PGN has been well characterized in both plant [10–14] and animal systems [9, 15] and recently plant PGN receptors have been identified in *Arabidopsis thaliana* and *Oryza sativa* [14, 16, 17]. For PGN binding, these plant receptors contain one or more lysin motifs (LysM) in their extracellular domain. The LysM is an ancient motif that confers the interaction of proteins with GlcNAc-containing glycans such as PGN, chitin and chitin-related symbiosis signals [18] and can be found across all kingdoms [19].

Whereas some of the plant LysM proteins have indeed glycan-binding activity, others merely function as co-receptors, for instance by contributing a kinase domain for signal transduction to a protein complex [20]. Importantly, binding preferences to certain glycans are also determined by the LysM domain. However, the exact mechanism how a given LysM domain specifically recognizes its ligand is still not known. One possibility suggested is the concerted action of several LysMs present within one protein or in complex-forming proteins to bind a certain ligand [21–23].

When studying receptors, analyzing their interaction with the corresponding ligands is of great importance. There are various methods suitable to determine binding activities and affinities of LysM proteins to GlcNAc-containing ligands such as isothermal titration calorimetry (ITC), surface plasmon resonance (SPR), or microscale thermophoresis (MST) analysis. For instance, MST was applied to determine PGN-binding affinities of the *Bacillus subtilis* PGN-hydrolyzing endopeptidase P60, which contains four LysMs [22]. Using ITC, the LysM domain-containing effector protein ECP6 from the fungal pathogen *Cladosporium fulvum* was shown to exhibit high binding affinity to chitin fragments, thus preventing their binding to plant receptors [21, 24]. ITC, MST, and SPR were also successfully applied to demonstrate direct binding of nodulation factors at high-affinity binding sites of the *Lotus japonicus* Nod factor receptor 5 (NFR5) and NFR1, two LysM-domain receptor kinases [25, 26]. Various binding studies also make use of labeled glycan ligands such as biotinylated chitooctaose [27], and labeling of PGN with for instance fluorescein isothiocyanate (FITC) has been described [28]. Biotin-labeled glycans and glycoconjugates have also been immobilized on microplates to create a fluorescence-based solid-phase assay to study protein-glycoconjugate interactions [29]. However, more complex glycoconjugates are often lacking from these glycan microarrays. For the investigation of LysM protein binding to different ligands a glycan microarray was recently developed using carbohydrate-linker conjugates [30]. However, whenever introducing structural alterations by these attachments care must be taken as to whether this affects ligand function or binding properties. Also, these approaches do not allow the determination of any binding affinities.

All methods mentioned above require specific equipment or the availability of defined soluble ligand structures. Alternatively, initial binding studies might get along with simple carbohydrate pull-down assays using insoluble glycans. This method can be applied to any complex structure such as PGN that is insoluble in water and thus serves as affinity matrix [31]. After co-incubation of proteins and the insoluble carbohydrate, glycan-binding proteins will co-precipitate and can be found in the glycan pellet after centrifugation (Fig. 1). This method has been used to demonstrate direct binding of PGN or chitin to their corresponding plant receptor proteins [14, 17, 32].

This chapter will thus focus on the description of simple carbohydrate-binding assays and the purification of the required PGN preparations and recombinant proteins.

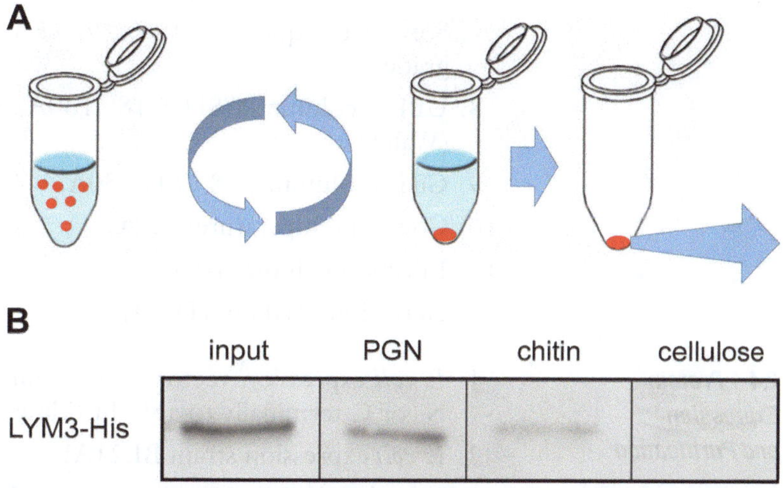

Fig. 1 The LysM protein LYM3 preferentially binds to peptidoglycan. (**a**) Cartoon of a simple carbohydrate-binding assay using insoluble carbohydrates (*red*) as affinity matrix to pull down binding proteins in solution (*blue*). After centrifugation the carbohydrate pellet is analyzed for the presence of epitope-tagged proteins using Western blot analysis. (**b**) The ectodomain of LYM3 was expressed in *E. coli* as 6×His-tagged fusion. 1 μg of purified protein (input) was incubated with 50 μg PGN, chitin, or cellulose for 10 min at 4 °C. After centrifugation, glycan-bound proteins in the pellet were analyzed by SDS polyacrylamide gel electrophoresis (SDS-PAGE) and immunoblotting using anti-His antibodies

2 Materials

2.1 Strains for PGN Preparation [7]

1. *S. aureus* strain SA113 [33].
2. Basic medium (BM; 10 g/L soy peptone, 5 g/L yeast extract, 5/L g NaCl, 1 g/L glucose, 1 g/L K_2HPO_4; adjust pH to 7.2).

2.2 Solutions for PGN Isolation

1. Solution A (1 M NaCl).
2. Tris/HCl buffer (0.1 M Tris/HCl buffer pH 6.8).
3. Solution B (15 μg/mL DNase and 60 μg/mL RNase in 0.1 M Tris/HCl, pH 6.8).
4. Solution C (50 μg/mL trypsin in ddH_2O).
5. Solution D (1 M HCl solution).

2.3 Solutions and Equipment for UPLC Analysis

1. Digestion buffer (12.5 mM NaH_2PO_4, pH 5.5).
2. Mutanolysin solution (5000 U/mL ddH_2O).
3. Borate buffer (0.5 M borax in ddH_2O, pH 9.0).
4. Reduction buffer (10 mg/mL sodium borohydrate in 0.5 M borate buffer).
5. UPLC: Acquity H-class (Waters).
6. Solvent A (aqueous 0.1% (v/v) trifluoroacetic acid (TFA), 5% (v/v) methanol MS grade).
7. Solvent B (aqueous 0.1% (v/v) TFA, 30% (v/v) methanol MS grade).
8. UPLC column: CSH C18, 130Å, 1.7 μm, 2.1 × 100 mm (Waters).
9. Guard column: C18 CSH 130Å, 1.7 μm, 2.1 × 5 mm (Waters).
10. Column temperature: 52 °C.
11. Injection volume: 10 μL.
12. Detection: 210 nm (DAD).

2.4 Protein Expression and Purification

1. *E. coli* expression vector fusing your protein of interest either N- or C-terminally to a six-histidine tag (e.g., pDEST17).
2. *E. coli* expression strain BL21AI.
3. Luria broth (LB; 10 g/L tryptone, 5 g/L yeast extract, 5 g/L NaCl) for *E. coli* cultivation, supplemented with antibiotics according to the expression vector used.
4. l-arabinose.
5. Lysis buffer (50 mM NaH_2PO_4, 300 mM NaCl, 10 mM imidazole, 0.1% (v/v) Tween 20, adjust pH to 8.0 using NaOH).
6. Wash buffer (50 mM NaH_2PO_4, 300 mM NaCl, 20 mM imidazole, adjust pH to 8.0 using NaOH).

7. Elution buffer (50 mM NaH_2PO_4, 300 mM NaCl, 250 mM imidazole, adjust pH to 8.0 using NaOH).
8. Lysozyme.
9. DNase I.
10. RNase A.
11. Ni-NTA agarose (e.g., Qiagen).
12. Bradford solution (e.g., with a commercial Bradford solution from Biorad or Carl Roth).

2.5 Carbohydrate Affinity Assay, SDS-PAGE, and Western Blotting

1. Carbohydrate suspension in water.
2. 100 mM Phosphate buffer, pH 7.
3. Loading buffer stock: 33.3% (v/v) glycerol, 6% (w/v) SDS, 1.5% (w/v) Bromphenol blue, 188 mM Tris/HCl pH 6.8. To obtain the 3× loading buffer freshly mix 200 μL 1M DTT + 500 μL loading buffer stock.
4. SDS-PAGE running buffer, 10× stock: 250 mM Tris base, 1.92 M glycine, 1% (w/v) SDS.
5. Protein marker, prestained.
6. Gel running apparatus.
7. Transfer buffer, 10× stock: 250 mM Tris base, 1.92 M glycine, add 20% (v/v) methanol to the diluted 1× buffer prior blotting.
8. Nitrocellulose membrane (e.g., Hybond, GE Healthcare).
9. Buffer-tank-blotting apparatus.
10. Ponceau S Red staining solution: 0.1% (w/v) Ponceau S Red, 5% (v/v) glacial acetic acid.
11. TBST: 20 mM Tris/HCl pH 7.5, 150 mM NaCl, 0.1% (v/v) Tween-20.
12. Nonfat dried milk powder.
13. Anti-His antibody (e.g., Sigma, product no. H1029).
14. Secondary anti-mouse Immunoglobin G from goat, coupled to alkaline phosphatase (e.g., Sigma, product no. A3562).
15. Alkaline Phosphatase (AP) buffer: 150 mM Tris/HCl pH 9.5, 5 mM $MgCl_2$, 100 mM NaCl.
16. 200× BCIP: 50 mg/mL 5-bromo-4-chloro-3-indolylphosphate in dimethylformamide, store at −20 °C.
17. 200× NBT: 50 mg/mL nitrobluetetrazolium-chloride in 70% (v/v) dimethylformamide, store at −20 °C.

3 Methods

In this section, we describe (a) the isolation of PGN from the gram-positive bacterium *S. aureus*, (b) the expression and purification of recombinant LysM proteins in *E. coli*, and (c) a carbohydrate-binding assay using insoluble PGN as affinity matrix.

3.1 PGN Isolation from S. aureus

1. Grow *S. aureus* in BM medium overnight (*see* **Note 1**).
2. Spin down 2 mL of culture in a 2 mL microcentrifuge tube (5 min at 10,000 × *g*).

   Alternatively, spin down 2 × 2 mL of a culture with a lower OD.
3. Resuspend the pellet in 1 mL solution A (*see* **Note 2**) and boil the suspension for 20 min at 100 °C in a heating block (*see* **Note 3**).
4. Spin down the suspension (5 min at 10,000 rpm) and wash at least two times with 1.5 mL ddH_2O, finally resuspend the pellet in 1 mL ddH_2O.
5. Put the suspension to a sonifier water bath for 30 min.
6. Add 500 μL of solution B and incubate for 60 min at 37 °C in a shaker.
7. Add 500 μL of solution C and incubate for 60 min at 37 °C in a shaker.
8. Boil the suspension for 3 min at 100 °C in a heating block.
9. Spin down the suspension (5 min at 10,000 rpm) and wash once with 1 mL ddH_2O.
10. Add 500 μL of solution D and incubate for 4 h at 37 °C in a shaker.
11. Spin down the suspension (5 min at 10,000 rpm) and wash with ddH_2O until the pH is 5–6.
12. Either use the suspension directly for binding experiments or lyophilize for concentration purposes or for storage. PGNs should finally be dissolved in water at a concentration of ~10 mg/mL and can be stored at −20 °C.

3.2 PGN Quality Control by UPLC

1. Spin down 1/10 of the PGN solution or take a small amount of the lyophilized PGN (*see* **Note 4**).
2. Resuspend the pellet in 50 μL digestion buffer and add 5 μL mutanolysin solution.
3. Incubate the sample for 16 h at 37 °C and 150 rpm.
4. Inactivate the enzyme by boiling at 100 °C for 3 min.
5. Spin down the samples (5 min at 10,000 rpm) and use the supernatant.

6. Add 10 µL of the reduction solution and incubate the samples for 20 min at room temperature (*see* **Note 5**).
7. Stop the reaction with 5 µL phosphoric acid (50%); pH must be between 2 and 3.
8. Wash the column 30 min with methanol (*see* **Note 6**).
9. Wash 30 min with ddH_2O water.
10. Wash 30 min with solvent B.
11. Wash 30 min with solvent A. A steady baseline is important.
12. Gradient conditions: 1 min 100% solvent A, then in 60 min to 100% solvent B; flow rate 0.176 mL/min.
13. Wash 5 min with solvent B.
14. Re-equilibrate for 10 min with solvent A.

3.3 Protein Expression in E. coli and Protein Purification Using Ni-NTA Matrix

1. Clone the coding sequence of the extracellular part of your desired LysM protein (excluding the regions coding for the predicted signal peptide, GPI-anchoring motif, transmembrane domain, or intracellular domain) into an *E. coli* expression vector for 6x His fusions such as pDEST17 (Gateway system, Invitrogen) (*see* **Note 7**).
2. When using the pDEST17 vector, finally transfer your expression clone into *E. coli* BL21AI cells (Invitrogen) (*see* **Note 8**).
3. Grow an overnight culture at 37 °C and 180 rpm in LB-medium supplemented with vector-specific antibiotic.
4. Use the overnight culture to inoculate a fresh 200 mL LB culture supplemented with vector-specific antibiotic (*see* **Note 9**) to an $O.D._{600}$ of 0.1 and grow cells for 2 h at 37 °C and 180 rpm.
5. Induce cells by adding L-arabinose at a final concentration of 0.2% (w/v) and further grow the cells at 28 °C for 2 h and 180 rpm (*see* **Note 10**).
6. Collect the *E. coli* cells by centrifugation for 20 min at 5000–13,000 × *g* and 4 °C, freeze in liquid N_2 (*see* **Note 11**). Batch purification using Ni^{2+}-NTA agarose (Qiagen) is performed according to the manufacturer's handbook "The QIAexpressionist" (Qiagen).
7. Thaw the cell pellet for 15 min on ice and resuspend the cells in 4 mL lysis buffer (*see* **Notes 12** and **13**).
8. Add lysozyme to 1 mg/mL, DNase I to 5 µg/mL, and RNase A to 10 µg/mL and incubate on ice for 30 min (*see* **Note 14**).
9. Sonicate on ice using a sonicator equipped with a microtip. Use six 10 s bursts at 200–300 W with a 10 s cooling period between each burst.
10. Centrifuge lysate at 10,000 × *g* for 30 min at 4 °C to pellet the cellular debris. Save supernatant.

11. Add 1 mL of the 50% Ni-NTA slurry to 4 mL cleared lysate and mix gently by shaking (200 rpm on a rotary shaker) at 4 °C for 60 min (*see* **Note 15**).
12. Load the lysate–Ni-NTA mixture into a column with the bottom outlet capped. Remove bottom cap and collect the column flow-through (can be discarded or analyzed by SDS-PAGE).
13. Wash twice with 4 mL wash buffer (can be discarded or analyzed by SDS-PAGE).
14. Elute the protein four times with 0.5 mL elution buffer. Collect the eluate in four tubes (can be analyzed by SDS-PAGE) (*see* **Note 16**).
15. Determine the protein concentration after Bradford in a photometer at 595 nm (*see* **Note 17**).

3.4 Carbohydrate-Binding Assay

1. Place 1 μg purified His_6 fusion protein into 250 μL 100 mM phosphate buffer, pH 7.
2. Add 50 μg of insoluble carbohydrate (*see* **Note 18**), mix well, and incubate at 4 °C for 10 min (*see* **Note 19**).
3. Spin down insoluble carbohydrate for 10 min at 4 °C and 13,000 × *g*.
4. Wash carbohydrate pellet twice with 1 mL 100 mM phosphate buffer, pH 7, each time collecting the pellet again by centrifugation for 10 min at 4 °C and 13,000 × *g* and carefully removing the supernatant without disturbing the pellet.

3.5 SDS-PAGE and Western Blotting

1. Add 30 μL SDS loading buffer to the pellet, heat for 5 min at 95 °C, spin briefly, and load onto a standard 10% SDS polyacrylamide gel.
2. Run the electrophoresis for 10 min at 130 V, than change to 200 V until your protein of interest approaches the middle of the gel as indicated by the prestained protein marker.
3. Following electrophoresis, transfer proteins from the polyacrylamid gel to a nitrocellulose membrane in a buffer tank-blotting apparatus for 1 h at 100 V (for gels of 1 mm thickness).
4. After transfer, check transfer efficiency and equal protein loading by staining the proteins on the membrane with Ponceau S-Red for 30 s, than destain briefly in water before the membrane can be scanned.
5. Block unspecific binding sites by incubating the membrane in TBST containing 5% (w/v) nonfat dried milk for at least 1 h at room temperature with gentle agitation.
6. Wash the membrane with TBST three times, each 5 min at room temperature with gentle agitation.

7. Incubate the membrane with primary antibody raised against the histidine tag (1:3000 dilution, from mouse) overnight at 4 °C in 10 mL TBST containing 5% (w/v) nonfat dried milk with gentle agitation.
8. Wash the membrane with TBST three times, each 5 min at room temperature with gentle agitation.
9. Incubate the membrane with secondary antibody raised against mouse immunoglobin G (from goat, coupled to alkaline phosphatase, 1:3000 dilution, *see* **Note 20**) for 2 h in 10 mL TBST at room temperature with gentle agitation.
10. Wash the membrane with TBST three times, each 5 min at room temperature with gentle agitation.
11. Equilibrate the membrane in AP-buffer for 2 min at room temperature with gentle agitation.
12. Add BCIP and NBT as a 1:200 dilution in AP buffer and let the color reaction proceed until clear bands are visible.
13. Wash the membrane with distilled water to stop alkaline phosphatase reaction.

4 Notes

1. Also PGN from other bacteria, for instance gram-negatives, can be purified according to Kühner et al. [7].
2. Sometimes, NaCl treatment is not sufficient for PGN isolation from the cells. Use 0.25% SDS solution in 0.1 M Tris/HCl (pH 6.8) instead. SDS has to be washed out thoroughly after boiling.
3. Make sure the samples are boiling at 100 °C. Bad isolation results are mostly caused by too low heat.
4. This step is to verify that the isolation was successful and to estimate the amount of PGN. The muropeptide peaks should be sharp and the highest should be about 150 mAU. Examples are given by Kühner et al. [7].
5. The reduction solution contains a lot of bubbles. The exact volume of 10 μL reduction solution is therefore impossible to obtain. Set your pipette to a volume of 100 μL and add 1 drop to the sample. This accords to the needed volume. Reduction of MurNAc to NAc-muraminitol prevents mutarotation resulting in double peaks during UPLC analysis.
6. Very important for good UPLC results is a long and intensive equilibration of the column and an exact column temperature. Use degassed solvents only.

7. The use of Glutathion-S-transferase (GST)-fusion proteins to increase protein solubility should be avoided as in our hands GST tends to bind to PGN itself.
8. For different vectors, e.g., pET28 (Novagen) or pRSET (Thermo Fisher Scientific), different *E. coli* expression strains such as, e.g., BL21(DE3) for IPTG induction might have to be chosen.
9. Choosing a different growth medium such as modified LB medium (2× YT) or Super Broth might increase protein expression.
10. For optimal protein expression different times and temperatures need to be tested prior to the main experiment.
11. At this point, cells can be stored at −80 °C for long term.
12. The lysis buffer contains 10 mM imidazole to minimize binding of untagged, contaminating proteins and increase purity with fewer wash steps. If the tagged protein does not bind under these conditions, the amount of imidazole should be reduced to 1–5 mM. With 6×His-tagged proteins exhibiting high binding affinities, the imidazole concentration can be increased to 20 mM ("The QIAexpressionist," Qiagen).
13. For proteins that are expressed at high levels (10–50 mg of 6×His-tagged protein per liter of cell culture) a 10× concentrated cell lysate (resuspend the pellet from a 40 mL culture in 4 mL lysis buffer) can be used. Four mL of a 10× concentrated cell lysate in lysis buffer will contain approximately 0.4–2 mg of 6×His-tagged protein. For much lower expression levels (1–5 mg/liter), 200 mL of cell culture should be used to obtain a 50× concentrated cell lysate (4 mL cell lysate = 0.2–1 mg of 6×His-tagged protein) ("The QIAexpressionist," Qiagen).
14. DNase and RNase need to be certified proteinase-free (preparations from bovine pancreas sometimes are not). Alternatively, draw the lysate through a narrow-gauge blunt-ended syringe needle several times ("The QIAexpressionist," Qiagen).
15. Ni-NTA residue is also suitable for FPLC purification of 6×His-tagged proteins from *E. coli* under native conditions.
16. The composition of the lysis, wash and elution buffers can be modified to suit the particular application, e.g., by adding 0.1 % Tween, 5–10 mM β-ME, or 1 mM PMSF, or increasing NaCl or glycerol concentrations ("The QIAexpressionist," Qiagen).
17. Instead of a fresh BSA standard curve the protein concentration can be estimated using the following formula: protein concentration [mg/mL] = $O.D._{595}/(0.0283 \times \text{used volume})$.
18. We routinely use 50 μg insoluble carbohydrate, but to roughly estimate affinities a dose-response with increasing glycan amounts can be performed.

19. An incubation time of 10 min should be sufficient, but depending on the affinity this time can be increased or decreased.
20. Alternatively, a secondary antibody coupled to horseradish peroxidase can be used and detected using the ECL detection system, secondary antibody dilutions then have to be increased to up to 1: 50,000.

Acknowledgments

We thank the Deutsche Forschungsgemeinschaft (SFB 766) for support to U.B. and A.A.G. Roland Willmann is acknowledged for preparing Fig. 1 and for helpful discussions on the manuscript.

References

1. Turner RD, Vollmer W, Foster SJ (2014) Different walls for rods and balls: the diversity of peptidoglycan. Mol Microbiol 91:862–874
2. Vollmer W, Seligman SJ (2010) Architecture of peptidoglycan: more data and more models. Trends Microbiol 18:59–66
3. Schleifer KH, Kandler O (1972) Peptidoglycan types of bacterial cell walls and their taxonomic implications. Bacteriol Rev 36:407–477
4. de Pedro MA, Cava F (2015) Structural constraints and dynamics of bacterial cell wall architecture. Front Microbiol 6:449
5. de Jonge BL, Chang YS, Gage D, Tomasz A (1992) Peptidoglycan composition of a highly methicillin-resistant *Staphylococcus aureus* strain. The role of penicillin binding protein 2A. J Biol Chem 267:11248–11254
6. Glauner B (1988) Separation and quantification of muropeptides with high-performance liquid chromatography. Anal Biochem 172:451–464
7. Kühner D, Stahl M, Demircioglu DD, Bertsche U (2014) From cells to muropeptide structures in 24 h: peptidoglycan mapping by UPLC-MS. Sci Rep 4:7494
8. Bertsche U, Mayer C, Götz F, Gust AA (2015) Peptidoglycan perception--sensing bacteria by their common envelope structure. Int J Med Microbiol 305:217–223
9. Dworkin J (2014) The medium is the message: interspecies and interkingdom signaling by peptidoglycan and related bacterial glycans. Annu Rev Microbiol 68:137–154
10. Felix G, Boller T (2003) Molecular sensing of bacteria in plants. The highly conserved RNA-binding motif RNP-1 of bacterial cold shock proteins is recognized as an elicitor signal in tobacco. J Biol Chem 278:6201–6208
11. Erbs G, Silipo A, Aslam S, De Castro C, Liparoti V, Flagiello A, Pucci P, Lanzetta R, Parrilli M, Molinaro A, Newman MA, Cooper RM (2008) Peptidoglycan and muropeptides from pathogens Agrobacterium and Xanthomonas elicit plant innate immunity: structure and activity. Chem Biol 15:438–448
12. Gust AA, Biswas R, Lenz HD, Rauhut T, Ranf S, Kemmerling B, Gotz F, Glawischnig E, Lee J, Felix G, Nurnberger T (2007) Bacteria-derived peptidoglycans constitute pathogen-associated molecular patterns triggering innate immunity in Arabidopsis. J Biol Chem 282:32338–32348
13. Millet YA, Danna CH, Clay NK, Songnuan W, Simon MD, Werck-Reichhart D, Ausubel FM (2010) Innate immune responses activated in Arabidopsis roots by microbe-associated molecular patterns. Plant Cell 22:973–990
14. Liu B, Li JF, Ao Y, Qu J, Li Z, Su J, Zhang Y, Liu J, Feng D, Qi K, He Y, Wang J, Wang HB (2012) Lysin motif-containing proteins LYP4 and LYP6 play dual roles in peptidoglycan and chitin perception in rice innate immunity. Plant Cell 24:3406–3419
15. Wheeler R, Chevalier G, Eberl G, Gomperts Boneca I (2014) The biology of bacterial peptidoglycans and their impact on host immunity and physiology. Cell Microbiol 16:1014–1023
16. Ao Y, Li Z, Feng D, Xiong F, Liu J, Li JF, Wang M, Wang J, Liu B, Wang HB (2014) OsCERK1 and OsRLCK176 play important roles in peptidoglycan and chitin signaling in rice innate immunity. Plant J 80:1072–1084
17. Willmann R, Lajunen HM, Erbs G, Newman MA, Kolb D, Tsuda K, Katagiri F, Fliegmann J,

Bono JJ, Cullimore JV, Jehle AK, Gotz F, Kulik A, Molinaro A, Lipka V, Gust AA, Nürnberger T (2011) Arabidopsis lysin-motif proteins LYM1 LYM3 CERK1 mediate bacterial peptidoglycan sensing and immunity to bacterial infection. Proc Natl Acad Sci U S A 108: 19824–19829

18. Gust AA, Willmann R, Desaki Y, Grabherr HM, Nürnberger T (2012) Plant LysM proteins: modules mediating symbiosis and immunity. Trends Plant Sci 17:495–502
19. Buist G, Steen A, Kok J, Kuipers OR (2008) LysM, a widely distributed protein motif for binding to (peptido)glycans. Mol Microbiol 68:838–847
20. Antolin-Llovera M, Petutsching EK, Ried MK, Lipka V, Nurnberger T, Robatzek S, Parniske M (2014) Knowing your friends and foes—plant receptor-like kinases as initiators of symbiosis or defence. New Phytol 204:791–802
21. Sanchez-Vallet A, Saleem-Batcha R, Kombrink A, Hansen G, Valkenburg DJ, Thomma BP, Mesters JR (2013) Fungal effector Ecp6 outcompetes host immune receptor for chitin binding through intrachain LysM dimerization. Elife 2:e00790
22. Wong JE, Alsarraf HM, Kaspersen JD, Pedersen JS, Stougaard J, Thirup S, Blaise M (2014) Cooperative binding of LysM domains determines the carbohydrate affinity of a bacterial endopeptidase protein. FEBS J 281:1196–1208
23. Wong JE, Midtgaard SR, Gysel K, Thygesen MB, Sorensen KK, Jensen KJ, Stougaard J, Thirup S, Blaise M (2015) An intermolecular binding mechanism involving multiple LysM domains mediates carbohydrate recognition by an endopeptidase. Acta Crystallogr D Biol Crystallogr 71:592–605
24. de Jonge R, van Esse HP, Kombrink A, Shinya T, Desaki Y, Bours R, van der Krol S, Shibuya N, Joosten MH, Thomma BP (2010) Conserved fungal LysM effector Ecp6 prevents chitin-triggered immunity in plants. Science 329: 953–955
25. Broghammer A, Krusell L, Blaise M, Sauer J, Sullivan JT, Maolanon N, Vinther M, Lorentzen A, Madsen EB, Jensen KJ, Roepstorff P, Thirup S, Ronson CW, Thygesen MB, Stougaard J (2012) Legume receptors perceive the rhizobial lipochitin oligosaccharide signal molecules by direct binding. Proc Natl Acad Sci U S A 109:13859–13864
26. Sorensen KK, Simonsen JB, Maolanon NN, Stougaard J, Jensen KJ (2014) Chemically synthesized 58-mer LysM domain binds lipochitin oligosaccharide. Chembiochem (A European journal of chemical biology) 15:2097–2105
27. Hayafune M, Berisio R, Marchetti R, Silipo A, Kayama M, Desaki Y, Arima S, Squeglia F, Ruggiero A, Tokuyasu K, Molinaro A, Kaku H, Shibuya N (2014) Chitin-induced activation of immune signaling by the rice receptor CEBiP relies on a unique sandwich-type dimerization. Proc Natl Acad Sci U S A 111:E404–E413
28. Maeda H (1980) A new lysozyme assay based on fluorescence polarization or fluorescence intensity utilizing a fluorescent peptidoglycan substrate. J Biochem 88:1185–1191
29. Leppänen A, Cummings RD (2010) Fluorescence-based solid-phase assays to study glycan-binding protein interactions with glycoconjugates. Methods Enzymol 478:241–264
30. Maolanon NN, Blaise M, Sorensen KK, Thygesen MB, Clo E, Sullivan JT, Ronson CW, Stougaard J, Blixt O, Jensen KJ (2014) Lipochitin oligosaccharides immobilized through oximes in glycan microarrays bind LysM proteins. Chembiochem (A European journal of chemical biology) 15:425–434
31. Biswas R, Voggu L, Simon UK, Hentschel P, Thumm G, Götz F (2006) Activity of the major staphylococcal autolysin Atl. FEMS Microbiol Lett 259:260–268
32. Petutschnig EK, Jones AM, Serazetdinova L, Lipka U, Lipka V (2010) The lysin motif receptor-like kinase (LysM-RLK) CERK1 is a major chitin-binding protein in Arabidopsis thaliana and subject to chitin-induced phosphorylation. J Biol Chem 285:28902–28911
33. Iordanescu S (1975) Host controlled restriction mutants of Staphylococcus aureus. Arch Roum Pathol Exp Microbiol 34:55–58

Chapter 2

Characterization of Plant Cell Wall Damage-Associated Molecular Patterns Regulating Immune Responses

Laura Bacete, Hugo Mélida, Sivakumar Pattathil, Michael G. Hahn, Antonio Molina, and Eva Miedes

Abstract

The plant cell wall is one of the first defensive barriers that pathogens need to overcome to successfully colonize plant tissues. Plant cell wall is considered a dynamic structure that regulates both constitutive and inducible defense mechanisms. The wall is a potential source of a diverse set of Damage-Associated Molecular Patterns (DAMPs), which are signalling molecules that trigger immune responses. However, just a few active wall ligands, such as oligogalacturonic acids (OGs), have been characterized so far. To identify additional wall-derived DAMPs, we obtained different plant wall fractions and tested their capacity to trigger immune responses using a calcium read-out system. To characterize the active DAMPs structures present in these fractions, we applied Glycome Profiling, a technology that uses a large and diverse set of specific monoclonal antibodies against wall carbohydrate ligands. The methods describe here can be used in combination with other biochemical approaches to identify and purify new plant cell wall DAMPs.

Key words Pectin, Hemicellulose, Cell wall, Arabidopsis, Immunity

1 Introduction

The cell wall is an essential component of plant cells that regulates diverse biological processes and directly determines the ability of a plant to grow and resist biotic and abiotic stress [1]. Plant cell wall composition and structure is constantly remodeled during development and in response to external influences. The alteration of plant cell wall integrity (CWI) has been hypothesized to be sensed by monitoring systems that initiate compensatory responses to restore wall integrity [2, 3]. This CWI system consists of a set of wall sensor/receptors that specifically bind wall-derived ligands, so-called Damage-Associated Molecular Patterns (DAMPs) that are released upon alteration of wall integrity. This plant monitoring system also functions during pathogen infection, since microbes modify wall composition to favor colonization by means of secreted cell wall degrading enzymes. The perception of wall DAMPs and microbial

Libo Shan and Ping He (eds.), *Plant Pattern Recognition Receptors: Methods and Protocols*, Methods in Molecular Biology, vol. 1578, DOI 10.1007/978-1-4939-6859-6_2, © Springer Science+Business Media LLC 2017

molecules (so-called Pathogen-Associated Molecular Patterns or PAMPs) by specific plant Pattern Recognition Receptors (PRRs) triggers immune responses (PAMP-triggered immunity, PTI) via the activation of several downstream molecular responses: (1) increase of cytoplasmic calcium (Ca^{2+}) concentration; (2) production of reactive oxygen species (ROS); (3) activation of calcium-dependent protein kinases (CDPKs) and mitogen-activated protein kinase (MAPK) cascades; (4) alteration of defensive gene expression; and (5) synthesis of antimicrobial biomolecules [4, 5]. Despite the increasing relevance of CWI signaling in the control of resource reallocation between growth and immunity, only a few wall-derived DAMPs (e.g., OGs) have been characterized. We describe here a combination of two experimental methods to identify potential DAMPs from cell walls of 25 day-old soil grown *Arabidopsis thaliana* wild-type plants (Col-0). Cell walls were fractionated by sequential extraction with increasingly harsh chemicals and four main fractions were obtained (Pectins (PC1 and PC2) and hemicelluloses (HC1 and HC2); [6, 7]). The immunogenic activity of these fractions was determined by using transgenic Col-0 plants expressing cytosolic Apoaequorin, a Ca^{2+} sensor protein (*35S::Apoaequorin*; [8, 9]), that measures the increase in cytosolic Ca^{2+} concentration, which is one of the earliest immune responses observed after PAMP/DAMP perception by plant PRRs [10]. The immune functionality of the selected active wall fractions was further demonstrated by MAPK phosphorylation, a downstream event in PTI signalling [11]. The active cell wall fractions were further characterized by comparative glycomics using ELISA-based technology and a collection of 155 glycan-directed monoclonal antibodies (mAbs) that recognize specific cell wall structures [12]. Here, we describe these combinatorial methodologies to identify novel wall-derived DAMPs.

2 Materials

Some of the chemicals used below are toxic; therefore, proper laboratory safety practices should be followed (gloves, coat, fume hood, etc.). Prepare all solutions using deionized water and analytical grade reagents. Prepare and store all reagents at room temperature (unless indicated otherwise). Follow waste disposal regulations when disposing waste materials.

2.1 AIR (Alcohol Insoluble Residue) Cell Wall Isolation

1. Liquid nitrogen (*see* **Note 1**).
2. Methanol (MeOH)/chloroform ($CHCl_3$) (1:1) solution: in a laboratory hood (*see* **Note 2**) add 500 mL of MeOH to a graduated cylinder and then fill up to 1 L with $CHCl_3$. Store in a laboratory glass bottle at 4 °C.
3. Equipment: fume hood, orbital shaker, rotary evaporator, and water bath that reaches at least 90 °C.

4. Miracloth (VWR, Millipore #CA80058-394) (*see* **Note 3**).
5. CW dehydration: 70% (v/v) EtOH in water and acetone.
6. Glass petri dishes. Never use plastic material since the solvents used may extract residues from the plastic and contaminate the samples.

2.2 AIR (Cell Wall) Fractionation

1. Centrifuge equipped with a rotor that can accommodate volumes of 50–100 mL at 13,900 × *g*, magnetic stirrer, stirring bars (12 × 3 mm) and freeze-dryer.
2. CDTA solution: 50 mM cyclohexane-trans-1,2-diamine-*N*,*N*,*N'*,*N'*-tetraacetic acid sodium salt (CDTA), pH 6.5. Add 18.21 g of CDTA to a glass beaker containing 900 mL of water. Adjust the pH to 6.5 with concentrated ammonia and make up to 1 L with water in a graduated cylinder (*see* **Note 4**).
3. Glacial acetic acid to neutralize the pH of the samples and nitrogen gas to remove the oxygen in the sample.
4. Dialysis tubes with a molecular weight cutoff 3.5 kDa (Spectra/Por® 3 Dialysis Membranes, Spectrum® Laboratories # 132724).
5. Na_2CO_3 solution: 50 mM Na_2CO_3, pH 10.8, containing 20 mM $NaBH_4$. Add 5.30 g of Na_2CO_3 and 756 mg of $NaBH_4$ to a glass beaker containing 700 mL of water. Adjust the pH to 10.8 with HCl and make up to 1 L with water in a graduated cylinder.
6. 4% KOH solution: 4% (w/v) potassium hydroxide (KOH) containing 20 mM $NaBH_4$. Add 40 g of KOH and 756 mg of $NaBH_4$ to a glass beaker containing 800 mL of water. Dissolve it and make up to 1 L with water in a graduated cylinder.
7. 24% KOH solution: 24% (w/v) KOH containing 20 mM $NaBH_4$. Add 240 g of KOH and 756 mg of $NaBH_4$ to a glass beaker containing 800 mL of water. Dissolve it and make up to 1 L with water in a graduated cylinder.

2.3 Total Sugar Quantification

1. Fractionated cell wall, deionized water, 96% sulfuric acid (*see* **Note 5**), 5% (w/v) phenol in deionized water, and 1 mg/mL glucose solution as standard.
2. 15 mL glass tubes (*see* **Note 6**) and plastic 96-well plate.
3. Plate reader spectrophotometer (Thermo Scientific, Varioskan Lux # N16045).

2.4 Ca^{2+}-Mediated Signalling in PTI–Apoaequorin Luminescence Measurements

1. Sterile 24-well plates and white 96-well plates (Sigma #BR781605).
2. Laminar flow hood and plant-growth chamber (16:8-h light:dark cycle at 22 °C).

3. MS medium: dissolve 2.5 g of sucrose in a glass beaker containing 900 mL of water. Then add 2.15 g of Murashige & Skoog medium [13] including Gamborg B5 vitamins (MS, Duchefa Biochemie) and 195.2 mg of morpholiethane sulfonic acid (MES). Adjust the pH to 5.7 with 1 M KOH and make up to 1 L with water in a graduated cylinder. Autoclave the medium and use it only in sterile conditions.
4. Coelenterazine solution: 2-(p-hydroxybenzyl)-6-(p-hydroxyphenyl)-8-benzyl-imidazol[1,2-a]pyrazin-3-(7H)-one (Coelenterazine-native, PJK GmbH #10271) is prepared as a 5 mM stock in MeOH. Add 470 μL of MeOH to 1 mg of coelenterazine and store it at −20 °C protected from light. Working solution is 40 μM (prepare just before use).
5. Microplate luminometer, ideally with an automatic injector (Thermo Scientific, Varioskan Lux # N16045).
6. Solution of the flagellin active PAMP peptide (flg22; [5]): make a 1 mM stock solution by diluting 5 mg of flg22 (EZBiolab #cp7201) in 2.2 mL of water and store at −20 °C in smalls aliquots. The flg22 working solution concentration is 3 μM.

2.5 Glycomic Profiling

1. Cell wall fractions (PC1, PC2, HC1, HC2): 10 μg of total sugar/mL and deionized water as a control.
2. Flat-bottom 96-well plates Costar 3598 (Sigma #CLS3598) (*see* **Note 7**).
3. Incubator (37 °C).
4. Filtering Wash buffer: 1X TBS (Tris Buffered Saline): 50 mM Tris–HCl, pH 7.6, and 100 mM NaCl in MilliQ water. Prepare a stock solution 8X: Add 2.224 g of Tris base, 9.696 g Tris–HCl, and 46.752 g of NaCl to a glass beaker containing 950 mL of water. The pH of the solution should be about 7.6 at room temperature. Dissolve it and make up to 1 L with MilliQ water. To prepare the 0.1 M Wash buffer, filter the solution with a 0.45 μm bottle top filter (VWR, Nalgene #16211-054) before diluting in MilliQ water.
5. Blocking solution: 1% (w/v) instant non-fat dry milk (Carnation) in 1X TBS.
6. Primary antibodies: CCRC series of antibodies were generated in mouse; JIM, MAC and LM series of antibodies were generated in rat. More information at WallMabDB (http://www.wallmabdb.net). The three main suppliers of plant glycan-directed antibodies are CarboSource (http://www.carbosource.net), PlantProbes (http://www.plantprobes.net), and BioSupplies (http://www.biosupplies.com.au/).
7. Peroxidase-conjugated secondary antibody diluted 1:5000. Peroxidase-conjugated, goat anti-mouse, IgG, whole molecule (Sigma #A4416) or goat anti-rat, IgG, whole molecule (Sigma #A9037).

8. Substrate 3,3′,5,5′-tetramethylbenzidine (TMB) solution (Vector Laboratories #SK4400). Prepare according to the manufacturer's directions: Add two drops "buffer pH 5.4," three drops of "substrate" and two drops of "hydrogen peroxide" to 15 mL of deionized water.
9. Stop Peroxidase Buffer: 0.5 N sulphuric acid. In a laboratory hood, add 1.36 mL of sulfuric acid (98%) carefully to 100 mL of water.

3 Methods

3.1 AIR (Alcohol Insoluble Residue) Cell Wall Isolation

1. Harvest 25-day-old Arabidopsis plants (grown at 16:8-h light:dark cycle, 22 °C) before the chamber lights turn on and quickly freeze them in liquid nitrogen. Frozen tissues can be stored at −20 °C until use.
2. Add 25 mL of MeOH/$CHCl_3$ solution per gram of fresh tissue and incubate the mixture in an orbital shaker for 4 h at 4 °C.
3. Remove the solvent by filtration through Miracloth filters. Add 15 mL of MeOH:$CHCl_3$ solution per gram of fresh tissue weight and incubate it overnight in the same conditions as in the previous step.
4. Repeat the filtration step and the incubation during 5 h.
5. Filtrate and incubate in the same conditions for 2 h.
6. Replace the MeOH:$CHCl_3$ by 70% EtOH, and heat the solution at 90 °C for 1 h. After the incubation, allow the cell wall suspension to stand overnight at room temperature and filter afterward through Miracloth.
7. Wash the tissues with 10 mL of acetone per gram of fresh tissue (*see* **Note 8**). Perform four washes of 30 min each at 4 °C 100 rpm. Filter the wall suspensions as before between each washing step, discarding the acetone according to waste disposal regulations.
8. After the final acetone wash and filtration, allow the insoluble residues to dry overnight in glass Petri dishes inside a fume hood. The dry residue is considered to be the Alcohol Insoluble Residue (AIR) (*see* **Note 9**).

3.2 AIR (Cell Wall) Fractionation

Cell walls (AIR) were sequentially extracted with increasingly harsh reagents essentially as described previously [14] with some modifications.

1. Start the fractionation with 300 mg of dry AIR. The incubations can be performed directly in centrifuge tubes containing a small stirring bar. In all cases, the shaking will be performed over magnetic stirrers at maximum speed. Add 30 mL of 0.05 M CDTA solution and incubate 24 h at room temperature

(*see* **Note 10**). Centrifuge the samples at 13,900 × *g* for 15 min. Keep the CDTA supernatant and repeat the incubation during 24 h by adding 30 mL of fresh CDTA solution to the pellet. Centrifuge as before and pool the CDTA supernatants. Dialyze the extracts against distilled water at 4 °C (molecular weight cutoff 3.5 kDa) and freeze-dry the contents of the dialysis tubes. The freeze-dried material is the CDTA or Pectin 1 (PC1) fraction (*see* **Note 11**).

2. Incubate the pellet after the CDTA extraction with 30 mL of Na_2CO_3 solution at 4 °C. Follow the same procedure as for the CDTA fraction; putting in fresh solvent after 24 h after centrifugation. The pooled Na_2CO_3 extracts are neutralized (pH 7.0) with glacial acetic acid, dialyzed and freeze-dried to obtain the Na_2CO_3 or Pectin 2 (PC2) fraction.
3. The pellet after CDTA and Na_2CO_3 extraction is then extracted with 4% KOH solution (bubbled with gaseous N_2, *see* **Note 12**) under the same conditions as for the CDTA extraction; 48 h at room temperature changing the KOH after 24 h by centrifugation. The KOH extracts are treated as the Na_2CO_3 fraction to produce the 4% KOH or Hemicellulose 1 (HC1) fraction.
4. Finally, the remaining pellet is treated with 24% KOH solution (bubbled with gaseous N_2) for 48 h. Repeat exactly as for the previous step to produce the freeze-dried 24% KOH or Hemicellulose 2 (HC2) fraction. The final residue is considered to be α-cellulose.

3.3 Total Sugar Quantification of All Cell Wall Fractions Was Done Using the Phenol-Sulphuric Acid Method [15]

1. Mix 400 μL of cell wall fraction or glucose solution (0–40 μg), and 400 μL of 5% phenol in a glass tube and vortex it.
2. Add 2 mL of concentrated sulphuric acid directly to the samples and vortex strongly (*see* **Note 13**).
3. Transfer 250 μL the samples to 96-well plate after cooling down to room temperature for 5 min.
4. Measure absorbance at 490 nm by spectrophotometer. Use the glucose solution as standard to quantify the total glucose equivalents in the samples.

3.4 Ca^{2+}-Mediated Signalling in PTI: Apoaequorin Luminescence Measurements

Arabidopsis transgenic calcium sensor lines expressing 35S::*Apoaequorin* cytoplasmic gene [8] were used for Ca^{2+} measurements using the method described [16] with minor modifications.

1. Grow Arabidopsis sterile seedlings expressing 35S::*Apoaequorin* in 24-well plates containing 2 mL of MS liquid medium during 8–10 days. Stratify during 3 days at 4 °C in the dark and then transfer the plates to a plant-growth chamber (16:8-h light:dark cycle, 22 °C).

2. Transfer the seedlings individually into the wells of a white 96-well plate containing 75 μL of sterile distilled water. Add 25 μL of 40 μM coelenterazine to reach a final concentration of 10 μM. Incubate overnight in the dark at room temperature.
3. Luminescence released by the activated form of colenterazine (colenteramide) is monitored by means of a microplate luminometer. The measurements should be preferentially taken by complete rows or columns of the 96-well plate. Before the addition of the PAMP/DAMP/wall fraction solution, the basal levels of luminescence are monitored during 1 min (basal luminescence step). Monitor the luminescence for each well during 20 min using 300 ms of integration time after the automatic application of the PAMP/DAMP/wall fraction (i.e., 50 μL of flg22 or cell wall fractions; elicited luminescence step). Figure 1a shows the comparative activity of the four wall fractions from Col-0 plants. The capacity of these fractions to regulate PTI was further corroborated by determining the phosphorylation of MAPK in Col-0 plants treated with the wall fractions (Fig. 1b).

3.5 Glycome Profiling by ELISA [12]

1. Add 50 μL of cell wall fractions solutions (10 μg/mL) to each well in alternating columns. Add water to the first and last columns as controls.
2. Place the coated plates (without lids) in an incubator at 37 °C and allow them to evaporate to dryness overnight. Dry plates can be stored covered at room temperature in the dark for up to 2 weeks.
3. Block wells with 200 μL of blocking solution for one hour at room temperature.
4. Remove blocking agent and add 50 μL of non diluted hybridoma supernatant, incubate for one hour at room temperature.
5. Remove supernatant and wash wells three times each with 300 μL/well (or 200 μL if washing by hand) of Wash buffer.
6. Add 50 μL of peroxidase-conjugated secondary antibody diluted 1:5000 in wash buffer and incubate for one hour at room temperature.
7. Remove secondary antibody and wash the wells five times with Wash buffer.
8. Add 50 μL of TMB substrate solution.
9. After 20 min, stop the reaction by adding 50 μL of Stop Peroxidase Buffer to each well. Do not remove the TMB substrate solution or the acid.
10. Read the optical density of the plate at 450–655 nm in spectrophotometer.

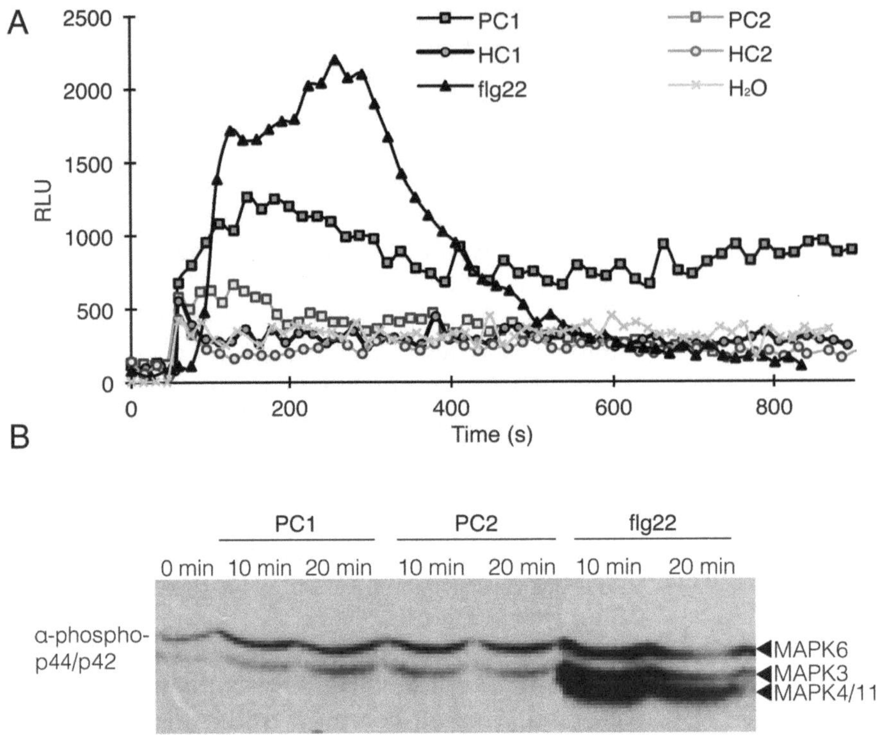

Fig. 1 PTI activation in Arabidopsis seedling by Col-0 cell wall fractions containing DAMPs. (**a**) Variation of intracellular Ca^{2+} concentration after treatment of 35S::*Apoaequorin* seedlings with 50 ng/μL of different cell wall fractions (PC1, PC2, HC1, and HC2) from wild-type (Col-0) plants. Ca^{2+} concentration is measured as Relative Luminescence Units (RLU) at different times (seconds, s) after treatment. flg22 (1 μM) and water (H_2O) treatments were used as positive and negative PTI controls, respectively. Treatments were performed by columns in a 96-well plate. Mean values (±SE; n = 8) are shown. (**b**) MAPK phosphorylation detected with anti-phosphop-44/p42 at 10 and 20 min after the treatment with 50 ng/μL of Pectin 1 (PC1) and Pectin 2 (PC2) fractions. MAPK phosphorylation (MAPK3/6/4/11) is a well-characterized event occurring at early PTI stages upon PAMPs/DAMPs perception [5, 11]. MAPK phosphorylation showed (**b**) corroborates the results obtained with the calcium sensor lines (**a**)

11. Transfer the OD readings to Excel and subtract the reading of the control well from that of each test well on the same plate that contained the same primary and secondary antibodies but not immobilized cell wall fractions. Figure 2 shows a heat-map of the results obtained with the four Col-0 wall fractions (PC1, PC2, HC1, and HC2). These data demonstrate the differential composition of the fractions tested.

4 Notes

1. Wear cryogenic gloves and appropriate protection goggles when working with liquid nitrogen.
2. MeOH:$CHCl_3$ solution should only be opened in a fume hood. Make sure that bottles containing it are properly closed

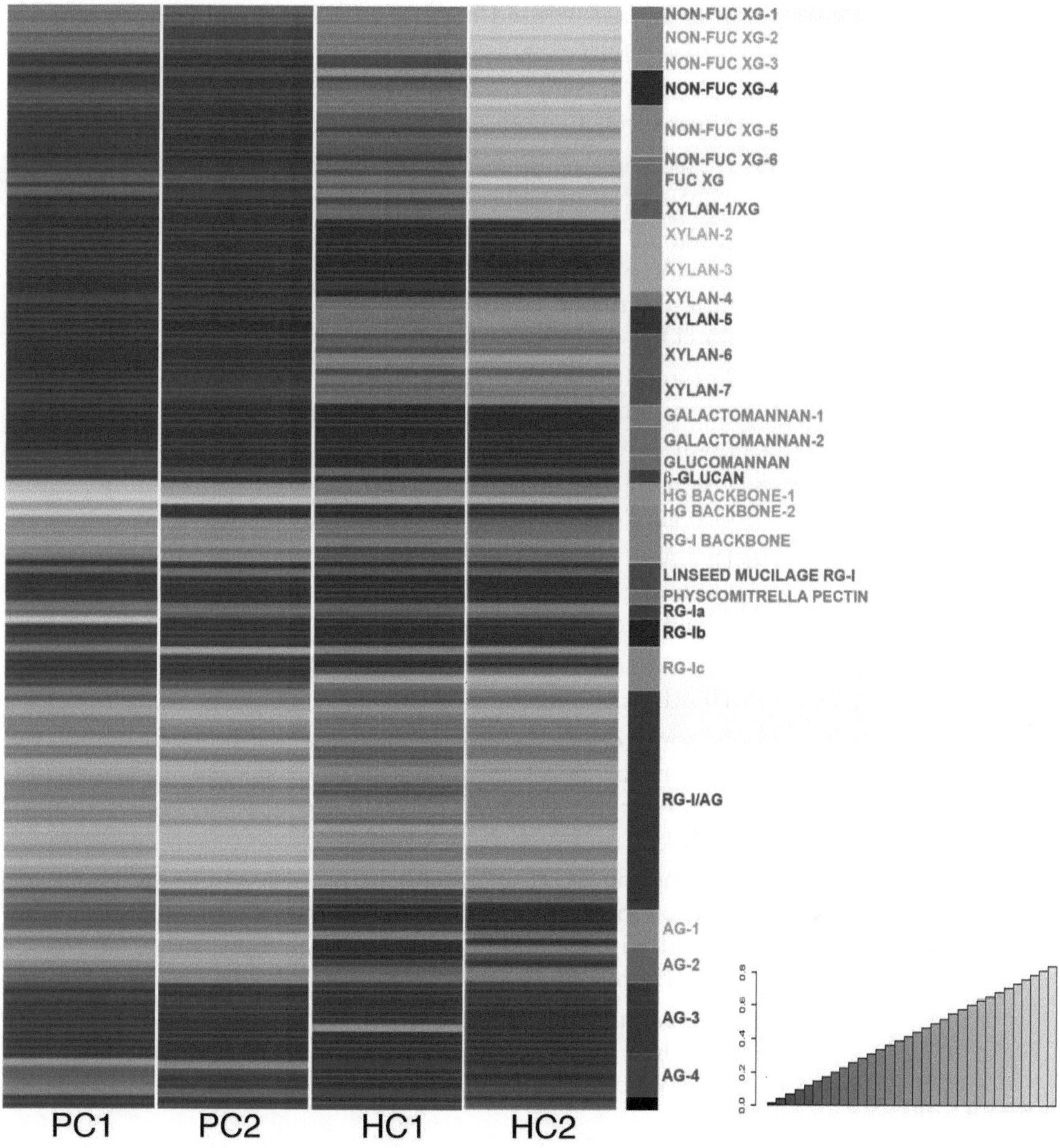

Fig. 2 Heat map analyses of relative abundance of plant cell wall epitopes in Arabidopsis Col-0 wall fractions (Pectin 1 (PC1), Pectin 2 (PC2), Hemicellulose 1 (HC1), and Hemicellulose 2 (HC2). The panel on the right-hand side of the heat map shows the groups of mAbs based on the class of cell wall epitopes recognized by each group. Response values are depicted as heat maps with *black-gray-white color* scheme, where white represents the highest antibody-binding signal and black represents no binding to the extract by the mAb

during the incubation steps. Most of the chemicals used during the procedure are quite harmful, so laboratory gloves and coat are mandatory.

3. Prepare the filters making a double layer of Miracloth. If Miracloth is not available, a plain weave such as synthetic Organdy (Organza) can be used. However, a paper-based

filter *cannot* be used because it will contaminate the samples with cellulose residues.

4. CDTA will only dissolve when the pH gets close to 6.5. Alternatively to ammonia, KOH can be used to adjust the pH.
5. We recommend using a ceramic dispenser for adding sulfuric acid, increasing safety and reproducibility of the measurements.
6. It should be noted that glass tubes are necessary for proper development of the reaction, which must be carried out with a heat treatment (the plastic plates are not stable to boiling water).
7. Among the ELISA plates tested, Costar 3598 gave the highest mean signal across all of the mAb/polysaccharide pairs tested [12].
8. Use adequate protection when working with acetone. Fume hood and gloves (double layer) are mandatory.
9. Dry AIR can be stored indefinitely, but a dry environment is necessary. We recommend adding silica gel bags to the container where AIR is stored.
10. The extraction solutions during the fractionation are applied at 10 mL per milligram of AIR dry weight.
11. In some cases, traces of CDTA may be retained complexed with pectic polymers even after long dialysis, so we recommend the use of additional methods to completely remove it [17, 18].
12. During the KOH treatments we recommend to bubble N_2 gas to the tubes before closing, which minimizes oxidation during the alkali extraction. Use a low N_2 flow to avoid sample dispersion.
13. The addition of concentrated sulfuric acid should be done in the fume hood and with a total protection of hands and arms to avoid burning with acid.

Acknowledgment

Research in the A. M. lab was financially supported by grants BIO2012-32910 and BIO2015-64077-R from Spanish Ministry of Economy and Competitiveness (MINECO), and in the M. G. H. lab (CCRC) by the United States NSF Plant Genome Program (DBI-0421683 and IOS-0923992). L. B. was a PhD fellow (BES-2013-065010) from MINECO and H. M. was supported by an IEF-SignWALLINg-624721 grant from the European Union.

References

1. Miedes E, Vanholme R, Boerjan W, Molina A (2014) The role of the secondary cell wall in plant resistance to pathogens. Front Plant Sci 5:358. doi:10.3389/fpls.2014.00358
2. Thorsten H (2012) Plant cell wall integrity maintenance as an essential component of biotic stress response mechanisms. Front Plant Sci 23:77. doi:10.3389/fpls.2012.00077
3. Thorsten H (2015) The plant cell wall integrity maintenance mechanism—concepts for organization and mode of action. Plant Cell Physiol 56:215–223
4. Boller T, Felix G (2009) A renaissance of elicitors: perception of microbe-associated molecular patterns and danger signals by pattern-recognition receptors. Annu Rev Plant Biol 60:379–406
5. Monaghan J, Zipfel C (2012) Plant pattern recognition receptor complexes at the plasma membrane. Curr Opin Plant Biol 15:349–357
6. Miedes E, Lorences EP (2004) Apple (*Malus domestica*) and tomato (*Lycopersicum esculentum*) fruits cell-wall hemicelluloses and xyloglucan degradation during *Penicillium expansum* infection. J Agric Food Chem 52: 7957–7963
7. Miedes E, Lorences EP (2006) Changes in cell wall pectin and pectinase activity in apple and tomato fruits during *Penicillium expansum* infection. J Sci Food Agric 86:1359–1364
8. Knight MR, Campbell AK, Smith SM, Trewavas AJ (1991) Transgenic plant aequorin reports the effects of touch and cold-shock and elicitors on cytoplasmic calcium. Nature 352:524–526
9. Ranf S, Gish N, Schäffer M, Illig T, Westphal L, Knirel YA et al (2015) A lectin S-domain receptor kinase mediates lipopolysaccharide sensing in *Arabidopsis thaliana*. Nat Inmunol 16:426–433
10. Ranf S, Eschen-Lippold L, Pecher P, Lee J, Scheel D (2011) Interplay between calcium signalling and early signalling elements during defence responses to microbe- or damage-associated molecular patterns. Plant J 68: 100–113
11. Meng X, Zhang S (2013) MAPK cascades in plant disease resistance signaling. Annu Rev Phytopathol 51:245–266. doi:10.1146/annurev-phyto-082712-102314
12. Pattathil S, Avci U, Baldwin D, Swennes AG, McGill JA, Popper Z et al (2010) A comprehensive toolkit of plant cell wall glycan- directed monoclonal antibodies. Plant Physiol 153:514–525
13. Murashige T, Skoog F (1962) A revised medium for rapid growth and bio assays with tobacco tissue cultures. Physiol Plant 15: 473–497
14. Pattathil S, Avci U, Miller JS, Hahn MG (2012) Immunological approaches to plant cell wall and biomass characterization: glycome profiling. In: Himmel M (ed) Biomass conversion: methods and protocols. Methods in molecular biology, vol 908. Springer Science+Business Media, LLC, New York, NY, pp 61–72
15. Dubois M, Gilles KA, Hamilton JK, Rebers PA, Smith F (1956) Colorimetric method for the determination of sugars and related substances. Anal Chem 28:350–356
16. Ranf S, Grimmer J, Poschl Y, Pecher P, Chinchilla D, Scheel D, Lee J (2012) Defense-related calcium signaling mutants uncovered via a quantitative high-throughput screen in *Arabidopsis thaliana*. Mol Plant 5:115–130
17. Mort AJ, Moerschbacher BM, Pierce ML, Maness NO (1991) Problems encountered during the extraction, purification, and chromatography of pectic fragments, and some solutions to them. Carbohydr Res 215: 219–227
18. Coimbra MA, Waldron KW, Selvendran RR (1994) Isolation and characterisation of cell wall polymers from olive pulp (*Olea europaea* L). Carbohydr Res 252:245–262

Chapter 3

Methods of Isolation and Characterization of Oligogalacturonide Elicitors

Manuel Benedetti, Benedetta Mattei, Daniela Pontiggia, Gianni Salvi, Daniel Valentin Savatin, and Simone Ferrari

Abstract

Oligogalacturonides (OGs) are pectic fragments derived from the partial degradation of homogalacturonan in the plant cell wall and able to elicit plant defence responses. Recent methodological advances in the isolation of OGs from plant tissues and their characterization have confirmed their role as *bona fide* plant Damage-Associated Molecular Patterns. Here, we describe the methods for the isolation of OGs from Arabidopsis leaf tissues and for the characterization of OG structure and biological activity.

Key words Oligogalacturonides, Damage-associated molecular patterns, Innate immunity, Pectin, Elicitors

1 Introduction

Oligogalacturonides (OGs) are oligomers of alpha-1,4-linked galacturonosyl residues derived from the partial hydrolysis of homogalacturonan (HGA), a major component of plant cell wall pectin. Their ability to activate plant defence responses was described more than 30 years ago [1]. It was therefore speculated that the degradation of HGA by microbial polygalacturonases (PGs) may cause the accumulation, in the apoplast, of OGs that trigger defence responses [2]. Subsequent studies, mostly conducted using OGs prepared in vitro and applied exogenously to plants, confirmed their ability to trigger an immune response in different plant species, including accumulation of phytoalexins [3], glucanase and chitinase [4, 5], deposition of callose, production of reactive oxygen species (ROS) [6, 7], and nitric oxide [8], indicating that these elicitors are *bona fide* Damage-Associated Molecular Patterns. Furthermore, plant Polygalacturonase-Inhibiting Proteins (PGIPs) have been shown to limit PG action and therefore favor the accumulation of OGs with eliciting

Libo Shan and Ping He (eds.), *Plant Pattern Recognition Receptors: Methods and Protocols*, Methods in Molecular Biology, vol. 1578, DOI 10.1007/978-1-4939-6859-6_3, © Springer Science+Business Media LLC 2017

activity [9]. Most studies aiming to define the mode of action of OGs, including the molecular elements involved in their perception and transduction [10], were based on the use of OGs prepared in vitro by partial digestion of purified polygalacturonic acid (PGA) using pectic enzymes [11], mild acid treatments [12], or high temperature [13]. OGs prepared by these methods usually comprise a fraction with a degree of polymerization (DP) between 10 and 15, which is the elicitor-active range of oligomers, that can be enriched by precipitation with ethanol, followed by dialysis and lyophilization [14].

We have recently reported that OGs with DP 10–15 can be isolated from the cell walls of Arabidopsis plants expressing a chimeric protein obtained by fusing the *Fusarium phyllophilum* FpPG with its bean inhibitor PvPGIP2 [15]; the protocol for this method of isolation is described elsewhere [14]. The chromatographic profile of these OGs, obtained by HPAEC-PAD, was comparable to that of OGs prepared in vitro from PGA. Here, we will provide details for a method for the direct isolation of OGs released from plant tissues incubated in a liquid medium, rather than from cell walls. This method, which we named "leaf strip" isolation of OGs, is much simpler and faster, since it does not require the preparation of cell wall fractions.

Differences in the molecular weight distribution and in the degree and pattern of methylation/acetylation of OGs can produce molecules with different biological activities. Therefore, a complete structural characterization of OGs aimed at obtaining the full details of their structure is needed, and mass spectrometry has been greatly useful to this purpose. Mass spectra of OGs can be obtained using either matrix-assisted laser desorption/ionization (MALDI) or electrospray ionization (ESI) ion sources. MALDI has often been preferred due to its high sensitivity for glycans and its tolerance toward contaminants; protocols for the analysis of OGs using different matrices have been described [14, 16]. ESI-MS and MS/MS methods for structural analysis of OGs are presented in this chapter. The structural analysis of OGs in complex samples is also possible with liquid chromatography coupled to mass spectrometry (LC-MS/MS) using hydrophilic interaction chromatography (HILIC) [17] or porous graphitized carbon HPLC columns [18] for the separation of different oligosaccharide structures. Biological activity of these OGs, as well as of OGs prepared by other methods, can be tested by a variety of approaches, which differ in terms of time of response (from minutes to days), type of biological material (seedlings, leaves, entire plants, cell cultures), cost and time required to complete the analysis. Most of these assays are summarized in Table 1; please refer to the indicated publications for more details.

Table 1
Assays for biological activity of OGs

Timing	Response	Tissue	Species	References
Very early (1–5 min)	Ion fluxes	Cultured cells	*Nicotiana tabacum*	[20, 21]
		Leaf mesophyll cells	*Solanum lycopersicon*	[22]
	Calcium spike	Cultured cells	*Daucus carota*	[23]
		Hypocotyl-derived cells	*A. thaliana*	[24]
	Nitric oxide production	Leaf discs	*A. thaliana*	[8]
	Oxidative burst	Seedlings	*A. thaliana*	[7]
		Leaf discs		[25, 26]
		Cultured cells	*Glycine max*	[27]
		Cultured cells	*N. tabacum*	[28]
		Leaf mesophyll cells	*S. lycopersicon*	[29]
	Activation of MAPKs	Seedlings	*A. thaliana*	[25, 30]
		Leaves	*A. thaliana*	[31]
	Changes in protein phosphorylation	Leaves and seedlings	*S. lycopersicon*, *S. tuberosum* and *A. thaliana*	[32, 33]
Early (15–60 min)	Ethylene biosynthesis	Fruit peel	*Citrus sinensis*	[34, 35]
		Seedlings	*A. thaliana*	[31]
	Gene expression reprogramming	Seedlings	*A. thaliana*	[36]
		Cultured cells	*A. thaliana*	[24]
	Inhibition of auxin-regulated gene induction	Seedlings and leaf explants	*A. thaliana*	[37]
Late (hours–days)	Phytoalexin production	Cotyledon	*G. max*	[3]
		Leaves	*A. thaliana*	[25, 38]
	Callose deposition	Leaves	*A. thaliana*	[7, 31]
	Protection to *B. cinerea*	Plantlets	*Vitis vinifera*	[39]
		Leaves	*A. thaliana*	[40]
	Regulation of plant morphogenesis	Leaf explants	*A. thaliana*	[37]
		Thin cell layer	*N. tabacum*	[41]
		Stem segments	*Pisum sativum*	[42]
	Inhibition of light-induced opening of stomata	Leaves	*Commelina communis*	[43]
	Fruit ripening	fruit pericarp	*S. lycopersicon*	[44]

2 Materials

2.1 Preparation of Standard OGs

1. Sodium acetate 50 mM, pH 5.0.
2. Ice-cold absolute ethanol.
3. PGA solution: 2 g high-molecular weight demethylated polygalacturonic acid (PGA; Alfa Aesar) in 100 mL sodium acetate

50 mM, pH 5.0. The mixture is irradiated in a microwave oven for 1 min at 800 W power to completely dissolve the PGA. The solution is cooled down at room temperature.

4. AnPGII: endo-polygalacturonase II from *Aspergillus niger* [19], purified from Pectinase from *A. niger* (Sigma). Other PGs can be used, though digestion conditions must be adjusted for each different enzyme.

2.2 Isolation and Quantification of Endogenous OGs from Cell Wall Material

Prepare and store all reagents at room temperature (unless otherwise indicated).

1. CDTA solution: 50 mM CDTA (trans-1,2-Diaminocyclohexane-*N,N,N′,N′*-tetraacetic acid monohydrate). Weigh 18.2 g CDTA and dissolve it in 1 L of ultrapure water in a graduated cylinder or a glass beaker. Add NaOH anhydrous pellets up to complete solubilization of CDTA.
2. ChA solution: 50 mM ammonium acetate pH 5.0, 50 mM CDTA, 50 mM oxalate. Weigh 7.1 g ammonium oxalate monohydrate, 3.8 g ammonium acetate. Dissolve the salts in 1 L CDTA solution. Adjust the pH to 5.0 by adding glacial acetic acid. Filter-sterilize the solution.
3. 1:1 (v/v) chloroform:methanol.
4. 70% (v/v) ethanol.
5. Plant material: rosette leaves of soil-grown 4-week-old *Arabidopsis thaliana* plants.

2.3 Isolation of OGs with the Leaf Strip Method

1. Sterilizing solution: 1% NaClO. Add about 14 mL of 7% NaClO to 86 mL of ultrapure water and filter-sterilize the solution.
2. CDTA solution: as described in Subheading 2.2.
3. OG extraction solution: ChA solution (see the above paragraph 2.2.3) supplemented with 10 mM Na_2SO_3.
4. Plant material: rosette leaves from 25-day-old *Arabidopsis thaliana* plants.

2.4 Analysis of OGs by HPAEC-PAD

The HPAEC analysis was carried out using a ICS3000 (Dionex Thermo Fischer) HPLC: Ion Chromatography System with pulsed amperometric detection (PAD), equipped with an Analytical anion exchange column Carbopac PA200 (3 × 250 mm Dionex Thermo Fischer) with a guard column (*see* **Note 1**).

All solutions are prepared using ultrapure water and HPLC grade reagents.

1. Eluent A: 50 mM potassium hydroxide (*see* **Note 2**).
2. Eluent B: 50 mM potassium hydroxide, 1 M potassium acetate (*see* **Note 2**).

2.5 Characterization of OGs by Mass Spectrometry

All solutions are prepared using LC-MS grade water and methanol.

1. BioRex MSZ 501 cation exchange resin beads (BIO-RAD) suspended in ultrapure water.
2. DHB solution: 5% 2,5-dihydroxybenzoic acid (DHB) in 0.2% trifluoroacetic acid (TFA). Weigh 5 mg DHB and dissolve in 100 µL of 0.2% (v/v) TFA. Centrifuge at 30,000 × *g* for 10 min and use the supernatant.

3 Methods

3.1 Preparation of Standard OGs

This method, optimized to obtain a preparation enriched in elicitor-active OGs (DP10-15), is based on a previously described protocol [14] with few modifications.

1. Incubate a PGA solution (100 mL in 500 mL flasks; at least five flasks should be processed in a single experiment) with AnPGII (0.1 RGU/mL) for 180 min at 30 °C in a water bath under gentle shaking (a representative profile of the preparation at the end of this step is shown in Fig. 1a) (*see* **Note 3**).
2. Inactivate the enzyme by boiling the digests for 10 min in a water bath.
3. Cool samples at 4 °C on ice.
4. Dilute samples with cold 50 mM sodium acetate and add ice-cold ethanol to a final concentration of 0.5% (w/v) PGA and 17% (v/v) ethanol. Ethanol is added to the PGA solution and mixed slowly (*see* **Note 4**).
5. Incubate samples overnight at 4 °C.
6. Centrifuge samples at 30,000 × *g* for 30 min. At the end of the centrifugation, an abundant white pellet should be visible.
7. Solubilize the pellet in a small amount of ultrapure water (about 200 mL).
8. Centrifuge samples at 30,000 × *g* for 30 min, recover the supernatant, and discard the brown pellet, containing insoluble material.
9. Dialyze samples against ultrapure water in a dialysis tube with a molecular mass cutoff of 1000 Da (Spectra/Por®).
10. Lyophilize samples.

The expected yield is approximately 20% (about 2 g OGs, when starting with 10 g PGA). The samples can be directly analyzed by HPAEC-PAD (a representative chromatogram of the OGs prepared with this method is shown in Fig. 1b).

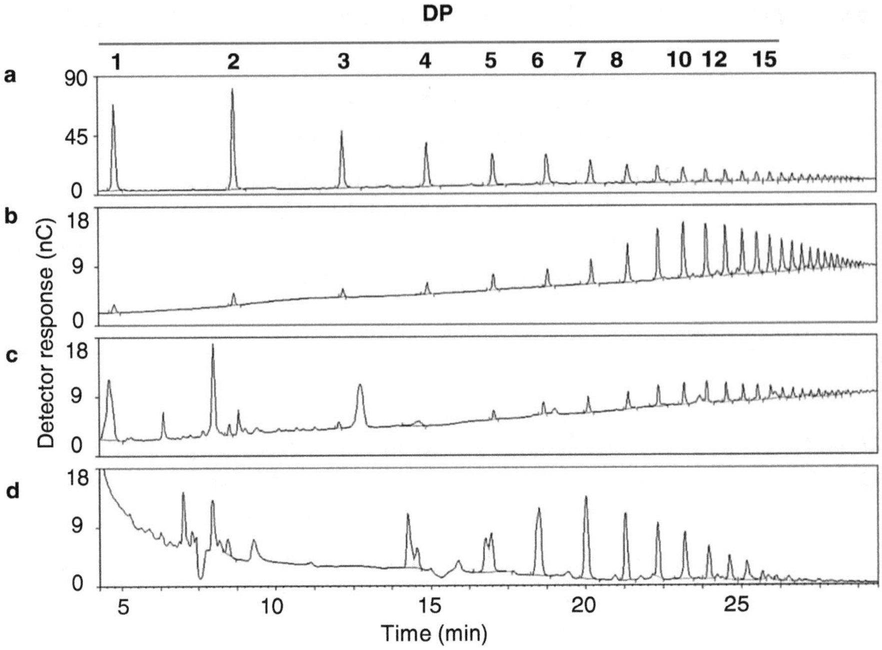

Fig. 1 HPAEC-PAD analysis of OGs. Numbers on top of the figure indicate the DP of the different oligomers. (**a**) Typical profile of OGs obtained from PGA after 210 min digestion with AnPGII. The sample was diluted 1:50. (**b**) OGs enriched in the DP range 10–15, prepared as described. The loaded amount was 2.5 μg. (**c**) OGs obtained from cell walls of 25-day-old Arabidopsis leaves. The sample was diluted 4:10. (**d**) OGs obtained with the leaf strip method from 25-day-old Arabidopsis leaves. The sample was diluted 2:10

3.2 Isolation and Quantification of Endogenous OGs from Cell Wall Material

Endogenous OGs isolated from cell wall material were prepared according to previously described protocols [14, 15], with some modifications.

Carry out all procedures at room temperature unless otherwise specified.

1. Harvest Arabidopsis rosette leaves (about 50 mg) using a razor blade.
2. Immediately freeze leaves in liquid nitrogen and homogenize for 2 min at 30 Hz in a mixer mill MM301 (Retsch), using 6-mm steel beads.
3. Add 1 mL 70 % ethanol to each sample and mix well.
4. Centrifuge at 12,000 × *g* for 10 min.
5. Repeat **steps 3** and **4**.
6. Solubilize the pellet in 1 mL of 1:1 chloroform:methanol (vol/vol).
7. Centrifuge at 14,000 × *g* for 10 min.
8. Repeat **steps 6** and 7.
9. Solubilize the pellet in 1 mL of acetone, mixed and vortexed.
10. Centrifuge at 14,000 × *g* for 10 min.

11. Repeat **steps 9** and **10**.
12. Dry the pellets containing the alcohol insoluble solids (AIS) overnight at room temperature under a chemical hood.
13. Solubilize pellets in 100 μL of ChA solution.
14. Incubate for 4 h at 70 °C under gentle shaking (*see* **Note 5**).
15. Centrifuge the sample at 18,000 × *g* for 20 min.
16. Collect the clarified supernatants; about 80 μL for each sample should be recovered.
17. The samples can be directly analyzed by HPAEC-PAD (Fig. 1c) or can be stored at −20 °C for subsequent analyses, i.e., using ESI-MS (see below).

3.3 Isolation of OGs with the Leaf Strip Method

Carry out all procedures at room temperature, unless otherwise specified.

1. Harvest leaves and cut a large piece for each leaf (about 90–100 mg) with a razor blade, removing the tip and the petiole.
2. Put each leaf piece in a well of a 6-well tissue culture plate.
3. Add 3 mL of sterilizing solution to each well and incubate at room temperature for 3 min.
4. Discard the sterilizing solution and wash leaf pieces at least four times with ultrapure water. Leave the leaf pieces in the water of the last washing step.
5. Cut each leaf piece in 2-mm-wide strips with a razor blade and transfer all the strips obtained from a single leaf to a well of a new sterile tissue culture plate.
6. Add 2 mL of OG extraction buffer (*see* **Note 6**) to each well. Incubate for 16 h at 30 °C under gentle shaking.
7. Collect the medium from each well and discard the leaf strips. Keep the samples from each well separated for the duration of the entire procedure. At this point, the medium can be directly processed (see **step 8**) or stored at −20 °C for subsequent analyses.
8. Transfer the samples to an Oak Ridge polypropylene centrifuge tube and add absolute ethanol to a final concentration of 80% (v/v).
9. Centrifuge at 15,000 × *g* for 30 min. At the end of the centrifugation a white pellet should be visible. Before proceeding to **step 10**, air dry the pellet (*see* **Note 7**).
10. Solubilize the pellet in 0.2 mL ultrapure water (*see* **Note 8**), and incubate at 65 °C for 20 min to eliminate residual enzymatic activities.
11. Centrifuge at 3000 × *g* for 3 min. The oligosaccharides will remain in solution.

12. Collect the clarified supernatant; you should be able to collect about 150 μL for each sample.
13. The sample can be directly analyzed by HPAEC-PAD (Fig. 1d) and by MS (*see* **Note 9**) or can be stored at −20 °C for subsequent analyses (*see* **Note 10**).

3.4 Analysis of OGs by HPAEC-PAD

1. Prepare a 5 mg/mL OG standard solution in ultrapure water as indicated above.
2. Prepare a 0.25 mg/mL dilution in ultrapure water of the OG standard.
3. Prepare a series of dilutions of each sample (obtained either from the cell wall or using the "leaf-strip method") in ultrapure water.
4. Filter standard OGs and samples with non-sterile Spin-X® centrifuge tubes with cellulose acetate filter membranes (0.45 μm cutoff) (Corning Costar).
5. Equilibrate the column for 10 min with a 90% Eluent A and 10% Eluent B before injecting the sample.
6. Inject 10 μL of sample (about 2.5 μg OGs, in the case of the standard).
7. Separate oligosaccharides with a flow-rate of 0.3 mL/min applying a linear gradient from 10 to 80% of eluent B for 30 min.
8. At the end of the run, wash the column with 100% eluent B for 5 min.
9. OGs are detected by a pulsed amperometric detector (PAD) using a gold electrode with waveform A, according to the manufacturer's instructions (Fig. 1).

3.5 Characterization of OGs by Mass Spectrometry

OG samples can be analyzed by MALDI-TOF-MS or by ESI-MS/MS, e.g., using an ESI-LTQ-Orbitrap system. Purification of OGs from salts and contaminants is necessary prior analysis, in particular when an ESI source is used.

1. OG samples, either from the cell wall or obtained with the leaf strip method, are precipitated overnight with 80% (v/v) ethanol at −20 °C.
2. Centrifuge at 18,000 × *g* for 10 min.
3. Wash pellets with 80% ethanol.
4. Centrifuge at 18,000 × *g* for 10 min.
5. Repeat **steps 3** and **4**.
6. Dry 10 min under a flow hood.
7. Dissolve samples in water and pretreat for 10 min with BioRex MSZ 501 beads.

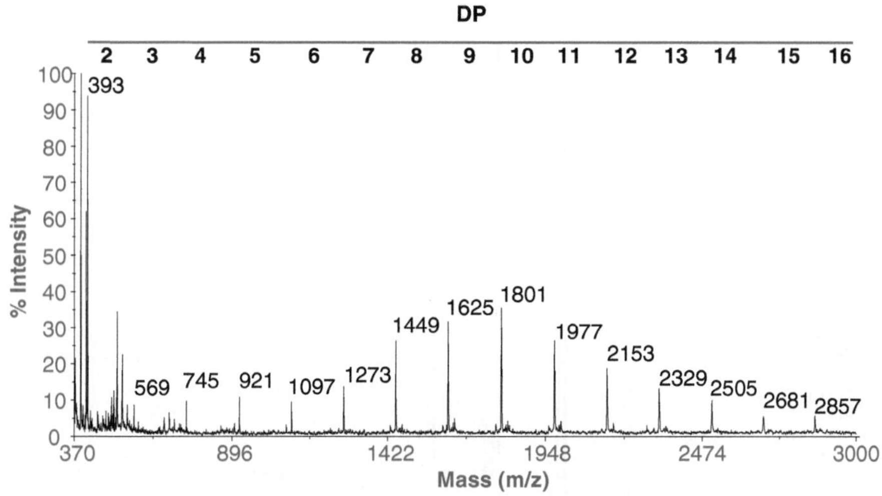

Fig. 2 MALDI-TOF MS analysis of OGs. Positive-ion mass spectrum of sodium-cationized standard OGs. Numbers on the *top* of the figure indicate the DP of the different oligomers

8. For MALDI-TOF-MS analysis, mix 1 μL of sample with 1 μL of DHB solution, spot onto a MALDI plate (Bioscience), and quickly crystallize under vacuum.
9. Acquire MS spectra using a Voyager DE-STR MALDI-TOF instrument (Applied Biosystems, Beverly, MA, USA) or similar, in positive ion, reflector mode, with an acceleration voltage of 20,000 V and a delay time of 350 ns (Fig. 2).
10. For ESI-MS/MS analysis, OG samples are mixed with methanol 80:20 (v/v) and directly infused at 5 μL/min in an ESI LTQ-Orbitrap mass spectrometer in positive ion mode using a capillary voltage of 4.2 kV, capillary temperature of 240 °C, sheath gas flow rate of 20 (arbitrary units), and an auxiliary gas flow rate of 10 (arbitrary units). MS is acquired at 30,000 FWHM (full width at half maximum) resolution in the FTMS (Fourier Transform Mass Spectrometer) and MS/MS is carried out in the linear ion trap, performing CID (collision-induced dissociation) at a collision energy of 35 keV to optimize the signal and obtain the maximum structural information for the ion of interest (Fig. 3).

4 Notes

1. As an alternative, the analytical anion exchange column PA1 or PA100 (Thermo Fischer) can be used with the same buffers, adjusting the gradient length.

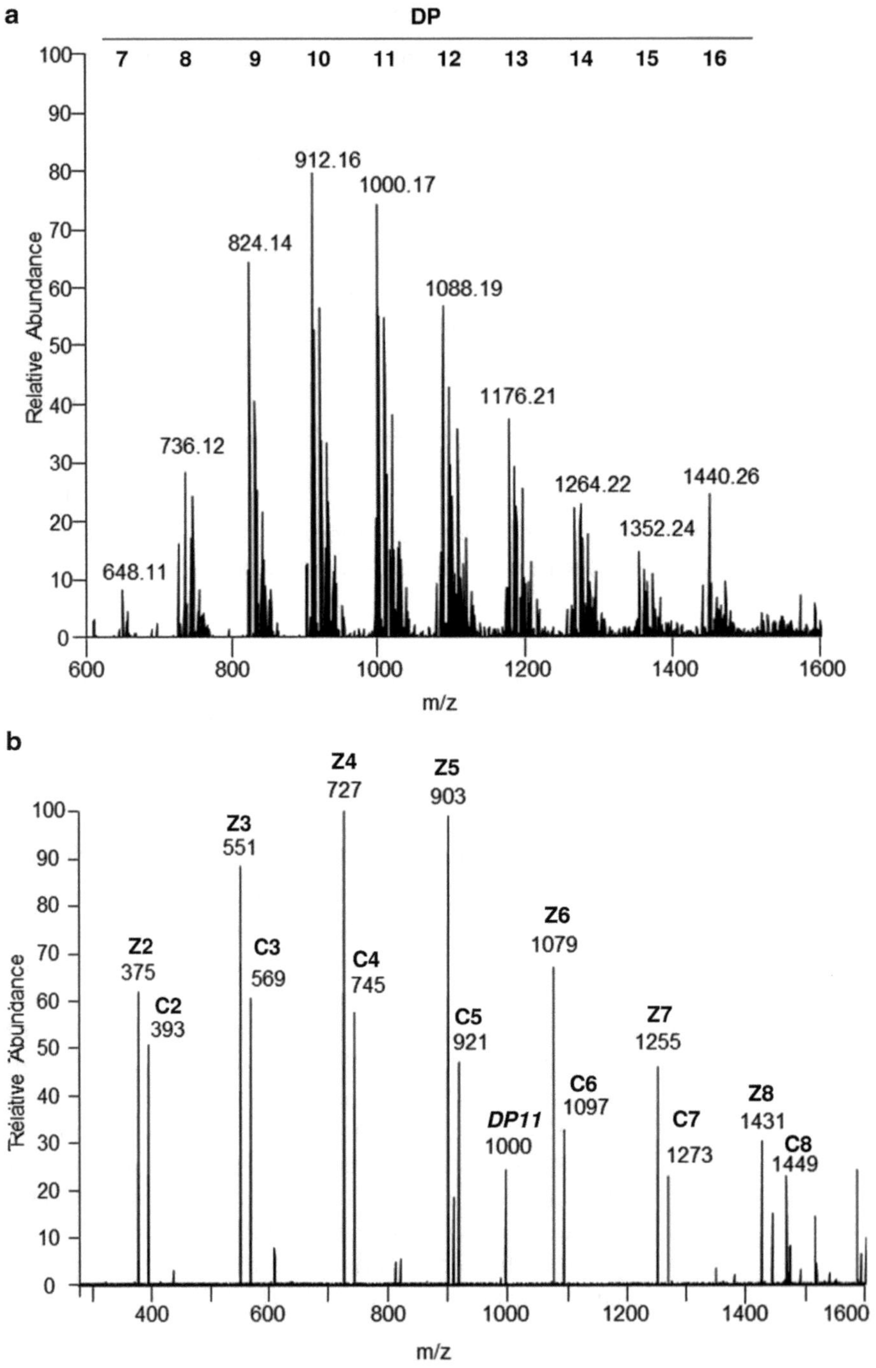

Fig. 3 Positive ion MS and MS/MS spectra of OGs by ESI-MS. (**a**) Full-scan mass spectrum of the $[M+2Na]^{2+}$ of standard OGs (0.8 μg/μL) in 20 % methanol. Numbers on the top of the figure indicate the DP of the different oligomers. (**b**) MS/MS of the sodiated ion at m/z 1000, corresponding to DP 11. Fragments $[M+Na]^{+}$ are identified according to the nomenclature of Domon and Costello [45]

2. Eluents must be filtered to eliminate biological contamination and particulate matter, and must be vacuum degassed to minimize carbonate contamination, since it may cause a loss of carbohydrate retention and consequently reduce PAD resolution. After the addition of hydroxide, eluents must be blanketed with helium.
3. Digestion time varies typically between 150 and 210 min. To determine the optimal digestion time, it is recommended to perform a time course experiment. The reaction should be stopped when the summed areas of the peaks corresponding to OGs with DP 10–12 reach a maximum and the amount of undigested material is minimized. Figure 1a shows as representative HPAEC-PAD profile of the OG preparation at this step.
4. Precipitation with 17% EtOH is aimed at maximizing the oligomers with DP around 11.
5. Incubation at 70 °C was introduced to increase the recovery of endogenous OGs, since this step inactivates enzymes that could degrade them during the extraction procedure.
6. The OG-Extraction buffer contains calcium chelating agents (CDTA and oxalate) that favor the release of pectin fragments in the medium, and sulfite as a preservative agent.
7. The pellet can be dried under the flux of a laboratory fume hood or with a vacuum dryer (Savant or similar). If a vacuum concentrator is used, do not dry the samples for more than 20 min.
8. This step can be laborious and time-consuming as the pellet may be recalcitrant to the solubilization. Even if complete solubilization is not achieved, proceed to **step 11**.
9. For MS analysis, incubate the sample with BioRex MSZ 501 beads for 10 min. Then proceed as described in **step 8** or **step 10** of Subheading 3.5 for MALDI-TOF-MS or ESI-MS/MS analysis, respectively.
10. The sample volume to be used for the HPAEC-PAD and MS analyses should be empirically determined.

Acknowledgment

This work was supported by ERA-Net (grant ERA-CAPS 2014 "SIPIS—DECODING LIGAND-RECEPTOR specificities of LysM-Proteins IN PLANT IMMUNITY AND SYMBIOSIS) and by "Sapienza Università di Roma (grant "Ricerche UNIVERSITARIE" 2015).

References

1. Hahn MG, Darvill AG, Albersheim P (1981) Host-pathogen interactions. XIX. The endogenous elicitor, a fragment of a plant cell wall polysaccharide that elicits phytoalexin accumulation in soybeans. Plant Physiol 68: 1161–1169
2. Cervone F, De Lorenzo G, Salvi G, Bergmann C, Hahn MG, Ito Y, Darvill A, Albersheim P (1989) Release of phytoalexin elicitor-active oligogalacturonides by microbial pectic enzymes. In: Lugtenberg BJJ (ed) Signal molecules in plants and plant-microbe interactions, NATO ASI Series, vol H36. Springer Verlag, Heidelberg, FRG, pp 85–89
3. Davis KR, Darvill AG, Albersheim P, Dell A (1986) Host-pathogen interactions. XXIX. Oligogalacturonides released from sodium polypectate by endopolygalacturonic acid lyase are elicitors of phytoalexins in soybean. Plant Physiol 80:568–577
4. Broekaert WF, Pneumas WJ (1988) Pectic polysaccharides elicit chitinase accumulation in tobacco. Physiol Plant 74:740–744
5. Davis KR, Hahlbrock K (1987) Induction of defense responses in cultured parsley cells by plant cell wall fragments. Plant Physiol 85:1286–1290
6. Bellincampi D, Dipierro N, Salvi G, Cervone F, De Lorenzo G (2000) Extracellular H_2O_2 induced by oligogalacturonides is not involved in the inhibition of the auxin-regulated *rolB* gene expression in tobacco leaf explants. Plant Physiol 122:1379–1385
7. Galletti R, Denoux C, Gambetta S, Dewdney J, Ausubel FM, De Lorenzo G, Ferrari S (2008) The AtrbohD-mediated oxidative burst elicited by oligogalacturonides in Arabidopsis is dispensable for the activation of defense responses effective against *Botrytis cinerea*. Plant Physiol 148:1695–1706
8. Rasul S, Dubreuil-Maurizi C, Lamotte O, Koen E, Poinssot B, Alcaraz G, Wendehenne D, Jeandroz S (2012) Nitric oxide production mediates oligogalacturonide-triggered immunity and resistance to *Botrytis cinerea* in *Arabidopsis thaliana*. Plant Cell Environ 35:1483–1499
9. Cervone F, Hahn MG, De Lorenzo G, Darvill A, Albersheim P (1989) Host-pathogen interactions. XXXIII. A plant protein converts a fungal pathogenesis factor into an elicitor of plant defense responses. Plant Physiol 90:542–548
10. Ferrari S, Savatin DV, Sicilia F, Gramegna G, Cervone F, De Lorenzo G (2013) Oligogalacturonides: plant damage-associated molecular patterns and regulators of growth and development. Front Plant Sci 4:49. doi:10.3389/fpls.2013.00049
11. Spiro MD, Kates KA, Koller AL, O'Neill MA, Albersheim P, Darvill AG (1993) Purification and characterization of biologically active 1,4-linked α-D-oligogalacturonides after partial digestion of polygalacturonic acid with endopolygalacturonase. Carbohydr Res 247:9–20
12. Hahn MG, Cheong J-J, Alba R, Enkerli J, Cote F (1993) Oligosaccharide elicitors: structures and recognition. In: Mechanisms of plant defense responses. Kluwer Academic Publishers (Dordrecht. The Netherlands), pp 99–116
13. Hotchkiss AT Jr, Hicks KB (1990) Analysis of oligogalacturonic acids with 50 or fewer residues by high-performance anion-exchange chromatography and pulsed amperometric detection. Anal Biochem 184(2):200–206
14. Pontiggia D, Ciarcianelli J, Salvi G, Cervone F, De Lorenzo G, Mattei B (2015) Sensitive detection and measurement of oligogalacturonides in Arabidopsis. Front Plant Sci 6:258
15. Benedetti M, Pontiggia D, Raggi S, Cheng Z, Scaloni F, Ferrari S, Ausubel FM, Cervone F, De Lorenzo G (2015) Plant immunity triggered by engineered in vivo release of oligogalacturonides, damage-associated molecular patterns. Proc Natl Acad Sci USA 112: 5533–5538
16. Korner R, Limberg G, Mikkelsen JD, Roepstorff P (1998) Characterization of enzymatic pectin digests by matrix-assisted laser desorption/ionization mass spectrometry. J Mass Spectrom 33:836–842
17. Remoroza C, Cord-Landwehr S, Leijdekkers AGM, Moerschbacher BM, Schols HA, Gruppen H (2012) Combined HILIC-ELSD/ESI-MS n enables the separation, identification and quantification of sugar beet pectin derived oligomers. Carbohydr Polym 90:41–48
18. Westphal Y, Schols HA, Voragen AGJ, Gruppen H (2010) Introducing porous graphitized carbon liquid chromatography with evaporative light scattering and mass spectrometry detection into cell wall oligosaccharide analysis. J Chromatogr A 1217:689–695
19. Kester HCM, Visser J (1990) Purification and characterization of polygalacturonases produced by the hyphal fungus *Aspergillus Niger*. Biotechnol Appl Biochem 12:150–160

20. Mathieu Y, Kurkdijan A, Xia H, Guern J, Koller A, Spiro M, O'Neill M, Albersheim P, Darvill A (1991) Membrane responses induced by oligogalacturonides in suspension-cultured tobacco cells. Plant J 1:333–343
21. Mathieu Y, Guern J, Spiro MD, O'Neill MA, Kates K, Darvill AG, Albersheim P (1998) The transient nature of the oligogalacturonide-induced ion fluxes of tobacco cells is not correlated with fragmentation of the oligogalacturonides. Plant J 16:305–311
22. Thain JF, Doherty HM, Bowles DJ, Wildon DC (1990) Oligosaccharides that induce proteinase inhibitor activity in tomato plants cause depolarization of tomato leaf cells. Plant Cell Environ 13:569–574
23. Messiaen J, Read ND, Van Cutsem P, Trewavas AJ (1993) Cell wall oligogalacturonides increase cytosolic free calcium in carrot protoplasts. J Cell Sci 104:365–371
24. Moscatiello R, Mariani P, Sanders D, Maathuis FJM (2006) Transcriptional analysis of calcium-dependent and calcium-independent signalling pathways induced by oligogalacturonides. J Exp Bot 57:2847–2865
25. Savatin DV, Gigli BN, Marti L, Fabbri C, Cervone F, De Lorenzo G (2014) The Arabidopsis NPK1-related protein kinases ANPs are required for elicitor-induced oxidative burst and immunity. Plant Physiol 165:1188–1202
26. Gigli BN, Gravino M, Savatin DV (2015) Luminol-based assay for detection of immunity elicitor-induced hydrogen peroxide production in *Arabidopsis thaliana* leaves. Bio-protocol 5(24):e1685http://www.bio-protocol.org/e1685
27. Legendre L, Rueter S, Heinstein PF, Low PS (1993) Characterization of the oligogalacturonide-induced oxidative burst in cultured soybean (*Glycine max*) cells. Plant Physiol 102:233–240
28. Binet MN, Bourque S, Lebrun-Garcia A, Chiltz A, Pugin A (1998) Comparison of the effects of cryptogein and oligogalacturonides on tobacco cells and evidence of different form of desensitization induced by these elicitors. Plant Sci 137:33–41
29. Stennis MJ, Chandra S, Ryan CA, Low PS (1998) Systemin potentiates the oxidative burst in cultured tomato cells. Plant Physiol 117:1031–1036
30. Galletti R, Ferrari S, De Lorenzo G (2011) Arabidopsis MPK3 and MPK6 play different roles in basal and oligogalacturonide- or flagellin-induced resistance against *Botrytis cinerea*. Plant Physiol 157:804–814
31. Gravino M, Savatin DV, Macone A, De Lorenzo G (2015) Ethylene production in Botrytis cinerea- and oligogalacturonide-induced immunity requires calcium-dependent protein kinases. Plant J 84:1073–1086
32. Farmer EE, Moloshok TD, Saxton MJ, Ryan CA (1991) Oligosaccharide signaling in plants: specificity of oligouronide-enhanced plasma membrane protein phosphorylation. J Biol Chem 266:3140–3145
33. Mattei B, Spinelli F, Pontiggia D, De Lorenzo G (2016) Comprehensive Analysis of the Membrane Phosphoproteome Regulated by Oligogalacturonides in Arabidopsis thaliana. Front Plant Sci 7:1107. doi: 10.3389/fpls.2016.01107
34. Baldwin EA, Pressey R (1988) Tomato polygalacturonase elicits ethylene production in tomato fruit. J Am Soc Hort Sci 113:92–95
35. Baldwin EA, Biggs RH (1988) Cell-wall lysing enzymes and products of cell-wall digestion elicit ethylene in citrus. Physiol Plant 73:58–64
36. Denoux C, Galletti R, Mammarella N, Gopalan S, Werck D, De Lorenzo G, Ferrari S, Ausubel FM, Dewdney J (2008) Activation of defense response pathways by OGs and Flg22 elicitors in Arabidopsis seedlings. Mol Plant 1:423–445
37. Savatin DV, Ferrari S, Sicilia F, De Lorenzo G (2011) Oligogalacturonide-auxin antagonism does not require posttranscriptional gene silencing or stabilization of auxin response repressors in Arabidopsis. Plant Physiol 157:1163–1174
38. Savatin DV, Gigli BN, Gravino M, Fabbri C, Pontiggia D, Mattei B (2015) Camalexin quantification in *Arabidopsis thaliana* leaves infected with *Botrytis cinerea*. Bio-protocol 5(2):e1379 http://www.bio-protocol.org/e1379
39. Aziz A, Heyraud A, Lambert B (2004) Oligogalacturonide signal transduction, induction of defense-related responses and protection of grapevine against *Botrytis cinerea*. Planta 218:767–774
40. Ferrari S, Galletti R, Denoux C, De Lorenzo G, Ausubel FM, Dewdney J (2007) Resistance to *Botrytis cinerea* induced in Arabidopsis by elicitors is independent of salicylic acid, ethylene, or jasmonate signaling but requires PHYTOALEXIN DEFICIENT3. Plant Physiol 144:367–379
41. Marfà V, Gollin DJ, Eberhard S, Mohnen D, Darvill A, Albersheim P (1991) Oligogalacturonides are able to induce flowers to form on tobacco explants. Plant J 1:217–225
42. Branca C, De Lorenzo G, Cervone F (1988) Competitive inhibition of the auxin-induced

elongation by α-D-oligogalacturonides in pea stem segments. Physiol Plant 72:499–504

43. Lee S, Choi H, Suh S, Doo IS, Oh KY, Choi EJ, Schroeder Taylor AT, Low PS, Lee Y (1999) Oligogalacturonic acid and chitosan reduce stomatal aperture by inducing the evolution of reactive oxygen species from guard cells of tomato and Commelina communis. Plant Physiol 121:147–152

44. Brecht JK, Huber DJ (1988) Products released from enzymically active cell wall stimulate ethylene production and ripening in preclimacteric tomato (*Lycopersicon esculentum* Mill.) fruit. Plant Physiol 88:1037–1041

45. Domon B, Costello CE (1988) A systematic nomenclature for carbohydrate fragmentations in FAB-MS/MS spectra of glycoconjugates. Glycoconj J 5:397–409

Chapter 4

Quantitative Analysis of Ligand-Induced Endocytosis of FLAGELLIN-SENSING 2 Using Automated Image Segmentation

Michelle E. Leslie and Antje Heese

Abstract

Plants are equipped with a suite of plant pattern recognition receptors (PRRs) that must be properly trafficked to and from the plasma membrane (PM), which serves as the host-pathogen interface, for robust detection of invading pathogenic microbes. Recognition of bacterial flagellin, or the derived peptide flg22, is facilitated by the PM-localized PRR, FLAGELLIN SENSING 2 (FLS2). Upon flg22 binding, FLS2 is rapidly internalized from the PM into endosomal compartments and subsequently degraded. To understand better the integration of FLS2 endocytosis and signaling outputs, we developed methods for the quantitative analysis of FLS2 trafficking using freely available bioimage informatic tools. Emphasis was placed on robust recognition of features and ease of access for users. Using the free and open-source software Fiji (Fiji is just ImageJ) and Trainable Weka Segmentation (TWS) plug-in, we developed a workflow for the automated identification of green fluorescent protein (GFP)-tagged FLS2 in endosomal puncta. Fiji-TWS methods can be adapted with ease for the analysis of FLS2 trafficking in various genetic backgrounds as well as for the endocytic regulation of diverse plant PRRs.

Key words FLAGELLIN SENSING 2 (FLS2), Endocytosis, Pattern recognition receptor (PRR), Spinning disc confocal microscopy, Bioimage informatics, ImageJ, Trainable Weka Segmentation

1 Introduction

FLAGELLIN-SENSING 2 (FLS2) is a plasma membrane (PM)-localized pattern-recognition receptor (PRR) present in both mono- and dicotyledonous plants [1, 2]. FLS2 recognizes bacterial flagellin (or its derived peptide flg22), thereby triggering a network of immune responses contributing to restriction of bacterial growth [1, 3]. For our studies, we utilize the *Arabidopsis thaliana* FLS2-flg22 model system [4, 5], for which microscopically observed events in response to flg22 include dynamic changes in FLS2 localization and abundance at the PM [6]. In confocal images of untreated plants, green fluorescence protein (GFP)-tagged FLS2 is primarily localized to the PM poised for flg22-recognition [7–9],

Libo Shan and Ping He (eds.), *Plant Pattern Recognition Receptors: Methods and Protocols*, Methods in Molecular Biology, vol. 1578, DOI 10.1007/978-1-4939-6859-6_4, © Springer Science+Business Media LLC 2017

with a low level of constitutive endocytosis of inactivated receptors [8]. However, after 30+ min of flg22-treatment, FLS2-GFP is rapidly removed from the PM, visualized as (1) increasing abundance of FLS2-GFP containing intracellular puncta that co-localize with markers of early and late endosomal compartments en route for degradation, likely in the lytic vacuole, and (2) decreasing abundance of FLS2-GFP at the PM [7–11] (Fig. 1a, b). While removal of flg22-activated FLS2 from the PM and subsequent degradation serves to desensitize cells to the flg22-stimulus [12] and correlates with the down-regulation of a subset of early signaling responses [9], it remains to be determined whether FLS2 endocytosis may also contribute to the activation of a subset of flg22-responses.

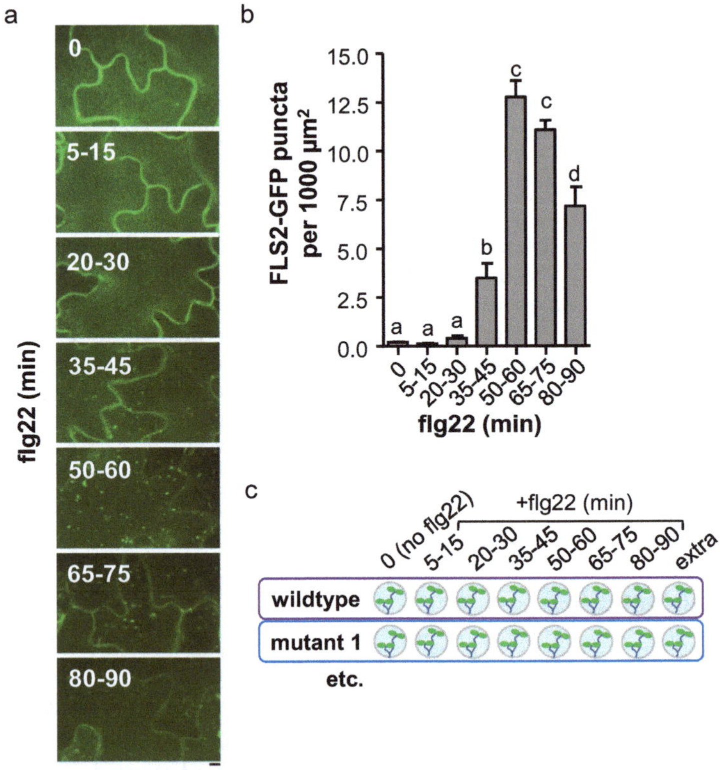

Fig. 1 Experimental setup for quantitative imaging of FLS2-GFP in endosomal puncta. Wild-type (Col-0 ecotype) seedlings expressing FLS2-GFP are treated with 1 μM flg22 for the indicated amount of time (min). (**a**) Representative maximum-intensity projection (MIP) images captured using spinning disc confocal microscopy. Scale bar, 10 μm. (**b**) Quantification of number of FLS2-GFP containing puncta per pavement cell area. This experiment was repeated three independent times with similar results. N = 14–25 images per timepoint. Values are mean ± SE. *Different letters* indicate significant differences ($P < 0.05$) while the *same letter* indicates no significant differences between samples ($P > 0.05$) based on Two-tailed Student's *t*-test. (**c**) Plate setup for treatment of FLS2-GFP seedlings with flg22 for quantitative live-cell imaging

In this chapter, we provide methods for quantitative live-cell imaging of FLS2-GFP endocytosis in intact *Arabidopsis* cotyledons using spinning disc confocal microscopy (SDCM) and the free and open-source image analysis software Fiji (Fiji is just ImageJ) [13]. SDCM provides high spatial and temporal resolution, allowing for the rapid capture of Z-series of images through the epidermal cell surface over a timecourse of flg22 treatment [8, 9, 11, 14]. Using Fiji Macros for the automated preprocessing of entire image datasets, Z-series are converted into single-plane maximum-intensity projections (MIPs) and subsequently analyzed for FLS2-GFP localization. The dynamic localization of FLS2-GFP to the PM and internal endosomal compartments has been effectively measured using both commercially and freely available bio-image analysis software packages [8–11, 14].

To aid researchers in realizing the benefits of enhanced image acquisition, we provide here detailed methods for utilizing the freely available Fiji Trainable Weka Segmentation (TWS) plug-in (http://fiji.sc/Trainable_Weka_Segmentation) for the detection of FLS2-GFP endosomal puncta. The Fiji-TWS plug-in provides a straightforward interface for application of machine learning algorithms (Waikato Environment for Knowledge Analysis; Weka) [15] for the automated identification of image features in large datasets. Supervised machine learning algorithms, as applied with Fiji-TWS, proceed in two phases: (1) the model Training Phase, in which the user identifies features of interest (i.e., FLS2-GFP endosomal puncta) in representative example images for the generation of a set of classification rules, termed the *Classifier*, and (2) the Classification Phase, in which the *Classifier* is automatically applied to a dataset for the segmentation of images into distinct feature classes (i.e., FLS2-GFP endosomal puncta versus background) [16]. Following image segmentation, FLS2-GFP endosomal puncta are automatically detected and quantified using the Fiji *Analyze Particles* function.

The Fiji-TWS workflow described herein has been previously applied to identify novel components functioning in ligand-induced endocytosis of FLS2 by comparing stimulated endocytosis of FLS2-GFP between wild-type and mutant backgrounds. In particular, *Arabidopsis* plants lacking the vesicular trafficking protein DYNAMIN-RELATED PROTEIN 2B (DRP2B), but not the closely related DRP2A, display a significant decrease in flg22-induced internalization of FLS2 in comparison to wild-type plants, which in turn correlates with enhanced Ca^{2+}-dependent signaling [9]. Using Fiji-TWS methods, additional genetic and/or environmental factors that contribute to the regulation of FLS2 trafficking may be identified. Furthermore, Fiji-TWS can be easily adapted for the quantitative analysis of diverse PRRs under the regulation of previously uncharacterized endocytic events. Beyond PRR trafficking, this method can be applied to measuring other flg22-responses observed at macro- and microscopic levels such as cell-wall modifications.

2 Materials

1. Six-day-old *pFLS2::FLS2-3xMyc-EGFP*(*FLS2-GFP*) *Arabidopsis thaliana* seedlings (*see* **Notes 1** and **2**) [8, 9].
2. Sterile, deionized water (dH_2O) at room temperature (*see* **Note 3**).
3. Sterile, 48-well tissue culture dish with transparent lid.
4. Forceps.
5. flg22 stock solution (1 mM): Dissolve 1–2 mg of lyophilized flg22 peptide (QRLSTGSRINSAKDDAAGLQIA; purity ≥90%) in an appropriate volume of DMSO for a final flg22 stock concentration of 1 mM (adjusted for purity). Store at −20 °C in 10–20 μL aliquots. For seedling treatments, the 1 mM flg22 stock solution is serially diluted in dH_2O to a final working concentration of 1 μM flg22.
6. Sterile microcentrifuge tubes (1.7 mL).
7. Sterile 15 mL conical tubes.
8. Microscope slides, precleaned.
9. Glass coverslips, 24 × 50 mm.
10. Timer.
11. Spinning disc confocal microscope (SDCM) equipped for rapid, live-cell imaging (*see* **Note 4**).
12. Computer for image analysis.
13. Fiji (Fiji is just ImageJ v1.50a or higher; NIH, Bethesda, MD) software with Trainable Weka Segmentation (TWS) plug-in installed (v2.20 or higher) (*see* **Note 5**).
14. Graphing and statistical analysis software (*see* **Note 6**).

3 Methods

3.1 Live-Cell Imaging

Spinning disc confocal microscopy (SDCM) is used to rapidly capture Z-series of images of cotyledons expressing FLS2-GFP over a timecourse of flg22 treatment (e.g., 0–90 min flg22) for quantification of FLS2-GFP endocytosis.

1. On the afternoon before imaging, prepare a 48-well sterile, tissue culture plate with 600 uL of sterile dH_2O per well (Fig. 1c). Using clean forceps, gently place groups of two to four healthy 6-day-old FLS2-GFP seedlings into wells, submerging the seedlings in the dH_2O (*see* **Note 7**). For quantitative imaging, prepare six or more seedlings per flg22 elicitation time-point for each genetic background (wild-type, mutant 1, etc.) (*see* **Note 8**). Keep covered plate in growth chamber

at 22 °C with either continuous light or a long-day (16-h light/8-h dark) cycle overnight.

2. Just prior to imaging, prepare 10 μM flg22 solution by diluting 1 mM flg22 in sterile dH_2O in a microcentrifuge tube (e.g., 10 μL 1 mM flg22 + 990 μL dH_2O). Briefly vortex to mix. Keep 10 μM flg22 solution on ice.
3. Determine optimal SDCM imaging parameters for the detection of FLS2-GFP with minimal photo-bleaching (*see* **Note 9**).
4. Begin the experiment by imaging untreated (0 min) seedlings for both wild-type and mutant genetic backgrounds (Fig. 1c; first column of 48-well plate).
5. Pipette approximately 250 μL of dH_2O onto the center of a microscope slide.
6. Use forceps to gently remove all seedlings from a single well with a scooping motion and place into dH_2O on the microscope slide. Arrange the intact seedlings on the slide such that cotyledons lie flat, are nonoverlapping, and the adaxial (top) surfaces are orientated upward for imaging (*see* **Note 10**). Gently apply coverslip.
7. Begin capturing Z-stacks of images through the adaxial epidermal cell layer of each cotyledon. Each Z-series starts at the outer surface of the pavement cells and ends at the point where the PMs of adjacent pavement cells are juxtaposed against one another (Fig. 2a). For 7-day-old seedlings, a Z-series spanning 20 μm is usually sufficient (*see* **Note 11**). Capture 2–4 fields of view per cotyledon. For rapid imaging, auto-save images during capture, preferably as .tif format.
8. After all wells of untreated (0 min) seedlings have been imaged, prepare the final working solution of 1 μM flg22 by diluting 10 μM flg22 solution (kept on ice) in room temperature dH_2O in a 15 mL conical tube (e.g., 400 μL of 10 μM flg22 solution + 3.6 mL of dH_2O). Vortex or invert to mix. Prepare fresh 1 μM flg22 solution immediately prior to each treatment.
9. For a timecourse of flg22 treatments, remove dH_2O from all six "+flg22" wells for the first row of wild-type seedlings in the 48-well plate (Fig. 1c). Immediately pipette 400 μL of 1 μM flg22 solution into each of these wells. Swirl seedlings briefly and gently with a pipette tip to ensure that all cotyledons are exposed to flg22 solution. Start a timer counting up from 0 min.
10. Within the first two minutes after the start of the flg22-treatment in **step 9**, pipette approximately 250 μL of 1 μM flg22 solution onto the center of a microscope slide. Using forceps, gently remove seedlings from the first flg22-treated well (Fig. 1c; wild-type, 5–15 min) and arrange cotyledons on microscope slide as for untreated seedlings in **step 6**.

11. After 5 min of flg22 treatment, begin capturing Z-series of images through the epidermal cell layer as done for untreated cotyledons in **step** 7. Continue capturing Z-series of images, 2–4 per cotyledon, over a 10-min period (i.e., 5–15 min of continuous flg22 treatment).
12. After 15 min of flg22 treatment, discard current slide and prepare a new slide with the next set of flg22-treated seedlings in the timed series (Fig. 1c; wild-type, 20–30 min). Again, image these cotyledons over a 10-min period (i.e., 20–30 min of continuous flg22 treatment).
13. Continue preparing and imaging slides over the entire flg22 treatment timecourse (i.e., 35–45, 50–60, 65–75, and 80–90 min) (*see* **Note 12**).
14. Treat additional sets of seedlings (Fig. 1c; mutant 1, etc.) with freshly prepared 1 μM flg22 and image over the timecourse until a sufficient number of images are captured (*see* **Note 13**).

3.2 Image Preprocessing

All image Z-series are converted to single-plane maximum-intensity projections (MIPs) (Fig. 2a, b) and uniformly adjusted for brightness and contrast in Fiji. These preprocessing steps are recorded as a script (Fiji Macro) (Fig. 2c). The saved Fiji Macro is automatically applied to all of the images in the dataset using Batch Processing.

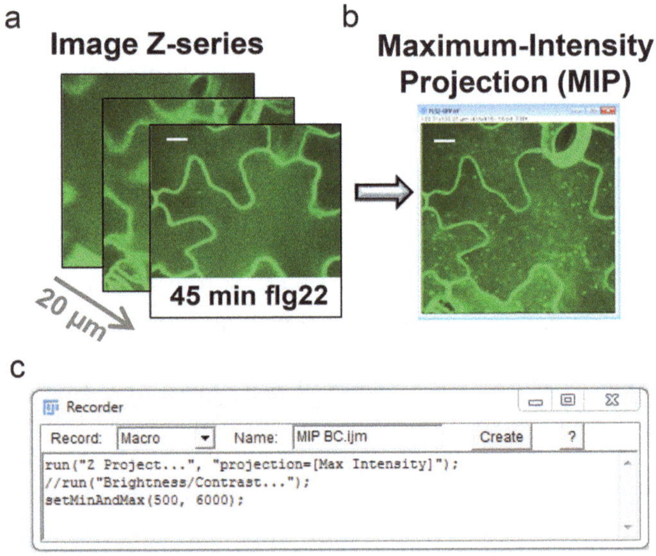

Fig. 2 Automated preprocessing of image Z-series using Fiji Batch Processing. (**a**) An example 20-μm-depth Z-series image captured by spinning disc confocal microscopy. (**b**) Brightness/contrast adjusted maximum-intensity projection (MIP). Scale bars, 10 μm. (**c**) An example recorded Fiji Macro script for automated generation of MIPs followed by brightness/contrast adjustment

1. Open a single image Z-series (.tif format) in Fiji and set image scale (*Analyze* ►*Set Scale…*) (*see* **Note 14**) using known calibration for the microscope objective that was used for imaging (*see* **Note 15**).
2. Turn on the Fiji Macro Recorder (*Plugins* ► *Macros* ► *Record…*), which will automatically write a script for each Fiji action.
   (a) First, convert the Z-series (Fig. 2a) to a single-plane MIP (*Image* ► *Stacks* ► *Z Project… Projection Type: Max Intensity*) (Fig. 2b).
   (b) Second, open the brightness and contrast adjustment window (*Image* ► *Adjust* ► *Brightness/Contrast…*).
   (c) Third, choose minimum and maximum pixel intensities to remove background from the MIP and enhance MIP brightness (*see* **Note 16**).
3. Within the Recorder window, check recorded script and select *Create* to generate the Macro (Fig. 2c). Save the Macro file (.ijm format) (*see* **Note 17**). Close the Recorder window.
4. Automatically apply the Macro saved in **step 3** to all Z-series images for the experiment using Fiji Batch Processing (*Process* ► *Batch* ► *Macro…*).
   (a) Within the Batch Process window, first select the Input folder with the Z-series images (.tif format).
   (b) Next, create and select a new Output folder (*see* **Note 18**) for the brightness/contrast-adjusted MIPs (.tif format) to be created.
   (c) Open the Macro file saved in **step 3**. Select *Process.*
5. Optional: Additional preprocessing steps may be necessary including cropping of images to focus on a particular cell type (pavement vs. guard cells) (*see* **Note 19**).

3.3 Automated Segmentation of Images

Following the preprocessing of all images, the TWS plug-in is utilized for the automated identification of FLS2-GFP-containing endosomal puncta. In the user-guided TWS Training Phase, features of interest (FLS2-GFP puncta) and background features are selected for the generation of a classification model, termed the *Classifier* (Fig. 3a). *Classifier* development is a reiterative process, in which the user visually inspects the training result (Fig. 3b) to determine whether or not it may be necessary to adjust feature selection and repeat training with one or more additional training images. The resulting *Classifier* is automatically applied to the entire data set for the segmentation of images into two classes, FLS2-GFP endosomal puncta versus background.

1. To begin the TWS Training Phase, select and copy a set of two or more brightness/contrast-adjusted MIP images (generated

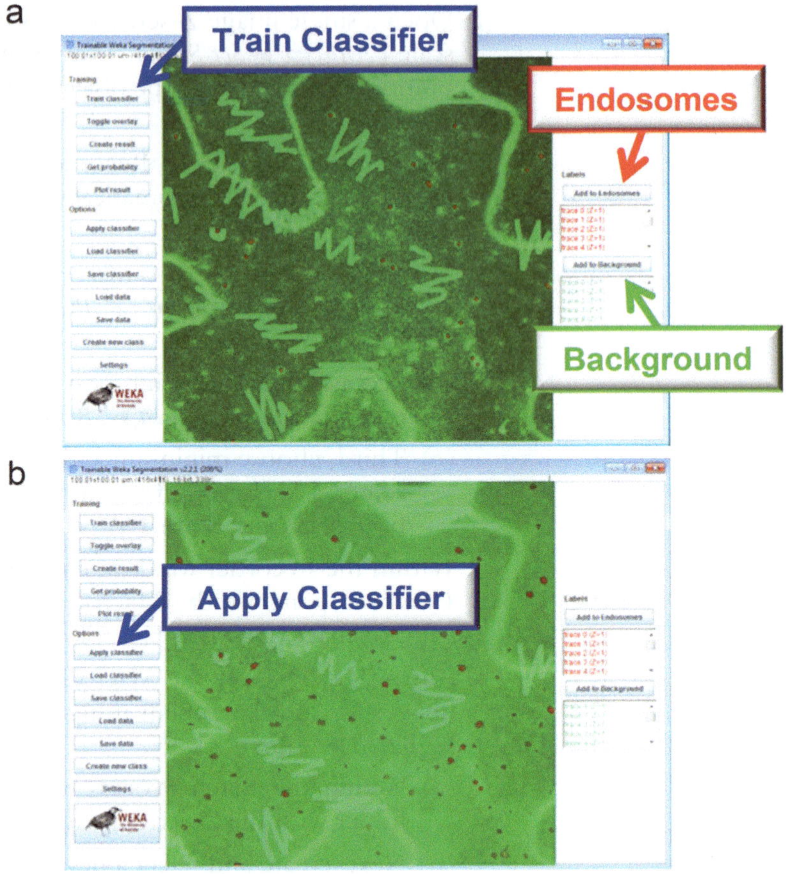

Fig. 3 Development of a Trainable Weka Segmentation *Classifier*. (**a**) Features are manually selected and added to one of two classes, Endosomes (*red*) or Background (*green*) for automated *Classifier* training. (**b**) The classification result is visually inspected, and additional rounds of feature selection and training are used to improve feature detection. The final *Classifier* (.model file) is saved and applied to the entire dataset

in Subheading 3.2, **step 4**) with clearly defined FLS2-GFP endosomal puncta into a folder for TWS training MIPs (.tif format).

2. Open training MIP #1 in Fiji. Load the TWS plug-in (*Plugins Segmentation* ►*Trainable Weka Segmentation*). The MIP will be visible within the TWS plug-in window (Fig. 3a).
3. Within the TWS plug-in Settings, define features in two classification groups: Class 1 = Endosomes (FLS2-GFP in endosomal puncta, i.e., feature of interest) and Class 2 = Background (cytosol, PM-localized FLS2-GFP, chloroplast autofluorescence) (Fig. 3a) (*see* **Note 20**).
4. Use *Freehand Selection*, *Freehand Line*, and *Selection Brush* tools in the Fiji main toolbar to mark features. After marking

an FLS2-GFP puncta or background feature, select *Add to Endosomes* or *Add to Background*, respectively. Assign approximately 10–20 features to each class (Fig. 3a).

5. Select *Train classifier* for automated classification of the image (Fig. 3b). Features identified as belonging to Class 1 (Endosomes) and Class 2 (Background) will be highlighted in red and green, respectively (Fig. 3b).
6. Visually inspect the classified image (Fig. 3b) for proper identification and assignment of features to each class.
7. If necessary and as needed, perform additional rounds of feature selection, classifier training, and visual inspection (**steps 4–6**) until features are correctly classified.
8. Save *Classifier* (.model) and *Data* (.arff) files.
9. To improve feature detection across the image dataset, additional images (training MIP #2, 3, etc.) can be loaded in succession into the TWS plug-in. First, apply the *Classifier* and *Data* (.model and .arff) files from the previous training session using the TWS *Load classifier* and *Load data* functions, followed by the *Train Classifier* function. Follow **steps 4–6** to select features of interest and/or background that were improperly classified and add to Class 1 (Endosomes) or Class 2 (Background), respectively, for further training (*see* **Note 21**).
10. Upon proper classification of the image, save the final *Classifier* (.model) and *Data* (.arff) files.
11. With the final *Classifier* (.model file) loaded into the TWS plug-in, apply the *Classifier* to the entire preprocessed image dataset (*see* **Note 22**).
    (a) Select *Apply Classifier*, and Open folder with all brightness/contrast-adjusted MIPs (.tif format; generated in Subheading 3.2, **step 4**).
    (b) Next, create and select a new Output folder for the classification results (.tif format) to be generated.
    (c) When prompted to create probability maps (see http://fiji.sc/Trainable_Weka_Segmentation), select *No*.

3.4 Quantification of Image Features

Automated TWS classification results in a binary, segmented image for each MIP, in which features of interest (FLS2-GFP endosomal puncta) are black and background features (cytosol, PM, chlorophyll) are white (Fig. 4a). Endosomal puncta are detected and quantified within the segmented, binary images using Fiji *Analyze Particles* functions (Fig. 4b). As for image preprocessing steps (*see* Subheading 3.2), feature detection is further automated using Fiji Batch Processing.

1. Open a single segmented, binary image (32-bit .tif format).

2. Turn on the Fiji Macro Recorder (*Plugins Macros Record…*).
   (a) First, change the image file type from 32-bit to 8-bit (*Image Type 8-bit*), which is a necessary step for particle analysis.
   (b) Second, open the *Analyze Particles* window to set parameters for detecting endosomal puncta within the 8-bit binary image (*Analyze* ►*Analyze Particles…*) (Fig. 4b).
   (c) Third, within the *Analyze Particles* window, set particle size to 0.25–2.5 μm^2 with a circularity index of 0.25–1.00 (a perfect circle = 1.00) to detect endosomal puncta (*see* **Note 23**). Select *Show: Nothing* from the drop-down menu since results will be automatically saved without visualization in this step.
   (d) Finally, select the *Summarize* function to create a table of particle analysis results. Select *OK*.
3. Within the Recorder window, check recorded script and select *Create* to generate the Macro. Save the Macro file (.ijm format) (*see* **Note 24**). Close the Recorder window.
4. Automatically apply the Macro saved in **step 3** to all 32-bit segmented, binary MIPs for the experiment using Fiji Batch Processing (*Process Batch Macro…*).
   (a) Within the Batch Process window, first select the Input folder with the segmented, binary MIP images (.tif format) created in Subheading 3.3, **step 11**.
   (b) Next, create and select a new Output folder for the 8-bit segmented, binary MIPs (.tif format) to be created.
   (c) Open the Macro file saved in **step 3**. Select *Process*. Files will be automatically converted, saved, and analyzed, resulting in a Summary table including each file name with the Count, Total Area, Average Size, and % Area of the puncta detected within the image (*see* **Notes 25** and **26**).
5. Calculate and graph the number of FLS2-GFP puncta per 1000 μm^2 for each image (Fig. 1b) (*see* **Note 27**) and select representative brightness/contrast-adjusted MIPs for each timepoint (Fig. 1a).

3.5 Determine Efficiency of FLS2-GFP Puncta Detection

Visually inspect a subset of images to determine the efficiency of FLS2-GFP puncta detection for each experiment.

1. Randomly choose five or more images per timepoint of the flg22 treatment.
2. Open both the brightness/contrast-adjusted MIP and the segmented, binary 8-bit MIP for each image.
3. Open the *Analyze Particles* window to set parameters for particle analysis within the 8-bit binary MIP (*Analyze* ►*Analyze*

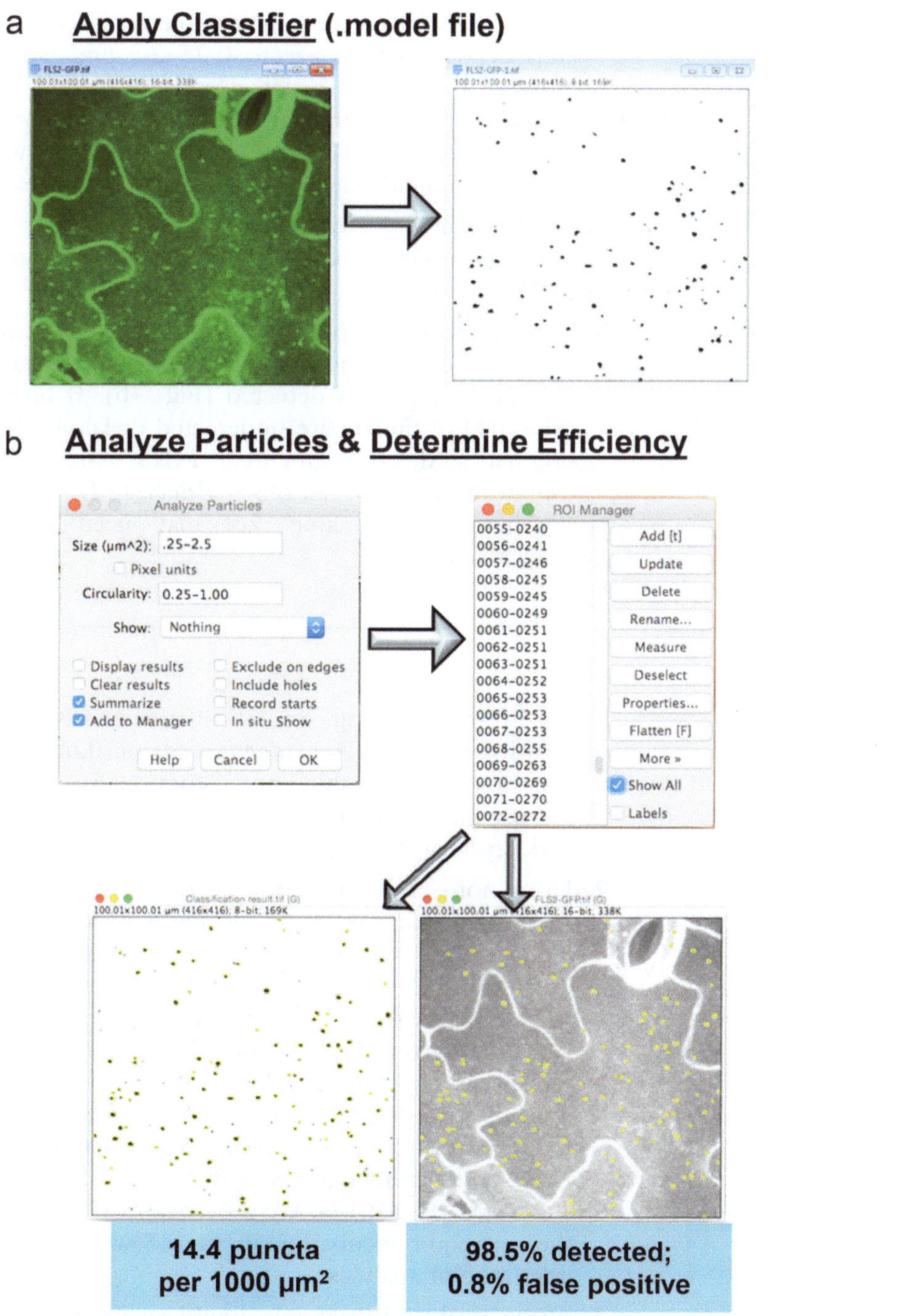

Fig. 4 Schematic for automated identification and quantification of FLS2-GFP endosomal puncta. (**a**) Brightness/contrast-adjusted MIPs are automatically segmented using the developed *Classifier* (.model). In the classification result, features of interest (Endosomes) are *black*, and all other features (Background) are *white*. (**b**) FLS2-GFP endosomal puncta are automatically detected and added to the Region of Interest (ROI) Manager using the *Analyze Particles* function. Detected puncta are displayed (*yellow outlines*) by selecting *"Show all"* within the ROI Manager. The efficiency of puncta detection is determined by visually inspecting an overlay of the puncta over the original and segmented MIPs. For Batch Processing, select "*Summarize*" within the *Analyze Particles* window to generate a table of results (not shown) used to calculate number of puncta per 1000 μm^2

Particles...). Using the same parameters as for Subheading 3.4, **step 2c**, set particle size to 0.25–2.5 μm² with a circularity index of 0.25–1.00 to detect endosomal puncta. Within the *Analyze Particles* window, select *Add to Manager* function to open the *Region Of Interest (ROI) Manager* with all of the detected puncta. Select *OK* (Fig. 4b).

4. Select *Show All* within the *ROI Manager* to highlight detected puncta (ROIs) within the original and segmented MIPs (Fig. 4b).
5. Through visual inspection of the detected puncta for each image, determine the percentage of total puncta detected and false-positive puncta detected (Fig. 4b). If high percentages (>5–10%) of puncta are undetected or falsely detected in the original and/or segmented MIPs, the TWS *Classifier* (Subheading 3.3) and/or *Analyze Particles* parameters (Subheading 3.4, **step 2c**) may need to be adjusted, respectively.

4 Notes

1. For live-cell imaging, our lab obtains consistent results when growing sterilized *Arabidopsis* seeds on 0.6% agar plates with half-strength Murashige and Skoog (0.5× MS) salts and 1% sucrose under continuous light or a long-day (16-h light/8-h dark) cycle at 22 °C [9].
2. Using non-segregating, homozygous lines is critical for quantitative analysis of FLS2 trafficking between mutant and wild-type genetic backgrounds, as any variability in FLS2-GFP expression will alter FLS2 abundance at the PM and in endosomal compartments. When crossing *FLS2-GFP* (wild-type Col-0 background) plants to mutant backgrounds, it is therefore necessary to propagate the plants until non-segregating, homozygous lines are identified by screening for FLS2-GFP fluorescence, PCR genotyping, and immunoblot analyses of FLS2-GFP protein levels with anti-FLS2 and/or anti-GFP antibodies [9, 12, 17]. We use extra *FLS2-GFP* seedlings from the overnight incubation in the 48-well plate (Fig. 1c) for immunoblot analyses.
3. Alternatively, 0.5× MS salts liquid media at room temperature may be used for all *FLS2-GFP* seedling incubations, flg22 treatments, and mounting on slides for imaging.
4. For live-cell imaging of trafficking events, we use a custom Olympus IX-71 inverted microscope (Center Valley, PA) equipped with a Yokogawa CSU-XI 5000 rpm spinning disc unit (Tokyo, Japan), PZ-2000 XYZ series automated stage

with Piezo Z-axis top plate (Applied Scientific Instrumentation; Eugene, OR), 60×-silicon oil objective (Olympus UPlanSApo 60×/1.30 Sil), and Andor iXon Ultra 897 High Speed EMCCD camera with Andor iQ3 software (Belfast, United Kingdom). GFP is excited with a 488-nm diode laser (Spectra Physics; Santa Clara, CA), and fluorescence emission is collected through a series of 488 nm single edge dichroic and 500–550 nm bandpass filters (Semrock; Rochester, NY) [9].

5. Fiji, a comprehensive and freely available software program for biological image analysis [13], can be downloaded with the TWS plug-in preinstalled (http://fiji.sc/Fiji). Additional information pertaining to the TWS plug-in is also available (http://fiji.sc/Trainable_Weka_Segmentation).
6. For organizing data and performing statistical tests, our lab uses Microsoft Excel (Redmond, WA) and Graph Pad Prism4 (La Jolla, CA) software, respectively.
7. Submerging seedlings in dH_2O overnight (16–20 h) helps to facilitate a rapid, and more uniform exposure of the flg22 solution to the cotyledons during the imaging experiment. In addition, the overnight incubation diminishes any wounding response(s) to the handling of seedlings when they are placed in dH_2O.
8. Imaging FLS2-GFP over the entire timecourse (5–90 min flg22) is lengthy, and it may be necessary to focus on one or two specific timepoints (e.g., 35–45 and 50–60 min) for quantitative comparison of ligand-induced endocytosis of FLS2 between wild-type and mutant backgrounds [9].
9. For FLS2-GFP SDCM imaging, we obtain consistent results using 30–40% laser power, 1500-rpm Yokagawa spinning disk speed, and 150 ms camera exposure time. However, optimal settings will need to be determined for each microscope. Be aware that increasing laser power and/or camera exposure time will increase the rate of photobleaching.
10. Higher FLS2-GFP expression is observed on the top (adaxial) surface of the cotyledons in comparison to the bottom (abaxial) surface.
11. Z-series depth will vary depending upon the thickness of the epidermal cell layer. For 7-day-old *Arabidopsis* Col-0 seedlings, we consistently use a Z-series depth setting of 20 μM (66 *z*-planes; 0.31 μm step) with a total acquisition time of approximately 12 s.
12. By imaging different groups of flg22-treated seedlings over 10-min periods, versus the same seedlings continuously over the entire flg22 treatment timecourse, seedlings are more uniformly exposed to the flg22 solution and mechanical stress from the coverslip is minimized.

13. For quantification of FLS2-GFP trafficking between genetic backgrounds, we typically capture 36–48 Z-series of images (from six or more seedlings) for each flg22 treatment timepoint and each genetic background.
14. Fiji actions are indicated throughout the methods in italics. A series of actions such as "*Analyze* ► *Set Scale...*" indicate to first select *Analyze* within the Fiji main menubar. Next, within the *Analyze* drop-down menu, select *Set Scale...* and enter information for number of pixels per μm.
15. To apply the same scale to all images taken with the same microscope objective and zoom settings, select *Global*.
16. To allow for proper image analysis, the same brightness/contrast adjustment must be applied to every image within the same experimental dataset such as images obtained for each timepoint for wild-type and mutant plant lines.
17. An example Macro script for generating a MIP followed by adjusting brightness/contrast is: run("Z Project...", "projection = [Max Intensity]"); //run("Brightness/Contrast..."); setMinAndMax(500, 6000);
18. Be careful to save brightness/contrast-adjusted MIPs in a separate folder to avoid writing over original data files. It is important to always keep unprocessed images for future reference and/or separate analyses.
19. We typically crop brightness/contrast-adjusted MIP images to focus on epidermal pavement cells. Whenever cropping an image, it is necessary to measure the adjusted image area (*Analyze Measure*) for accurate calculation of number of FLS2-GFP puncta per 1000 μm^2 (Fig. 1b).
20. Default TWS plug-in settings work well for the identification of endosomal puncta; see http://fiji.sc/Trainable_Weka_Segmentation for information on adjusting settings.
21. While multiple rounds of *Classifier* (.model) training with several example MIPs usually improve feature recognition, there are instances where feature recognition is not improved and/or the *Classifier* (.model) becomes too complex and significantly slows image processing. In these instances, you may need to restart training or use one of the intermediate *Classifier* (.model) files with good feature recognition.
22. Within the same experiment, a single *Classifier* (.model file) must be used. In addition, for images from different experiments that were acquired with identical microscope settings, and with similar fluorescence intensity of features versus background, the same *Classifier* (.model file) may often be used for feature detection.

23. We have found that the described parameters work best for detecting FLS2-GFP in endosomal puncta [9]. However, it may be necessary to optimize the parameters for different imaging systems and *Classifier* (.model) files.
24. An example Macro script for detecting endosomal puncta is: run ("Analyze Particles...", "size = .25–2.5 circularity = 0.25–1.00 summarize");
25. Results for measuring the size of detected puncta may vary depending upon the *Classifier* (.model) developed and applied to the dataset, thus making it difficult to determine absolute puncta size with TWS methods. However, within a single experiment, trends in size can be observed that can be compared between genetic backgrounds.
26. For images with less than two puncta per 1000 μm^2, visually inspect and correct for false positives, which occur more frequently in images with very few features of interest.
27. Results for the quantification of FLS2-GFP in endosomal puncta can be followed up with pixel intensity measurements of FLS2-GFP at the PM using established methods [9]. After flg22 treatment, the removal of FLS2 from the PM correlates with decreasing FLS2-GFP fluorescence intensity at the PM [8, 9, 11].

Acknowledgments

This work was supported by a grant received from the National Science Foundation to A.H. (NSF-IOS-1147032).

References

1. Boller T, Felix G (2009) A renaissance of elicitors: perception of microbe-associated molecular patterns and danger signals by pattern-recognition receptors. Annu Rev Plant Biol 60:379–406
2. Trdá L, Boutrot F, Claverie J et al (2015) Perception of pathogenic or beneficial bacteria and their evasion of host immunity: pattern recognition receptors in the frontline. Front Plant Sci 6:219
3. Tena G, Boudsocq M, Sheen J (2011) Protein kinase signaling networks in plant innate immunity. Curr Opin Plant Biol 14:519–529
4. Robatzek S, Wirthmueller L (2013) Mapping FLS2 function to structure: LRRs, kinase and its working bits. Protoplasma 250:671–681
5. Macho A, Zipfel C (2014) Plant PRRs and the activation of innate immune signaling. Mol Cell 54:263–272
6. Ben Khaled S, Postma J, Robatzek S (2015) A moving view: subcellular trafficking processes in pattern recognition receptor-triggered plant immunity. Annu Rev Phytopathol 53:379–402
7. Robatzek S, Chinchilla D, Boller T (2006) Ligand-induced endocytosis of the pattern recognition receptor FLS2 in *Arabidopsis*. Genes Dev 20:537–542
8. Beck M, Zhou J, Faulkner C et al (2012) Spatio-temporal cellular dynamics of the *Arabidopsis* flagellin receptor reveal activation status-dependent endosomal sorting. Plant Cell 24:4205–4219
9. Smith J, Leslie M, Robinson S et al (2014) Loss of *Arabidopsis thaliana* Dynamin-Related Protein 2B reveals separation of innate immune signaling pathways. PLoS Pathog 10:e1004578
10. Choi S, Tamaki T, Ebine K et al (2013) RABA members act in distinct steps of subcellular

trafficking of the FLAGELLIN SENSING2 receptor. Plant Cell 25:1174–1187

11. Spallek T, Beck M, Ben Khaled S et al (2013) ESCRT-I mediates FLS2 endosomal sorting and plant immunity. PLoS Genet 9:e1004035
12. Smith J, Salamango D, Leslie M et al (2014) Sensitivity to flg22 is modulated by ligand-induced degradation and de novo synthesis of the endogenous flagellin-receptor FLAGELLIN-SENSING2. Plant Physiol 164:440–454
13. Schindelin J, Arganda-Carreras I, Frise E et al (2012) Fiji: an open-source platform for biological-image analysis. Nat Methods 9:676–682
14. Beck M, Zhou J, Faulkner C et al (2014) High-throughput imaging of plant immune responses. In: Birch P, Jones J, Bos J (eds) Plant-pathogen interactions: methods and protocols, methods in molecular biology, vol 1127. Springer, New York, pp 67–80
15. Hall M, Frank E, Holmes G et al (2009) The WEKA data mining software: an update. SIGKDD Explor Newsl 11:10–18
16. Sommer C, Gerlich DW (2013) Machine learning in cell biology—teaching computers to recognize phenotypes. J Cell Sci 126:5529–5539
17. Heese A, Hann D, Gimenez-Ibanez S et al (2007) The receptor-like kinase SERK3/BAK1 is a central regulator of innate immunity in plants. Proc Natl Acad Sci USA 104: 12217–12222

Chapter 5

Analysis for Protein Glycosylation of Pattern Recognition Receptors in Plants

Takaakira Inokuchi and Yusuke Saijo

Abstract

Recognition of molecules typical of microbes or aberrant cellular states, termed microbe- or danger-associated molecular patterns (MAMPs/DAMPs), respectively, provides an important step in plant and animal innate immunity. In plants, pattern recognition receptors (PRRs) identified to date are limited to membrane-associated proteins, of which the majority has an extracellular leucine-rich repeat (LRR) or lysine-motif (LysM) domain. These PRRs undergo quality control (QC) in the Endoplasmic Reticulum (ER) that is dependent on Asn (N)-linked glycosylation ($Glc_3Man_9GlcNAc_2$ conjugation) of their extracellular domain. In Arabidopsis, genetic studies have revealed that a subset of these PRRs require an intact N-glycosylation pathway in the ER for their biogenesis and function. Here, we describe methods for immunoblot-based detection of protein glycosylation states in plants.

Key words Plant immunity, Endoplasmic Reticulum quality control (ERQC), Pattern recognition receptors (PRRs), Immunoblot analysis, N-linked glycosylation, Glycosidase treatment

1 Introduction

The LRR receptor kinases (RKs) EFR and FLS2 recognize the bacterial MAMPs elf18 and flg22, derived from elongation factor-Tu (EF-Tu) and flagellin, respectively. Genetic studies have shown that EFR function, but not FLS2 function, is impaired in the absence of the following ERQC branches: The first involves the oligosaccharyltransferase (OST) complex subunits STT3A and OST3/6, glucosidases I and II (GI and GII), the ER chaperone calreticulin3 (CRT3), and UDP-glucose:glycoprotein glucosyltransferase (UGGT); the second involves the Hsp70 family member BiP that acts together with the Hsp40 family members ERdJ and stromal cell-derived factor 2 (SDF2) [1]. This indicates a model in which EFR is a client protein that strictly requires ERQC mediated by these protein folding pathways. However, it cannot be ruled out that the observed dysfunction of the receptor in these

Libo Shan and Ping He (eds.), *Plant Pattern Recognition Receptors:Methods and Protocols*, Methods in Molecular Biology, vol. 1578, DOI 10.1007/978-1-4939-6859-6_5, © Springer Science+Business Media LLC 2017

ERQC-related mutant plants is an indirect consequence of ERQC defects. A simple protein biochemical analysis can help determine whether the protein of interest directly undergoes N-glycosylation. N-glycosylation on the client proteins can be typically detected as a shift of the mobility on SDS-PAGE. In the case of EFR and FLS2, protein biochemical analysis revealed an increase of the apparent size in a manner dependent on the OST complex [2]. This leads to the view that both receptors undergo N-glycosylation, despite the aforementioned divergence in the requirement for OST-mediated ERQC. Here, we describe methods that are useful for the detection of N-glycosylated proteins from plant protein extracts.

2 Materials

Prepare all solutions using Milli-Q (MILLIPORE Q-guard).

2.1 Plants

Ten-day-old *Arabidopsis thaliana* seedlings grown on 1/2 × MS medium (pH 5.6) containing 25–34 mM sucrose and 0.8% agar were treated with or without elf18 at 0.5 μM for 24 h, and then subjected to immunoblot analysis.

2.2 Lysis Buffer

Prepare for a 10× stock solution of Lysis Buffer, containing 50 mM Tris–HCl (pH 7.5), 150 mM NaCl, 10% Glycerol, 2 mM EDTA, 0.5% (v/v) NP-40, which can be stored at 4 °C. Immediately before use, dilute it to the working concentration, and then freshly add 2 mM DTT and 1% (v/v) P9599 protease inhibitor cocktail (Sigma).

2.3 Immunoblotting Reagents

1. PVDF membranes.
2. For 1 L of Immunoblot transfer buffer, dissolve 3 g of Tris, 14 g of Glycine and 200 mL of 100% MeOH in water.
3. For 1 L of 10× Tris-buffered saline (TBS), dissolve 87.66 g of NaCl and 100 mL of 1 M Tris–HCl (pH 7.5) in water.
4. For 1 L of TBS containing 0.1% Tween-20 (TBST), dilute 100 mL of 10× TBS and 5 mL of 20% Tween-20 in water.
5. Blocking solution: dissolve 5% skim milk in 0.1% TBST. Store at 4 °C.
6. Criterion™ Blotter.
7. Medium binder clips.
8. Plastic container.
9. Whatman 3 MM Chr paper.
10. Anti-FLS2 antibodies.
11. For 10 mL of 6× SDS sample buffer, dissolve 3 g of sucrose, 1.2 g of SDS, and 10 mg of bromophenol blue (BPB) in 6 mL

of 0.5 M Tris–HCl (pH 6.8), and then top up with water (if necessary). Store at room temperature before adding a reducing agent.

3 Method

3.1 Extraction of PRRs

1. Grind tissues in liquid nitrogen (Fig. 1a).
2. Suspend tissues in Lysis Buffer (100 mg in 300 μL lysis buffer). If necessary, keep the lysates on ice until they thaw, and then mix well.
3. Centrifuge at 13,200 × *g* 4 °C for 15 min.
4. Recover the supernatant (crude lysates) to a clean tube (*see* **Note 1**).

3.2 Pull Down with ConA Sepharose

1. Dilute crude lysates with Lysis buffer up to ~1.3 mL in a 1.5 mL tube (Fig. 1a).
2. Add 50 μL of ConA Sepharose™ 4B (GE Healthcare) after vortex (*see* **Note 2**).
3. Rocking at 4 °C for 1–3 h.
4. Centrifuge at 800 × *g*- 4 °C for 1–2 min and discard the supernatant (*see* **Note 3**).
5. Wash the Sepharose pellet with Lysis buffer (*see* **Note 4**).

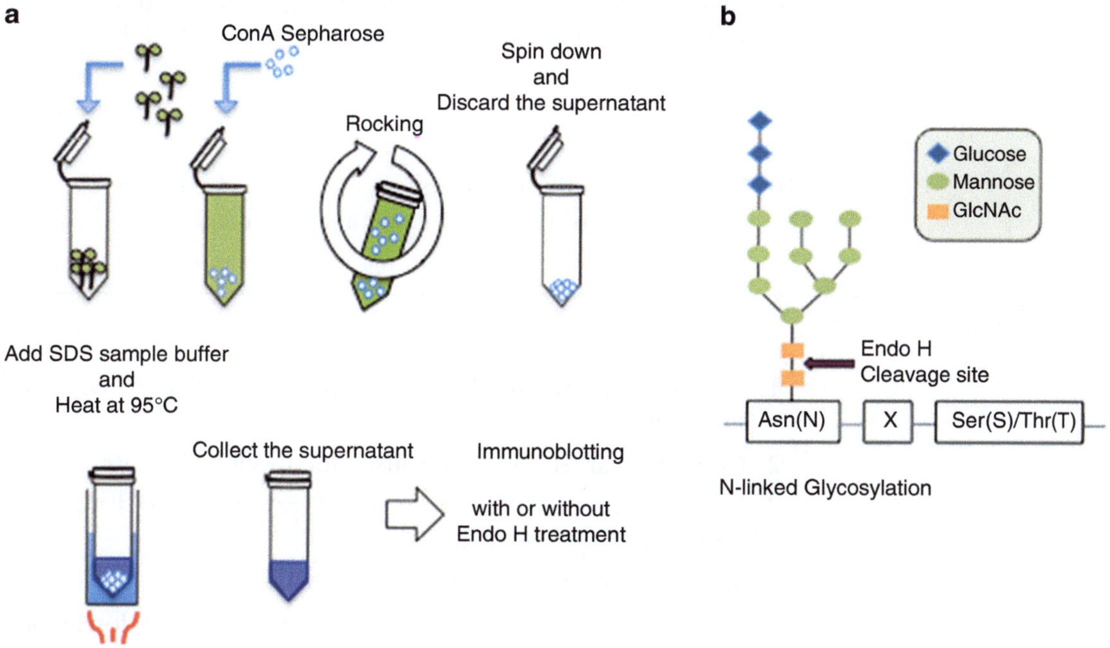

Fig. 1 (**a**) Schematic diagram for the isolation of ConA Sepharose-binding proteins. (**b**) Endo H cleavage site in N-linked glycan

6. Repeat three times the **steps 4** and **5**.
7. Add 50 μL of 1× SDS sample buffer (with DTT added before use to a final concentration of 10–20 mM).
8. Heat at 95 °C for 5 min.
9. Centrifuge at 800×*g* 4 °C for 1–2 min and collect the supernatant.

3.3 Cleavage of N-Linked Glycosylation of PRRs by Endo H

1. Add 2 μL of GlycoBuffer 3 (New England BioLabs) and 1 μL of Endo H (New England BioLabs) in 17 μL of the aforementioned ConA Sepharose-bound protein solution (*see* **Note 5**).
2. Incubate this mixture for 1 h at 37 °C (Fig. 1b).

3.4 Immunoblot Analysis for PRRs

1. SDS-PAGE on a 6.0–7.5% gel (*see* **Note 6**).
2. Transfer proteins to PVDF membrane for 1 h at 4 °C.
3. Stain the membrane with Ponceau S (*see* **Note 7**), to assess equal loading of the samples.
4. Destain the membrane with 1× TBS.
5. Block the membrane with blocking solution for 1 h at room temperature (*see* **Note 8**).
6. Incubate the membrane overnight at 4 °C with appropriate dilutions of the first antibodies of choice in Lysis buffer.
7. Wash the membrane with 0.1% TBST for 10 min. Repeat this step three times.
8. Incubate the membrane with appropriate dilutions of the second antibodies of choice in Lysis buffer for 1 h at room temperature.
9. Wash the membrane with 0.1% TBST three times for 10 min each.
10. Develop chemiluminescent signals with ECL reagent and detect them using a CCD camera-based imager (Fig. 2).

4 Notes

1. Floating debris should be filtered out of the recovered supernatant.
2. ConA Sepharose should be suspended well before use.
3. Do not exceed 3300 × *g* in centrifugation to avoid malfunctioning of ConA Sepharose.
4. Suspend gently by pipetting. Do not Vortex.
5. Suspend GlycoBuffer 3 and Endo H by pipetting. Endo H works in the presence of SDS, which is likely to increase the accessibility of Endo H to the cleavage sites on N-glycan chains.

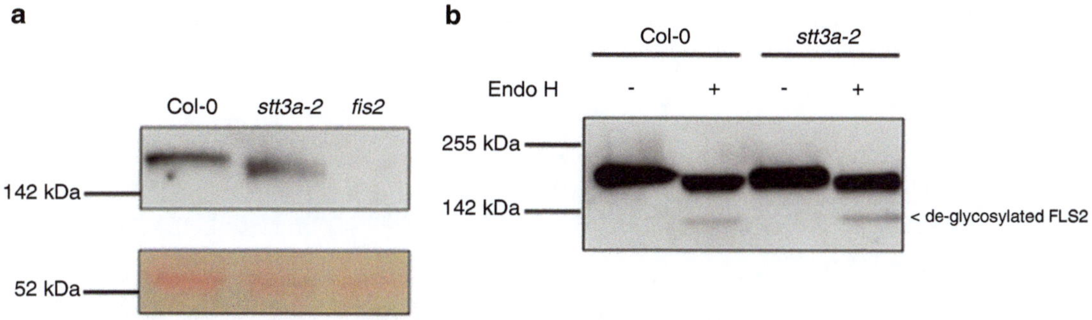

Fig. 2 Immunoblot analysis for an N-glycosylation-dependent mobility shift of the Arabidopsis pattern recognition receptor FLS2. (**a**) The position of the molecular weight marker is shown on the left. *Top*, immunoblot probed with anti-FLS2 antibodies; *Bottom*, immunoblot stained with Ponceau S for equal loading verification. (**b**) ConA Sepharose-binding proteins were digested with Endo H, and then subjected to immunoblot analysis with anti-FLS2 antibodies

6. According to the molecular weight of the proteins of interest, the gel percentage needs to be determined. Electrophoresis should be performed until the frontline of BPB reaches the bottom (Fig. 2).
7. Since the staining of Ponceau S is reversible, this staining does not affect immunoblotting.
8. When the background signal is high, blocking and washing should be extensively performed. Blocking and/or washing even overnight help enhance the signal/background ratio.

Acknowledgments

The researches in the Saijo Lab have been supported in part by JST PRESTO and Grant-in-Aid for Scientific Research to Y.S.

References

1. Saijo Y (2010) ER quality control of immune receptors and regulators in plants. Cell Microbiol 12:716–724
2. Saijo Y, Tintor N, Lu X, Rauf P, Pajerowska-Mukhtar K, Haweker H, Dong X, Robatzek S, Schulze-Lefert P (2009) Receptor quality control in the endoplasmic reticulum for plant innate immunity. EMBO J 28: 3439–3449

Chapter 6

Assays to Investigate the *N*-Glycosylation State and Function of Plant Pattern Recognition Receptors

Stacey A. Lawrence, Teresa Ceserani, and Nicole K. Clay

Abstract

The biogenesis and functionality of pattern recognition receptors (PRRs) are critical for robust plant immune responses. Here, we present methods to determine the *N*-glycosylation state and ligand-induced activity of these receptors for comparative quantitative analysis. These techniques can be used to identify mutants and chemical inhibitors affecting PRR biogenesis and functionality. When combined, these techniques can provide useful insights on biological processes necessary to synthesize a properly membrane-localized and ligand-responsive PRR.

Key words Asparagine-linked glycosylation, Protein quality control, Endoglycosidase H, Tunicamycin, MAPK activation, Seedling growth arrest, Anthocyanin inhibition

1 Introduction

1.1 N-Glycosylation of FLS2 and EFR

The addition of glycans to asparagine residues (*N*-glycosylation) is an essential, highly conserved co-translational modification. It targets newly synthesized membrane-localized and secreted proteins in the endoplasmic reticulum (ER) lumen in eukaryotic cells. The role of *N*-glycans in the ER-resident protein quality control (ERQC) network is to monitor the accurate folding of glycoproteins and mediate the distribution of correctly folded proteins within the secretory system (e.g., to the plasma membrane or vacuole) [1, 2]. Some of the best characterized proteins to undergo *N*-glycosylation are the *Arabidopsis* pattern recognition receptors (PRRs) FLS2 and EFR, which recognize microbe-associated molecular pattern molecules (MAMPs) bacterial flagellin and EF-Tu (as well as their bioactive epitopes flg22 and elf18/elf26), respectively [3, 4].

FLS2 and EFR require protein *N*-glycosylation for their maturation and transport to their final plasma membrane destination to mediate antibacterial immunity in response to their ligands [5]. EFR protein abundance and function are compromised when single *N*-glycosylation sites in the EFR protein, *N*-glycosylation

Libo Shan and Ping He (eds.), *Plant Pattern Recognition Receptors: Methods and Protocols*, Methods in Molecular Biology, vol. 1578, DOI 10.1007/978-1-4939-6859-6_6, © Springer Science+Business Media LLC 2017

enzymes, or single associated ERQC components are mutated [6–12]. Although individual *N*-glycan site mutants of FLS2 do not have the same effect, a significant portion of the mutated FLS2 protein is retained in the ER, reducing its ligand sensitivity and suggesting that it too is a client of the *N*-glycan-mediated ERQC mechanism [11, 12].

1.2 Quantitative Readouts of FLS2 and EFR Glycosylation State

FLS2 and EFR have 21 and 16 conserved *N*-glycosylation consensus sites—NX(S/T) motifs—on their respective LRR-containing ectodomains [11, 12]. In addition, the detected molecular masses of FLS2 and EFR are much larger than their predicted masses: ~175 vs. ~130 kDa for FLS2 and ~145 vs. ~110 kDa for EFR [3, 4, 13]. Two tools can be used to deglycosylate the FLS2 and EFR proteins: the *N*-glycosylation inhibitor tunicamycin and the deglycosylation enzyme endoglycosidase H (Endo H).

Tunicamycin treatment inhibits *N*-glycosylation and induces the unfolded protein response in the ER, a cellular stress response that initiates programmed cell death in plants [14–16]. In wild-type seedlings treated with tunicamycin, the FLS2 protein mass is reduced by ~40 kDa, whereas the EFR protein mass is reduced by ~5 kDa (Fig. 1a). Similar tunicamycin treatments were reported to produce two differently sized FLS2 proteins at ~175 and ~130 kDa and a single sized EFR protein at ~150 kDa [11]. Because plant *N*-glycans have an average mass of ~2 kDa, a mass shift of ~40 kDa for the FLS2 protein suggests that nearly all 21 potential *N*-glycan attachment sites are occupied, whereas a mass shift of ~5 kDa for the EFR protein indicates that only a handful of *N*-glycan sites may be occupied. The absence of an unglycosylated EFR protein at ~110 kDa suggests that EFR contains other substantial post-translational modifications or is more resistant to tunicamycin treatments.

Endo H has been used to functionally characterize ERQC components required for the biogenesis of the FLS2 and EFR proteins [8, 11, 12]. It cleaves *N*-glycans that are in the early stages of maturation before processing through the Golgi, and thus are mainly associated with ER-localized, immature glycosylated proteins. In contrast, mature *N*-glycans, which are only associated with membrane proteins that process through the Golgi, are largely insensitive to Endo H cleavage. Therefore, Endo H treatment combined with functional data allows the differentiation of plasma membrane-localized glycoproteins from their ER-localized counterparts. In wild-type seedling extracts treated with Endo H, FLS2, and EFR protein masses are reduced by ~5 kDa. This suggests that FLS2 and EFR carry mainly mature *N*-glycans and are mostly localized at the plasma membrane (Fig. 1b). These results are consistent with previous reports of similar treatments [8, 11, 12].

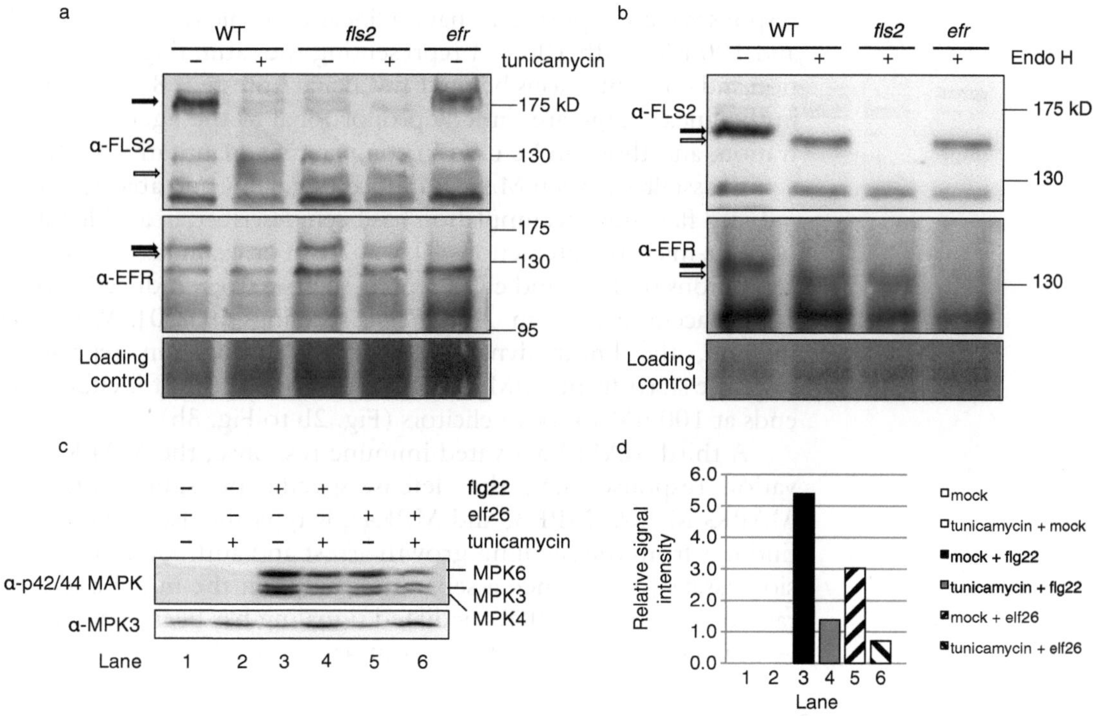

Fig. 1 (**a**) Immuno-detection of FLS2 and EFR proteins in total protein extracts from Col-0 (WT), *fls2*, and *efr* seedlings treated with 5 μg/mL tunicamycin and separated on an 8.5% SDS-PAGE gel. (**b**) Immuno-detection of FLS2 and EFR proteins from Col-0 (WT), *fls2*, and *efr* seedlings digested with Endo H and separated on a 6% SDS-PAGE gel. Black arrows indicate FLS2- or EFR-specific bands before treatment. White arrows indicate FLS2- or EFR-specific bands after treatment. Molecular mass sizes are indicated on the right. (**c**) Immuno-detection of phosphorylated MPK6, MPK3, and MPK4 proteins from flg22- or elf26-elicited WT total protein extracts separated on a 10% SDS-PAGE gel. (**d**) Data represents the combined signal intensities of phosphorylated MAPKs relative to that of MPK3 protein (loading control) in (**c**)

1.3 Quantitative Readouts of FLS2 and EFR Function in Genetic Studies

Two MAMP-induced immune responses have been used as high-throughput readouts in forward-genetic screens to identify new regulators of FLS2 and EFR function: the seedling growth arrest response and the anthocyanin repression response. Both responses require long-term exposure to elicitors (i.e., in days) and sustained activation of FLS2/EFR-mediated signaling. In addition, these assays were successfully used to identify ERQC components required for the biogenesis of the FLS2 and EFR proteins [6–9, 11, 12]. Here, we describe methods that adapt both MAMP-sensitive responses as robust quantitative readouts in small-scale assays.

Different MAMPs arrest seedling growth in *Arabidopsis* [3, 4]. With our method, flg22 and elf26 induce a seedling growth arrest response with different linear dynamic ranges: between 1 nM and 1 μM for flg22 (Fig. 2a) and between 0.1 and 100 nM for elf26 (Fig. 3a). Similar flg22/elf18-induced seedling growth arrest responses were reported to have a linear dynamic range between 1 and 100 nM, with 100 nM representing the saturating concentration and concentrations below 1 nM being untested [8]. Responses in the linear range are directly proportional to the ligand concentration, and thus can be used in comparative quantitative analysis.

Crosstalk between MAMP-induced signaling and abiotic stress-induced flavonoid accumulation has been described in a wide range of plant-microbe interactions [9, 17–20]. For example, 1 μM concentrations of flg22 and elf18 repress sucrose stress-induced anthocyanin accumulation in *Arabidopsis* seedlings [9, 20]. With our method, the linear dynamic range of anthocyanin repression response starts from 1 nM for flg22 and from 0.1 nM for elf26, and ends at 100 nM for both elicitors (Fig. 2b to Fig. 3b).

A third MAMP-activated immune response, the MAPK activation response (i.e., the defense-specific phosphorylation of MAPKs MPK6, MPK3, and MPK4), is typically used to confirm findings from the seedling growth arrest and anthocyanin repression assays [6–9], since a separation between the initial and sustained activation of PRR-mediated signaling has been reported in *Arabidopsis* [7, 21, 22]. In contrast to seedling growth and

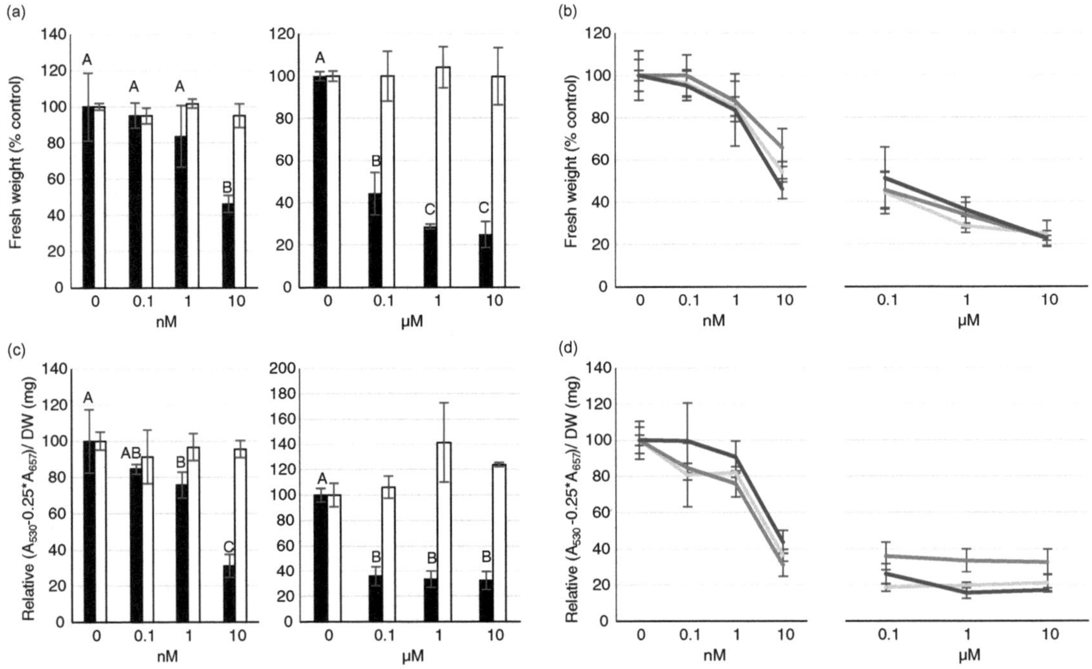

Fig. 2 Flg22-induced, dose-dependent seedling growth arrest (**a, b**) and anthocyanin repression (**c, d**) immune responses in Col-0 (WT) and *fls2* seedlings. The response to water is set to 100. (**a, c**) Data represent the mean ±1 standard deviation of $n = 5$ biological replicates in (**a**), $n = 10$ in (**c**, *left graph*), and $n = 16$ in (**c**, *right graph*). Black bars, WT; white bars, *fls2*. Different letters indicate statistically significant differences ($p \leq 0.01$, two-tailed *t*-test). (**b, d**) Three independent experiments, including those shown in (**a**) and (**c**), illustrate the variability inherent in the two MAMP immune response assays. DW, dry weight

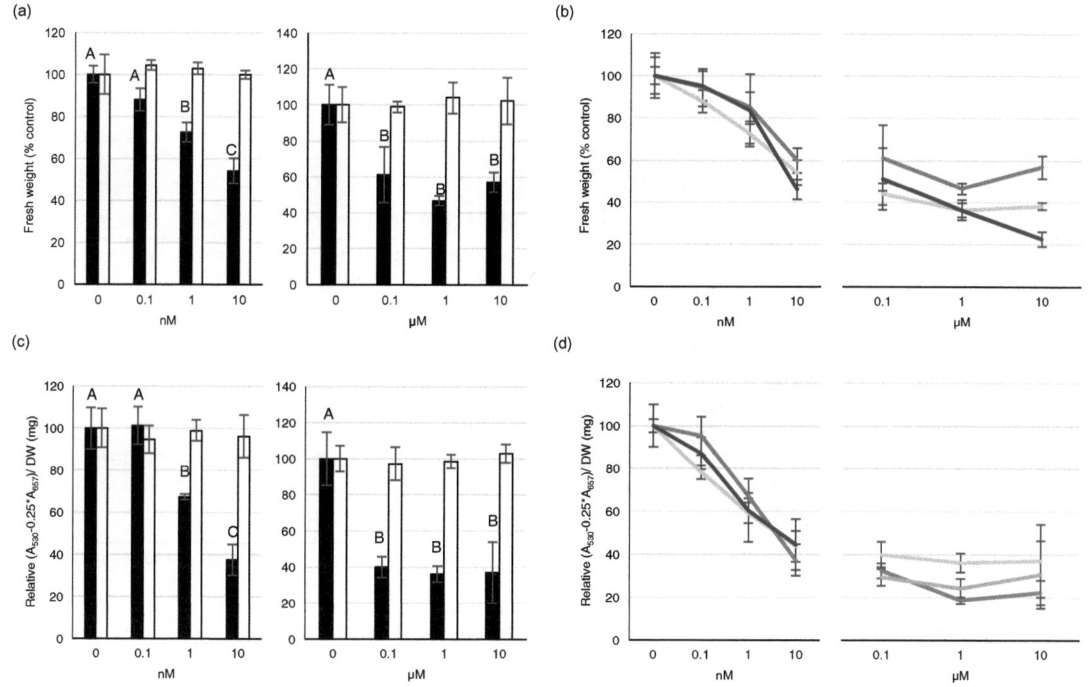

Fig. 3 Elf26-induced, dose-dependent seedling growth arrest (**a, b**) and anthocyanin repression (**c, d**) immune responses in Col-0 (WT) and *efr* seedlings. The response to water is set to 100. (**a, c**) Data represent the mean ±1 standard deviation of $n = 5$ biological replicates in (**a**), $n = 10$ in (**c**, *left graph*), and $n = 16$ in (**c**, *right graph*). *Black bars*, WT; *white bars*, *fls2*. Different letters indicate statistically significant differences ($p \leq 0.01$, two-tailed *t*-test). (**b, d**) Data from three independent experiments, including those shown in (**a**) and (**c**), illustrate the variability inherent in the two MAMP immune response assays. DW, dry weight

anthocyanin repression assays, whose MAMP exposures are in the order of days, the MAPK activation response occurs early in the FLS2/EFR-mediated signaling and thus requires a much shorter term of elicitor exposure (in the order of minutes) to flg22 and elf26. It also has two short linear dynamic ranges: the first between 1 and 10 nM and the second between 100 nM and 1 μM (Fig. 4). However, only the second linear range of the flg22-induced MAPK activation response appears to be significant, with a near-fold increase in signal intensity between 100 nM and 1 μM of flg22 (Fig. 4).

1.4 Qualitative Readout of FLS2 and EFR Function in Chemical Studies

The MAPK activation response is useful in pharmacological studies of FLS2 and EFR function, especially in the presence of toxic chemicals such as tunicamycin. In tunicamycin-treated seedlings, the deglycosylated FLS2 and underglycosylated EFR proteins exhibit reduced phosphorylation of MPK6, MPK3, and MPK4, in response to their respective MAMP ligands, demonstrating their decreased functionality (Fig. 1c, d). This result is consistent with the reported interference of tunicamycin treatment with FLS2 and EFR's ligand-binding ability and plasma membrane-localization [11].

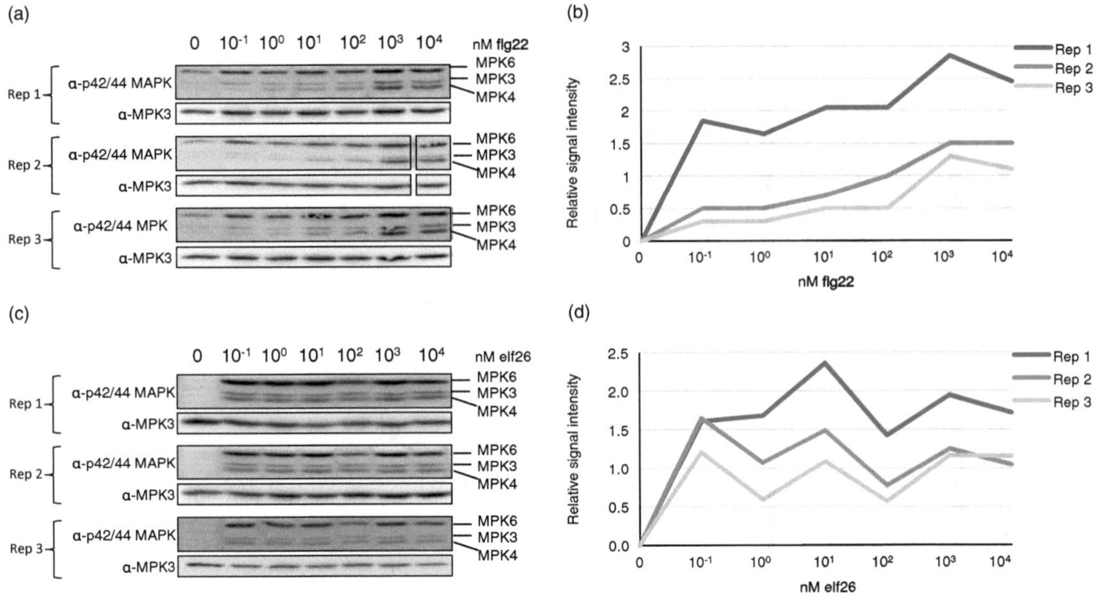

Fig. 4 Dose-dependent MAPK activation response in Col-0 (WT) seedlings elicited with increasing concentrations of flg22 (**a, b**) or elf26 (**c, d**). (**a, c**) Immuno-detection of phosphorylated MPK6, MPK3, and MPK4 (indicated on the *right*) from three technical replicates (Rep1–3) of WT total protein extracts separated on a 10% SDS-PAGE gel. (**b, d**) Data of the combined signal intensities of phosphorylated MAPKs in (**a**) and (**c**) relative to that of MPK3 protein (loading control) illustrate the variability inherent in the MAPK activation assay

2 Materials

2.1 Plant Growth

1. *Arabidopsis thaliana* seeds Col-0 (wild-type), *fls2* (SALK_062054C), and *efr* (SALK_044334) from Arabidopsis Biological Resource Center (Columbus, Ohio).
2. Seed sterilization solution: 5% (v/v) hypochlorite, 0.1% (v/v) Triton X-100.
3. Biosafety cabinet.
4. Autoclaved distilled water (dH_20).
5. MS media 1 (pH 5.7–5.8): 0.5× MS basal salts with vitamins (Phytotechnology labs, M519), 0.05% (w/v) MES, 0.5% (w/v) sucrose, filter-sterilized (0.45 μm).
6. MS media 2 (pH 5.7–5.8): 1× MS basal salts with vitamins, 0.05% (w/v) MES, 0.5% (w/v) sucrose, filter-sterilized (0.45 μm).
7. MS media 3 (pH 5.7–5.8): 0.5× MS basal salts with vitamins, 0.05% (w/v) MES, filter-sterilized (0.45 μm).
8. MS media 4 (pH 5.7–5.8): 0.5× MS basal salts with vitamins, 0.05% (w/v) MES, 3.42% sucrose, filter-sterilized (0.45 μm).
9. Sterile 12-well and 24-well tissue culture plates.

10. Light cart with adjustable light cycle and intensity in a 21 °C room and water trays to provide humidity.
11. Bacterial MAMP elicitors: 1 mM flg22 (Genscript, ≥85.1% purity) and 1 mM elf26 (Genscript, ≥96.6% purity) aqueous solutions.
12. Sterile 2-mL round-bottom and 1.7-mL microcentrifuge tubes.
13. Sterile 1-mL filter tips and 200 μL tips for pipetmen.
14. Sterile 15-mL conical tubes.

2.2 Tunicamycin Treatment

1. 0.5 mg/mL tunicamycin (Sigma, T7765) stock solution in methanol.

2.3 Total Protein Extraction

1. Total protein extraction buffer: 50 mM Tris-Cl (pH 7.5), 50 mM DTT, 4% (w/v) SDS, 10% (v/v) glycerol; add DTT prior to use.
2. 2-mL round-bottom and 1.7-mL microcentrifuge tubes.
3. Liquid nitrogen.
4. TissueLyzer homogenizer (Qiagen) and 5-mm stainless steel beads (Thomson, 440CSS).
5. Room-temperature benchtop microcentrifuge.
6. 95 °C heat block.

2.4 SDS-PAGE

1. Distilled H_2O (dH_2O).
2. Mini PROTEAN tetra cell and 1.5-mm plates (Bio-Rad).
3. 6% separating gel: 375 mM Tris-Cl (pH 8.8), 6% (w/v) 37.5:1 acrylamide to bisacrylamide, 0.1% (w/v) SDS, 0.05% (w/v) APS, 10% (v/v) TEMED.
4. 8.5% separating gel: 375 mM Tris-Cl (pH 8.8), 8.5% (w/v) 37.5:1 acrylamide to bisacrylamide, 0.1% (w/v) SDS, 0.05% (w/v) APS, 10% (v/v) TEMED.
5. 10% separating gel: 375 mM Tris-Cl (pH 8.8), 10% (w/v) 37.5:1 acrylamide to bisacrylamide, 0.1% (w/v) SDS, 0.05% (w/v) APS, 10% (v/v) TEMED.
6. 2-propanol or 0.1% SDS solution for gel overlay.
7. 4% stacking gel: 125 mM Tris-Cl (pH 6.8), 4% (w/v) 37.5:1 acrylamide to bisacrylamide, 0.1% (w/v) SDS, 0.05% (w/v) APS, 10% (v/v) TEMED.
8. Lint-free tissues (Kimwipes).
9. Prestained 10–175 kDa molecular-weight markers (Abcam, ab115832).
10. Running buffer: 25 mM Tris, 192 mM glycine, 0.1% (w/v) SDS.

2.5 Western Blotting

1. Two-gel tetra blotter or Criterion blotter (Bio-Rad).
2. Transfer buffer: 25 mM Tris, 192 mM glycine, freshly prepared and stored at 4 °C.
3. PVDF membrane, 0.45 μm pore size (Millipore, IPVH00010).
4. Methanol.
5. Precut chromatography paper.
6. 1% (w/v) Ponceau S (IBI Scientific, IB01100) in 3% (w/v) trichloroacetic acid.
7. 5% (w/v) acetic acid.
8. Sterile 10× phosphate-buffered saline (PBS) (pH 7.4): 1.369 M NaCl, 26.8 mM KCl, 1 mM Na_2HPO_4, 17.9 mM KH_2PO_4.
9. 5% (w/v) nonfat dry milk in 1× PBS.
10. Primary antibodies in 5% (w/v) nonfat dry milk in 1× PBS: anti-phospho-p44/42 MAPK (Thr202/Tyr204) (Cell Signaling Technology, 9101), 1:2000 dilution; anti-AtMPK3 (Sigma, A6979), 1:2000 dilution; anti-FLS2 (custom-made polyclonal antibody against the peptide CTKQRPTSLNDEDSQ, *A. thaliana* gene ID: At5g46330), 1:1000 dilution; anti-EFR (custom-made polyclonal antibody against the peptide CITESPRDAPQSSPQ, *A. thaliana* gene ID: At5g20480), 1:250 dilution.
11. Secondary antibody: Goat anti-rabbit IgG-HRP (Jackson Laboratories, 111-035-003), 1:20,000 dilution in 5% (w/v) nonfat dry milk in 1× PBS.
12. Enhanced chemiluminescence (ECL) prime western blotting detection reagent (GE Healthcare, RPN2232).
13. National Institutes of Health (NIH) ImageJ software (http://imagej.nih.gov/ij/).

2.6 Membrane Stripping

1. Stripping solution: 50 mM Tris-Cl (pH 7.5), 2% (w/v) SDS, 0.16% (v/v) β-mercaptoethanol.
2. 60 °C oven.
3. 1× PBS.

2.7 Endo H Treatment

1. Endoglycosidase H (Endo H; 1,000,000 units/mL; New England Biolabs, P0703L).
2. 10× citrate buffer (pH 5.4): 127.5 mM citric acid, 372.5 mM sodium citrate.
3. 37 °C incubator.
4. 80 °C heat block.

2.8 Seedling Growth Inhibition and Anthocyanin Assays

1. 18–25 G1½ hypodermal needle.
2. Bunsen burner.
3. Liquid nitrogen.

4. Lyophilizer (Labconco).
5. Balance (≥1 mg).
6. TissueLyzer homogenizer (Qiagen) and 5-mm stainless steel beads (Thomson, 440CSS).
7. Acidic methanol: 99% (v/v) methanol, 1% (v/v) 12.1 M HCl.
8. 70 °C incubator.
9. Visible light spectrophotometer (Beckman).

2.9 MAPK Activation Assay

1. MAPK extraction buffer: 25 mM β-glycerophosphate, 10% (v/v) glycerol, 50 mM Tris-Cl (pH 7.5), 200 mM NaCl, 0.1% (v/v) Tween-20, 5 mM NaF, 0.5 mM DTT, 1 mM EDTA (pH 8.0), 2 mM $NaVO_3$, 1 mini-complete protease inhibitor tablet (Roche, 4693159001), and 1 phosphatase inhibitor tablet (Roche, 4906845001), freshly prepared on ice (*see* **Note 1**).
2. 2× Laemmli buffer: 125 mM Tris-Cl (pH 6.8), 4% (w/v) SDS, 10% (v/v) β-mercaptoethanol, 20% (v/v) glycerol [23].
3. 2-mL round-bottom and 1.7-mL microcentrifuge tubes.
4. Liquid nitrogen.
5. TissueLyzer homogenizer (Qiagen) and 5-mm stainless steel beads (Thomson, 440CSS).
6. 4 °C benchtop microcentrifuge.
7. 95 °C heat block.
8. Room-temperature benchtop microcentrifuge.

3 Methods

3.1 Tunicamycin Treatment

1. Sterilize Col-0 *Arabidopsis* seeds in an excess of seed sterilization solution for 10 min. In biosafety cabinet, under sterile conditions, rinse seeds with autoclaved dH_20. Repeat three to five times.
2. Add 1 mL of MS media 2 and 12 seeds per well in a 12-well tissue culture plate.
3. Place sample-containing plate on a light cart, under a 16-h light cycle at 21 °C, 70–80 μE intensity, and 50–75% humidity.
4. On day 7, in biosafety cabinet, exchange media with 1 mL of MS media 2 (*see* **Note 2**).
5. On day 8, in biosafety cabinet, add 10 μL of tunicamycin to a final concentration of 5 μg/mL. Incubate the sample-containing plate under the same conditions as in **step 3** for 24 h.
6. On day 9, air-dry seedlings from each well on paper towels for 5 min, transfer seedlings to 2-mL round-bottom microcentrifuge tubes, and freeze samples in liquid nitrogen. Samples can be stored at −80 °C until processing (*see* **Note 3**).

3.2 Total Protein Extraction

1. Cool the homogenizer plates by pouring liquid nitrogen on them. Quickly transfer the sample tubes to the homogenizer plate and open the lids. Pour liquid nitrogen on top of the samples to keep them frozen. Add 80 μL of total protein extraction buffer and a liquid nitrogen-cooled 5-mm stainless steel bead to each sample.
2. Homogenize the samples for 2 min at 25 Hz.
3. Centrifuge at 3000 × *g* for 5 s to collect homogenate at the bottom of the tube.
4. Incubate samples at 95 °C for 10 min.
5. Centrifuge at 11,000 × *g* for 8 min at room temperature to precipitate insoluble material.

3.3 SDS-PAGE

1. Cast an 8.5% SDS polyacrylamide separating gel of 1.5-mm thickness (*see* **Note 4**).
2. Overlay gel with 0.1% SDS or 2-propanol and allow gel to polymerize for 20–30 min.
3. Gently rinse off the 0.1% SDS or 2-propanol layer with dH_20 and use a lint-free tissue to remove excess liquid on top of the gel.
4. Cast a 4% stacking gel on top of the separating gel, insert comb, and allow gel to polymerize for at least 10 min.
5. Prerun gel at 100 V for 10 min (*see* **Note 5**).
6. Load 35 μL of each sample per well of the gel. Load 7.5 μL of prestained molecular-weight markers in the first well.
7. Separate proteins at 150 V until the 62-kDa marker band reaches the bottom of the gel.

3.4 Western Blotting

1. Once the run is complete, incubate gel in dH_20 for 5 min with gentle agitation.
2. Equilibrate gel and sponges in cold transfer buffer for 20 min.
3. Incubate PVDF membrane in methanol for 1–5 min and equilibrate in ice-cold transfer buffer for 20 min.
4. Assemble blotting sandwich following manufacturer's instructions.
5. In cold-room, transfer proteins onto membrane at 100 V for 1 h or until current reaches 0.7 mA for a Tetra blotter apparatus, or at 100 V for 30 min or until current reaches 1 mA for a Criterion blotter apparatus.
6. Allow membrane to air dry completely (10–30 min).
7. Stain with Ponceau S for 5 min.
8. Destain using 5% acetic acid for 10 min and rinse three times in dH_20.

9. Image and rinse with 1× PBS on a shaker until Ponceau S staining has faded.
10. Block with 5% nonfat dry milk for 20–30 min at room temperature with gentle rocking.
11. Incubate with primary antibody anti-FLS2 overnight at 4 °C with gentle rocking.
12. Wash blot with 1× PBS for 5 min with gentle agitation. Repeat two times.
13. Incubate with secondary antibody at room temperature for 1 h with gentle agitation.
14. Wash blot with 1× PBS for 5 min with gentle agitation. Repeat two times.
15. Incubate with ECL solution per manufacturer's instructions.
16. Perform detection using digital imaging system or film.

3.5 Membrane Stripping

1. After detection, rinse membrane with 1× PBS and incubate in stripping solution at 60 °C for 30 min.
2. Rinse membrane with 1× PBS several times to remove all SDS and β-mercaptoethanol.
3. Proceed to **step 10** of Subheading 3.4, except incubate in primary antibody anti-EFR.

3.6 Endo H Treatment

1. Sterilize Col-0 *Arabidopsis* seeds in an excess of seed sterilization solution for 10 min. In biosafety cabinet, under sterile conditions, rinse seeds with autoclaved dH_20. Repeat three to five times.
2. Add 1 mL of MS media 2 and 12 seeds per well in a 12-well tissue culture plate.
3. Place sample-containing plate on a light cart, under a 16-h light cycle at 21 °C, 70–80 μE intensity, and 50–75% humidity.
4. On day 7, in biosafety cabinet, exchange media with 1 mL of MS media 2 (*see* **Note 2**).
5. On day 9, air-dry seedlings from each well on paper towels for 5 min, transfer seedlings to 2-mL round-bottom microcentrifuge tubes, and freeze samples in liquid nitrogen. Samples can be stored at −80 °C until processing (*see* **Note 3**).
6. Proceed to Subheading 3.2.
7. Set up two sample reactions:

   Reaction H: 34 μL of protein extract and 1000 U of Endo H in 1× citrate buffer.

   Reaction O: 34 μL of protein extract in 1× citrate buffer.
8. Incubate at 37 °C for 1–3 h (*see* **Note 6**).

9. Incubate the reaction at 80 °C for 10 min to heat-inactivate the enzyme.
10. Centrifuge samples at max speed for 1 min at room temperature.
11. Proceed to Subheading 3.3, except casting a 6% SDS-PAGE gel and separate proteins at 150 V until 95 kDa band reaches the bottom of the gel.
12. Proceed to Subheading 3.4.

3.7 Seedling Growth Inhibition Assay

1. Sterilize Col-0 and *fls2* (or *efr*) *Arabidopsis* seeds in an excess of seed sterilization solution for 10 min. In biosafety cabinet, under sterile conditions, rinse seeds with autoclaved $dH_2 0$. Repeat three to five times.
2. Label the first 12-well tissue culture plate with "Col-0" and the second plate with "*fls2*" (or "*efr*"). Add 1 mL of MS media 1 and five seeds of the appropriate genotype to each well. Each 12-well plate can accommodate the treatment of one genotype with a mock dilution and three tenfold serial dilutions of the MAMP elicitor flg22 (or elf26) in triplicate (*see* **Note 7**).
3. Synchronize seeds at 4 °C in the dark for a minimum of 48 h.
4. Place the sample-containing plate on a light cart, under a 16-h light cycle at 21 °C, 70–80 μE light intensity, and 50–75% humidity.
5. On day 3, in biosafety cabinet, prepare 100× working solution for each tenfold serial dilution (e.g., 1 mM, 100 μM, 10 μM) of the MAMP elicitor flg22 (or elf26) for a total of three concentrations and add 60 μL of each solution to a 15-mL conical tube containing 5.94 mL of MS media 1 for a total of three final concentrations (e.g., 10 μM, 1 μM, 100 nM).
6. Insert a 1-mL filter tip into a 200-μL tip and use it to remove the liquid media from the first column of wells of both plates using a pipetman.
7. Pipette 1 mL of MS media 1 (mock dilution) into each well in the first column of both plates. Repeat **steps 6** and **7** for the remaining columns of wells using the first serial dilution instead of the mock dilution. Repeat **steps 6** and **7** for the third and fourth columns of wells using the second and third serial dilutions, respectively (*see* **Note 8**).
8. Incubate the sample-containing plates under the same conditions as in **step 4** for 2 days, and then count the number of seedlings per well (*see* **Note 9**).
9. After an additional 3 days, air-dry the seedlings from each well on paper towels for 10 min and measure their collective fresh weight (FW) using a balance. Each 12-well plate generates 12 samples, three biological replicates per elicitor concentration for a given genotype (*see* **Note 10**).

10. Create a data sheet of the number of seedlings and their FW from each sample to calculate the FW per seedling. Set the mock treated value to 100% and calculate the percentage of inhibition of seedling growth induced by each MAMP elicitor concentration with respect to mock-treated samples. Repeat for each genotype.

3.8 Anthocyanin Accumulation Assay

1. Sterilize Col-0 and *fls2* (or *efr*) *Arabidopsis* seeds in an excess of seed sterilization solution for 10 min. In biosafety cabinet, under sterile conditions, rinse seeds with autoclaved dH_20. Repeat three to five times.
2. Label the first 24-well tissue culture plate with "Col-0" and the second with "*fls2*" (or "*efr*"). Add 0.5 mL of MS media 3 and five (or eight) seeds of the appropriate genotype to each well (*see* **Note 11**). Each 24-well plate can accommodate the treatment of one genotype with a mock dilution and three tenfold serial dilutions of the MAMP elicitor flg22 (or elf26) in triplicate (*see* **Note 12**).
3. Synchronize seeds at 4 °C in the dark for a minimum of 48 h.
4. Place the sample-containing plate on a light cart, under a 24-h light cycle at 21 °C, 100 μE light intensity, and 50–75% humidity.
5. On day 3, in biosafety cabinet, prepare 100× working solution for each tenfold serial dilution (e.g., 1 mM, 100 μM, 10 μM) of the MAMP elicitor flg22 (or elf26) for a total of three concentrations and add 60 μL of each solution to a 15-mL conical tube containing 5.94 mL of MS media 4 for a total of three final concentrations (e.g., 10 μM, 1 μM, 100 nM).
6. Insert a 1-mL filter tip into a 200-μL tip and use it to remove the liquid media from the first row of wells of both plates using a pipetman.
7. Pipette 0.5 mL of MS media 3 (mock dilution) into each well in the first row of both plates. Repeat **steps 6** and **7** for the second row of wells of both plates using the first serial dilution instead of the mock dilution. Repeat **steps 6** and **7** for the third and fourth rows of wells using the second and third serial dilutions, respectively (*see* **Note 13**).
8. Incubate plates under the same conditions as in **step 4** for 4 days (*see* **Note 14**).
9. Heat a needle using a Bunsen burner and poke two air holes in the lids of each 1.7-mL microcentrifuge tube for ventilation. Label tubes with an alcohol-resistant marker.
10. Combine seedlings from two wells for each row and air-dry them on paper towels for 2–5 min. Transfer the seedlings to the ventilated and labeled microcentrifuge tubes. Each 24-well

plate should generate 12 samples, three biological replicates per elicitor concentration (*see* **Note 15**).

11. Freeze samples in liquid nitrogen and lyophilize them overnight. Cover container with aluminum foil to prevent light-induced degradation of anthocyanins.
12. Calibrate the balance with the weight of an empty 2-mL round-bottom microcentrifuge tube, transfer the lyophilized contents from the 1.7-mL tube to the 2-mL tube, and record the sample dry weight (DW). Repeat for all samples.
13. Add one 5-mm stainless steel bead to each sample-containing 2-mL microcentrifuge tube and homogenize the sample to a fine powder for 1 min at 25 Hz.
14. Carefully open the microcentrifuge tubes so not to lose any homogenized contents and add 500 μL of acidic methanol to each sample.
15. Cap samples tightly and vortex until sample is completely dissolved in solution.
16. Incubate samples at 70 °C for 15 min to extract anthocyanin. *Caution*: this is the boiling point of methanol and can cause tubes to pop open when handled.
17. Centrifuge extracts at max speed for 5 min.
18. Transfer 200 μL of supernatant to a new 1.7-mL microcentrifuge tube containing 300 μL of acidic methanol.
19. Measure the absorbance (Abs) value of each sample with the spectrophotometer set at 530 and 657 nm.
20. Create a data sheet with DW, Abs_{530}, and Abs_{657} of each sample.
21. Determine the anthocyanin content using the following equation: $(Abs_{530}-0.25\times Abs_{657})/DW$ (mg) [24]. Set the anthocyanin content of the water-treated samples to 100% and calculate the relative anthocyanin repression elicited by each MAMP concentration (*see* **Note 16**).

3.9 MAPK Activation Assay

1. Sterilize Col-0 *Arabidopsis* seeds in an excess of seed sterilization solution for 10 min. In biosafety cabinet, under sterile conditions, rinse seeds with autoclaved dH_2O. Repeat three to five times.
2. Add 1 mL of MS media 2 and 12 seeds to each well of a 12-well tissue culture plate for a total of seven wells. Each 12-well plate can accommodate the treatment of a single genotype with a mock dilution and 3 serial dilutions of the MAMP elicitor flg22 (or elf26) in triplicate.
3. Place the sample-containing plate on a light cart, under a 16-h light cycle at 21 °C, 70–80 μE light intensity, and 50–75% humidity.

4. On day 7, in biosafety cabinet, exchange liquid media with 1 mL of MS media 2 (*see* **Note 2**). Place the sample-containing plate under the same conditions as in **step 3** (*see* **Note 17**).
5. On day 9, prepare 100× working solution for each tenfold serial dilution (e.g., 1 mM, 100 μM, 10 μM, 1 μM, 100 nM, 10 nM) of the MAMP elicitor flg22 (or elf26) for a total of six concentrations. Add 10 μL of mock dilution to the first well and repeat for each point of the serial dilution.
6. After 5 min, gently transfer seedlings to a paper towel to dry excess media (*see* **Note 18**).
7. After another 5 min, transfer seedlings to a 2 mL round-bottom sample tube (*see* **Note 19**).
8. Freeze samples in liquid nitrogen. Samples can be stored at −80 °C until processing.
9. Cool the homogenizer plates by pouring liquid nitrogen on them. Quickly transfer the sample tubes to the homogenizer plate and open the lids. Pour liquid nitrogen on top of the samples to keep them frozen. Add 100 μL of ice-cold MAPK extraction buffer and a liquid nitrogen-cooled 5-mm stainless steel bead to each sample (*see* **Note 1**).
10. Homogenize the samples for 2 min at 25 Hz.
11. Centrifuge at 4 °C for 3 min.
12. Transfer 90 μL of supernatant to a precooled 1.7-mL tube and add 90 μL of freshly prepared 2× Laemmli buffer. Repeat for each sample.
13. Incubate samples at 95 °C for 3 min.
14. Centrifuge for 1 min at room temperature at maximum speed.
15. Load 25 μL onto a 10% SDS-PAGE gel.
16. Continue from Subheading 3.3, except casting three 10% SDS polyacrylamide separating gels of 1.5-mm thickness, loading 25 μL of samples per well on triplicate gels, and separating proteins at 150 V until the 29-kDa marker band reaches the bottom of the gel.
17. Continue from Subheading 3.4, except incubating in primary antibody anti-phospho-p44/42 MAPK (*see* **Notes 20** and **21**).
18. After detection, quantify phosphorylated MPK6, MPK3, and MPK4 signal intensities *en bloc* using NIH ImageJ, following software instruction.
19. Continue from Subheading 3.5, except using primary antibody anti-AtMPK3 (*see* **Note 20**).
20. After detection, quantify MPK3 signal intensity using NIH ImageJ, following software instruction.

4 Notes

1. Due to the labile nature of the phosphate bond, short processing times and ice-cold temperature are a must for this technique. Therefore, keep all reagents, tubes, and samples on ice. We recommend processing no more than 12 samples at a time, although a maximum of 24 samples can be processed. The MAPK extraction buffer is stable on ice for two consecutive rounds of sample extractions.
2. Due to the evaporation, absorption, and alteration of the media by growing seedlings over the course of a week, the media in each well must be replenished with fresh media. To ensure enough time for the seedlings to recover from the mechanical stress caused by the media exchange, we recommend the media exchange to occur on day 7.
3. If performing a chemical treatment followed by MAPK activation assay after **step 6** proceed with **step 5** of Subheading 3.9.
4. We use different percentage (5–13.5%) SDS-PAGE separating gels to allow for differential separation of specific and nonspecific proteins for immuno-detection by a given primary antibody, especially for proteins whose molecular masses may be altered by chemical treatments.
5. Prerunning gels eliminates contaminants that may be present in the separating gel and thus can interfere with protein migration.
6. Although the incubation time can vary from 1 to 16 h, with 3 h as the ideal, longer incubation times will result in more protein degradation.
7. This plating scheme is used to test the elicitation strength of three serial MAMP dilutions. When comparing wild-type to mutant seedlings, label the first column of a 12-well tissue culture plate with "Col-0," the second with "m1" (for mutant number 1), the third with "m2" (for mutant number 2), and the last with "*fls2*" (or "*efr*"). Repeat this for two other 12-well plates to obtain three biological replicates per genotype.
8. The described exchange of media is valid when testing a dilution series of an elicitor. When comparing immune responses between wild-type, 1–2 mutants, and *fls2* (or *efr*) for a total of two serial MAMP dilutions, each column of a triplicate plate contains a different genotype. Exchange the media for each row of wells, so that the first row will receive the mock treatment; the second, the first serial dilution of elicitor; and the third, the second dilution of elicitor. Repeat for the second and third plates.

9. Total seedling count may be less than the total seed count due to reduced seed germination rates and seedling losses from media exchanges.
10. After 8 days of growth, we observed the growth rate of mock-treated seedlings to slow due to space limitations in the wells.
11. To ensure that all seedlings in a well are equally exposed to the MAMP elicitor, we recommend plating five seeds per well for MAMP treatments <100 nM, and plating eight seeds per well for MAMP treatments ≥100 nM.
12. The plating scheme provided is valid when testing an elicitor dilution series. When comparing immune responses between wild-type, 1–2 mutants, and *fls2* (or *efr*) for one concentration of elicitor, label the first row of a 24-well tissue culture plate with "Col-0," the second with "m1" (for mutant number 1), the third with "m2" (for mutant number 2), and the last with "*fls2*" (or "*efr*"). Repeat this for two other 24-well plates. Proceed by adding the media and five seeds of the appropriate genotype to each well.
13. The described exchange of media is valid when testing a dilution series of an elicitor. When comparing immune responses among wild-type, 1–2 mutants, and *fls2* (or *efr*) for one MAMP concentration, draw a vertical line on the lid for all three plates, between columns 3 and 4 and through all four rows. Exchange the media in the wells to the left of the line with media containing the mock dilution and the media in the wells to the right of the line with media containing the MAMP concentration to be tested.
14. To ensure that seedlings have dry weights of more than 1 mg, which is the lower limit of the balance, we recommend extending the length of elicitation to 5 days for MAMP elicitor treatments >100 nM. We also observed that increasing the length of elicitation can enhance differences in growth arrest or anthocyanin repression responses between wild-type and mutant seedlings.
15. When comparing immune responses among wild-type, 1–2 mutants, and *fls2* (or *efr*), combine seedlings from three wells per genotype. Each 24-well plate should generate one biological replicate of the mock-treated sample and one of the MAMP-elicited samples per genotype for a total of eight samples. Three plates are needed to generate the three biological replicates per elicitor concentration per genotype to test statistical significance.
16. Due to pipetting errors, plate-to-plate variability, and imprecise treatment-lengths, we highly recommend growing wild-type and *fls2* (or *efr*) seedlings on each 24-well plate to serve as normalization factors for comparative analysis.

17. When performing a tunicamycin treatment, add the desired amount of chemical to every other well on day 8 and place the sample-containing plate under the same conditions as in **step 3**.
18. Mechanical wounding can also activate MAPKs; therefore, gently handle the samples.
19. MAMP-induced MAPK activation is fast and dynamic, with a maximum signal intensity between 10 and 15 min.
20. Incubations with strong antibodies, like the anti-phospho-p44/42 MAPK and anti-MAPK3, can be carried out at room temperature for 1.5–2 h, with gentle rocking.
21. To minimize the technical variability due to transfer, it is possible to cut each of the 10% replicate gels between the 37 and 50 kDa molecular weight markers and transfer the three strips to a single PVDF membrane.

Acknowledgments

This work was supported by T32 GM007499 (to S.A.L.) and Yale University Elizabeth Brown Postdoctoral Fellowship (to T.C.).

References

1. Aebi M, Bernasconi R, Clerc S, Molinari M (2010) *N*-Glycan structures: recognition and processing in the ER. Trends Biochem Sci 35:74–82
2. Strasser R (2014) Biological significance of complex *N*-glycans in plants and their impact on plant physiology. Front Plant Sci 5:363
3. Gómez-Gómez L, Boller T (2000) FLS2: an LRR receptor-like kinase involved in the perception of the bacterial elicitor flagellin in *Arabidopsis*. Mol Cell 5:1003–1011
4. Zipfel C, Kunze G, Chinchilla D, Caniard A, Jones JDG, Boller T, Felix G (2006) Perception of the bacterial PAMP EF-Tu by the receptor EFR restricts *Agrobacterium*-mediated transformation. Cell 125:749–760
5. Saijo Y (2010) ER quality control of immune receptors and regulators in plants. Cell Microbiol 12:716–724
6. Li J, Zhao-Hui C, Batoux M, Nekrasov V, Roux M, Chinchilla D et al (2009) Specific ER quality control components required for biogenesis of the plant innate immune receptor EFR. Proc Natl Acad Sci U S A 106: 15973–15978
7. Lu X, Tintor N, Mentzel T, Kombrink E, Boller T, Robatzek S et al (2009) Uncoupling of sustained MAMP receptor signaling from early outputs in an *Arabidopsis* endoplasmic reticulum glucosidease II allele. Proc Natl Acad Sci U S A 106:22522–22527
8. Nekrasov V, Li J, Batoux M, Roux M, Chu Z-H, Lacombe S, Bougon A et al (2009) Control of the pattern-recognition receptor EFR by an ER protein complex in plant immunity. EMBO J 28:3428–3438
9. Saijo Y, Tintor N, Lu X, Rauf P, Pajerowska-Mukhtar K, Häweker H et al (2009) Receptor quality control in the endoplasmic reticulum for plant innate immunity. EMBO J 28: 3439–3449
10. von Numers N, Survila M, Aalto M, Batoux M, Heino P, Palva ET et al (2010) Requirement of a homolog of glucosidease II beta-subunit for EFR-mediated defense signaling in *Arabidopsis thaliana*. Mol Plant 3:740–750
11. Häweker H, Rips S, Koiwa H, Salomon S, Saijo Y, Chinchilla D, Robatzek S, von Schaewen A (2010) Pattern recognition receptors require *N*-glycosylation to mediate plant immunity. J Biol Chem 285:4629–4636
12. Sun W, Cao Y, Jansen KL, Bittel P, Boller T, Bent AF (2012) Probing the *Arabidopsis* flagellin receptor: FLS2-FLS2 association and the contributions of specific domains to signaling function. Plant Cell 24:1096–1113

13. Chinchilla D, Bauer Z, Regenass M, Boller T, Felix G (2006) The *Arabidopsis* receptor kinase FLS2 binds flg22 and determines the specificity of flagellin perception. Plant Cell 18:465–476
14. Elbein AD (1987) Inhibitors of the biosynthesis and processing of *N*-linked oligosaccharide chains. Annu Rev Biochem 56:497–534
15. Martínez IM, Chrispeels MJ (2003) Genomic analysis of the unfolded protein response in Arabidopsis shows its connection to important cellular processes. Plant Cell 15:561–576
16. Iwata Y, Koizumi N (2005) An *Arabidopsis* transcription factor, AtbZIP60, regulates the endoplasmic reticulum stress response in a manner unique to plants. Proc Natl Acad Sci U S A 102:5280–5285
17. Lozoya E, Block A, Lois R, Hahlbrock K, Scheel D (1991) Transcriptional repression of light-induced flavonoid synthesis by elicitor treatment of cultured parsley cells. Plant J 1:277–234
18. Lo SC, Nicholson RL (1998) Reduction of light-induced anthocyanin accumulation in inoculated sorghum mesocotyls. Implications for a compensatory role in the defense response. Plant Physiol 116:979–989
19. McLusky SR, Bennett MH, Beale MH, Lewis MJ, Gaskin P, Mansfield JW (1999) Cell wall alterations and localized accumulation of feruloyl-3'-methoxytyramine in onion epidermis at sites of attempted penetration by *Botrytis allii* are associated with actin polarisation, peroxidase activity and suppression of flavonoid biosynthesis. Plant J 17:523–534
20. Serrano M, Kanehara K, Torres M, Yamada K, Tintor N, Kombrink E et al (2012) Repression of sucrose/ultraviolet B light-induced flavonoid accumulation in microbe-associated molecular pattern-triggered immunity in Arabidopsis. Plant Physiol 158:408–422
21. Korasick DA, McMichael C, Walker KA, Anderson JC, Bednarek SY (2010) Novel functions of Stomatal Cytokinesis-Defective 1 (SCD1) in innate immune responses against bacteria. J Biol Chem 285:23342–23350
22. Smith JM, Leslie ME, Robinson SJ, Korasick DA, Zhang T, Backues SK et al (2014) Loss of *Arabidopsis thaliana* dynamin-related protein 2B reveals separation of innate immune signaling pathways. PLoS Pathog 10:e1004578
23. Laemmli UK (1970) Cleavage of structural proteins during the assembly of the head of bacteriophage T4. Nature 227:680–685
24. Mancinelli AL, Schwartz OM (1984) The photoregulation of the anthocyanin synthesis IX. The photosensitivity of the response in dark and light-grown tomato seedlings. Plant Cell Physiol 25:93–105

Chapter 7

Steady-State and Kinetics-Based Affinity Determination in Effector-Effector Target Interactions

André Reinhard and Thorsten Nürnberger

Abstract

Dissecting the functional basis of pathogenicity and resistance in the context of plant innate immunity benefits greatly from detailed knowledge about biomolecular interactions, as both resistance and virulence depend on specific interactions between pathogen and host biomolecules. While in vivo systems provide biological context to host-pathogen interactions, these experiments typically cannot provide quantitative biochemical characterization of biomolecular interactions. However, in many cases, the biological function does not only depend on whether an interaction occurs at all, but rather on the "intensity" of the interaction, as quantified by affinity. Specifically, microbial effector proteins may bind more than one host target to exert virulence functions, and looking at these interactions more closely than would be possible in a purely black-and-white qualitative assay (as classically based on plant or yeast systems) can reveal new insights into the evolutionary arms race between host and pathogen. Recent advances in biomolecular interaction assays that can be performed in vitro allow quantification of binding events with ever greater fidelity and application range. Here, we describe assays based on microscale thermophoresis (MST) and surface plasmon resonance (SPR). Using these technologies allows affinity determination both in steady-state and in kinetic configurations, providing two conceptually independent pathways to arrive at quantitative affinity data describing the interactions of pathogen and host biomolecules.

Key words Protein-protein interactions, Affinity, Kinetics, Microscale thermophoresis, Surface plasmon resonance

1 Introduction

1.1 Steady-State Affinity and Kinetics: Two Ways to Quantify an Interaction

Biomolecular interactions form the functional basis of most biological processes. Ligands bind to receptors, substrates bind to enzymes, and members of signaling cascades bind to one another to transmit signals. While resulting in large datasets, qualitative "black or white" screening approaches for interaction analysis, as commonly performed using two-hybrid systems or coaffinity purification followed by mass spectrometry, are notoriously error-prone; rates of false discoveries have been estimated to routinely reach about 50% [1, 2]. Quantitative description of an interaction

Libo Shan and Ping He (eds.), *Plant Pattern Recognition Receptors: Methods and Protocols*, Methods in Molecular Biology, vol. 1578, DOI 10.1007/978-1-4939-6859-6_7, © Springer Science+Business Media LLC 2017

provides an in-depth look that can drastically increase the confidence in a putative interaction [3]. A common way to quantify the intensity of biomolecular interactions is the determination of binding affinity—ultimately a description of the sum of attractive and repulsive intermolecular forces that govern the interaction. Fundamentally, biomolecular interactions can often be reasonably described by simple binding models that are governed by the law of mass action; formation of a biological complex can thus be described succinctly as the reversible formation of a product made up of two reactants A and B: $A + B \leftrightarrow AB$.

Steady-state affinity determination assays take advantage of the fact that such reactions eventually reach chemical equilibrium; that is, the association and dissociation rates of the complex are equal and there is no net change in the amount of complex. If the concentration of the reactants can be predetermined and the amount of complex can be observed, experimental titration approaches allow quantification of the interaction's binding affinity. The most common way to describe affinity in the context of biologically relevant interactions is that of the dissociation constant, KD, which is defined as the concentration of free A at which half of B is associated with A.

Kinetic affinity determination reaches the same information via a different pathway; instead of observing reactions that have reached chemical equilibrium, the rates of association and dissociation of the complex are measured directly. Formation of the complex AB is described by a forward reaction ($A + B \rightarrow AB$), which is governed by the association rate constant, ka. Conversely, dissociation of the complex AB is described by a backward reaction ($AB \rightarrow A + B$), which is governed by the dissociation rate constant, kd. The affinity of the interaction (KD) can then be determined directly from the association and dissociation rate constants: $KD = k\text{d}/k\text{a}$. Compared to steady-state analysis, kinetic analysis provides additional information to characterize an interaction, as two interactions with the same affinity can have different kinetic behaviors [4].

1.2 Microscale Thermophoresis (MST)

Microscale thermophoresis (MST) assays are based on the motion of particles in a temperature gradient (thermophoresis; also called thermodiffusion or Ludwig-Soret effect). This effect applies to particles ranging in size from macroscopic colloids and aerosols to single ions in solution. Although the existence of thermophoresis has been discovered in the mid nineteenth century, the precise molecular mechanisms governing this motion remain largely elusive. For typical biomolecules in aqueous solutions, important factors include size of the molecule, charge, and extent of the hydration shell [5]; the complex interplay of these and other factors makes accurate prediction of thermophoretic behavior of biomolecules difficult. Experimentally, biomolecules may exhibit positive (defined as movement from hot to cold) or negative (defined as movement from cold to hot) thermophoresis. Taking general molecular diffusion into account, a steady-state situation will be reached for any given temperature gradient when the thermophoretic force acting on a particle equals the diametric force of back-diffusion (Fig. 1a).

When biomolecular interactions take place, the complex that is being formed exhibits thermophoretic behavior that is different to that of its isolated constituent molecules. MST assays take advantage of this by tracking the movement of a specific biomolecule through fluorescence and using the differential thermophoretic behavior of its possible states (i.e., isolated vs. complex-bound) to infer the amount of complex formation (Fig. 1b). In practice, quantitative determination of binding affinities is possible for pairs of biomolecules by using a titration approach. While the concentration of the observed molecule is held constant, the second molecule is titrated over a broad range of concentrations (Fig. 1c). By setting the amount of complex being formed into relation with the concentration of the second molecule, the affinity of the interaction can be determined (Fig. 1d).

Instruments for MST assays use coupled optical systems to induce temperature gradients (infrared laser) and measure fluorescence (UV-vis detector, excitation by LED, emission filtering using RGB color channels) in glass capillaries of 10 μL total internal volume. The temperature gradients induced by the infrared laser are spheroid with a radius of ~200 μm between the center point (highest temperature) and a 70% temperature drop point [6]. Temperature gradients are typically on the order of 2–6 °C (depending on the user-adjustable instrument infrared laser power). Typical time scales for a single capillary measurement are about 60 s with a laser-heated phase of 30 s and continuous fluorescence signal recording. The intensity of the fluorescence signal can be manipulated by the adjustment of excitation LED power and the choice of a specific emission filter set. Measurements over multiple capillaries (typically 8–16 capillaries per interaction) are used to calculate thermophoretically induced signal shifts dependent on concentration of the unlabeled molecule.

1.3 Surface Plasmon Resonance (SPR) Spectroscopy

Surface Plasmon Resonance (SPR) spectroscopy uses an optical biosensor approach to quantify the total molecular mass accumulated on a sensor surface in real time with a maximum molecular weight resolution of ~200 Da [7]. Molecular interactions can be studied by fixing one of the binding partners (termed "ligand") on the sensor surface and injecting the other binding partner (termed "analyte") in a time-controlled manner. Binding events between these molecules then lead to accumulation of mass on the sensor surface which can be detected directly. In SPR assays, data is obtained as curves of system response (plotted as arbitrary reso-

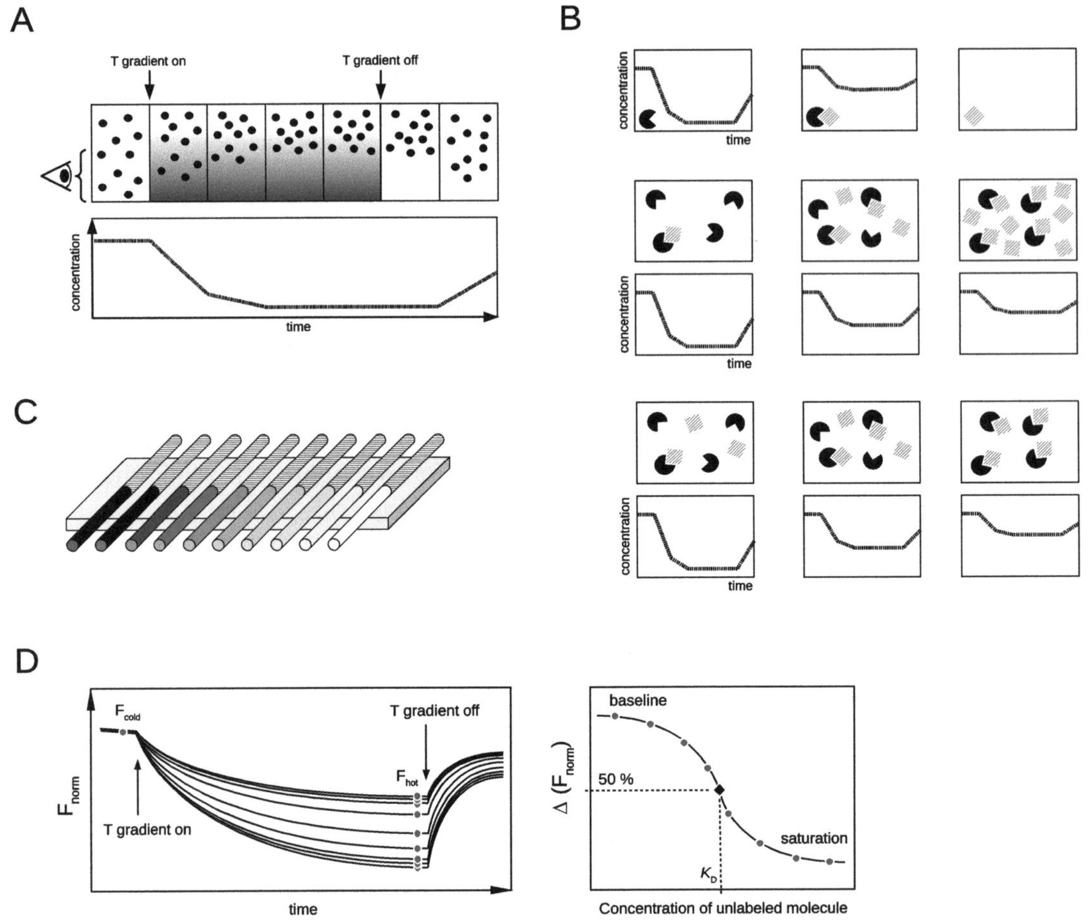

Fig. 1 Microscale thermophoresis. (**a**) Molecular principle for thermophoresis. *Top*: Compartment representation. Particles (*black dots*) are evenly and randomly distributed in a solution. Upon establishment of a temperature gradient (*gray*), particles move due to thermophoresis until a steady state has been reached. Upon abolishment of the temperature gradient, particles redistribute in the solution by diffusion. The eye symbolizes an external observer. *Bottom*: Apparent particle concentration over time (*dashed line*) as seen by the observer. (**b**) Complex formation and thermophoresis. Thermophoresis-induced concentration shifts (*dashed lines*) are depicted analogous to (**a**). One of the biomolecules (*black shape*) can be observed through fluorescence, while the other (*shaded diamond*) cannot. *Top*: Complex formation is reflected by a shift in thermophoretic mobility. *Center*: Titration of the nonobserved molecule leads to concentration-dependent shifts in apparent thermophoretic mobility for any given affinity of the interaction. *Bottom*: At a single constant concentration of the nonobserved molecule, differences in affinity are visible as differences in thermophoretic mobility. (**c**) Serial dilution capillary setup for MST. Equal volumes of two protein solutions are used. The labeled protein is used at the same concentration in each capillary (shaded part of capillary), while the unlabeled protein is diluted over the range of capillaries (*grayscale gradient*, with darker colors corresponding to higher concentration). Note: the two different protein solutions are in reality homogeneously distributed in each capillary; an artificial division in the center of each capillary was used for illustrative purposes. (**d**) MST affinity determination. *Left*: Fluorescence is traced before, during, and after the establishment of a temperature gradient (*arrows*) in multiple capillaries corresponding to one concentration of unlabeled molecule each. Fluorescence traces are plotted normalized to the initial ("cold") fluorescence for each capillary, yielding signal as F_{norm}. *Right*: The changes in normalized fluorescence ("hot" fluorescence vs. "cold" fluorescence, Δ F_{norm}) are plotted against the concentration of the unlabeled molecule, yielding a binding curve from which the affinity (K_D, *black diamond*) can be determined. Each data point (*red dot*) corresponds to a single fluorescent trace shown in the *left pane*

nance units, RU) over time. The signal is proportional to mass, with 1 RU corresponding to ~1 pg/mm^2 of molecules [8]. For a detailed description of the physical background of the method, see the relevant literature on SPR theory [9].

SPR instruments consist of a microfluidics system with automated sample injection mechanisms and constant flow regulated through a peristaltic pump (flow rates 5–30 μL/min). The reaction chamber (flow cells, Fig. 2a, left) is created by docking a sensor chip to the instrument, which closes the internal microfluidics system. The reaction chamber is enclosed in a temperature-controlled jacket (temperature accuracy ± 0.005 °C) to achieve a stable environment for kinetic measurements. The most commonly used type of sensor chip for protein-protein interactions (type CM5) consists of a thin gold surface on a glass substrate to which a carboxymethylated dextran hydrogel matrix has been covalently attached by the manufacturer to a thickness of ~100 nm, which corresponds to 0.2% of the flow cell height [10]. For immobilization of proteinaceous ligands, an amine coupling chemistry can be used which covalently links the N-terminus and lysine ε-amino groups of the ligand to the matrix carboxyl groups. This type of coupling results in a random, flexible, and three-dimensional distribution of ligand molecules within the dextran matrix (Fig. 2a, right), increasing the sensor's binding capacity compared to that of a purely planar surface and minimizing potential steric hindrance by decreasing the apparent ligand density in the flow system [11]. However, in the context of analyte diffusion across this third dimension, the thickness of the dextran layer can usually be neglected for practical experimental considerations [12, 13].

Injecting analyte into the system's flow cell (Fig. 2b) causes the total surface response signal to increase rapidly due to the presence of the analyte molecules in the laminar flow close to the surface. This does not necessarily indicate binding events. In the absence of binding the signal will return to the baseline value rapidly after the analyte sample flow has been switched off and the flow cell has been emptied of nonbound analyte (Fig. 2b, top). However, if a binding event takes place, the signal will initially remain at a high value after the analyte sample flow has been switched off and decrease more slowly. The following signal decrease is then due to dissociation of the complex, and the rate of this signal decrease is directly linked to the binding reaction's dissociation rate constant (Fig. 2b, bottom).

Under favorable circumstances, quantitative determination of binding parameters is possible using sensorgram data generated by analyte injection. Taking the predefined concentration of the analyte into account, an idealized binding event corresponding to a theoretical model (e.g., a simple 1:1 Langmuir binding of the analyte to the ligand) can be predicted. By varying flexible parameters in the model iteratively (e.g., association rate constant, *k*a, and

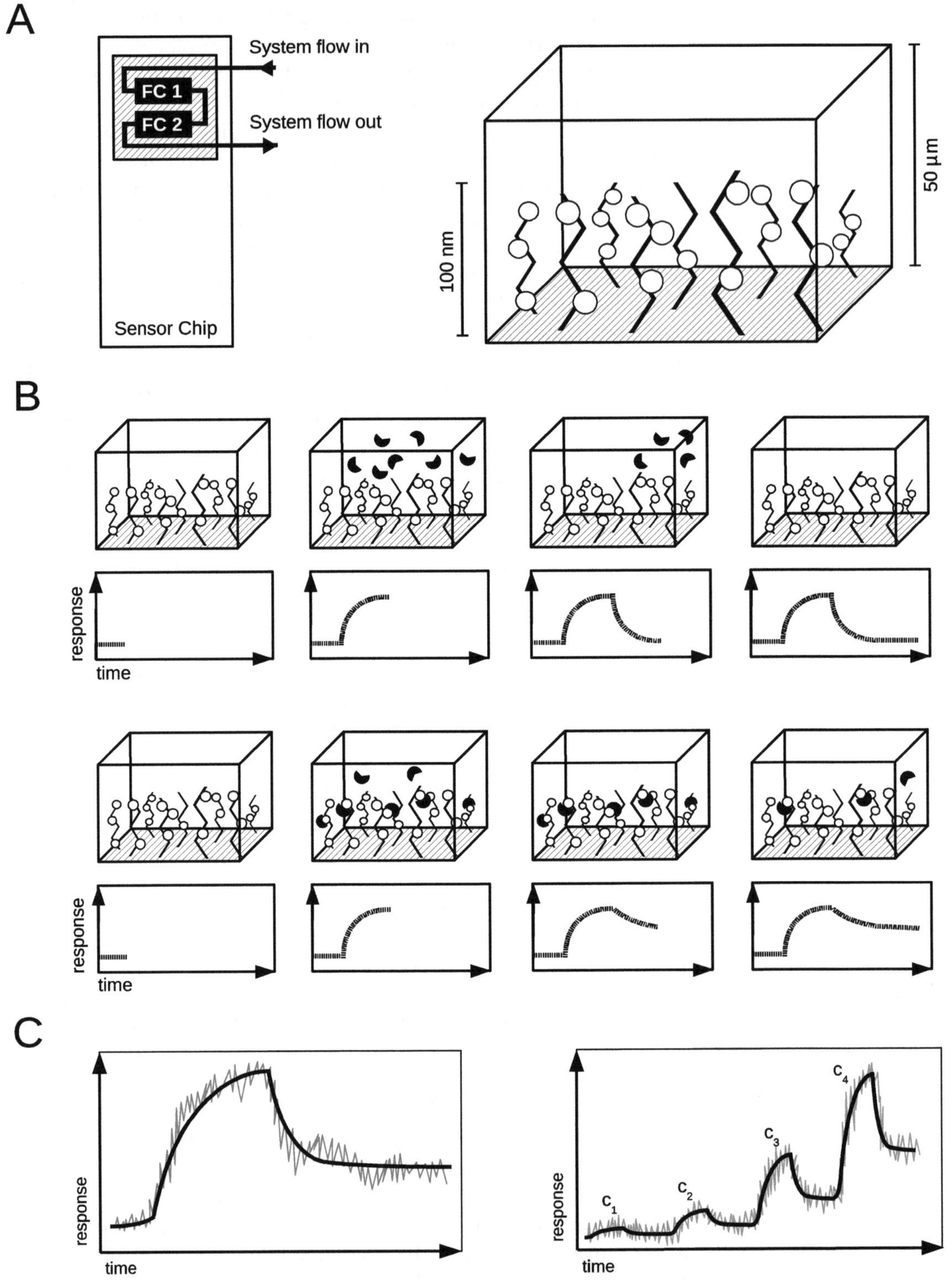

Fig. 2 Surface plasmon resonance. (**a**) Sensor chip for SPR measurements (not to scale). *Left*: Each sensor chip is made up of two flow cells (*black boxes*) that are gold-coated microscopic cavities etched into the glass substrate (*shaded*). Upon inserting the chip into the instrument, these cavities dock with corresponding cavities in the instrument to create the reaction volumes. System flow runs sequentially through both flow cells.

dissociation rate constant, *k*d, among others), the data can be fitted to the model until a minimum deviation is achieved (Fig. 2c).

Recent SPR platforms introduced kinetic analysis via single-cycle titration. In this work flow, the analyte is injected consecutively at increasing concentrations (Fig. 2c, right). In between injections, the analyte partially dissociates from the ligand. Although one injection of analyte at a known concentration would yield enough data to determine the kinetic constants, using a titration series increases the robustness of the fit by canceling out artifacts potentially arising at certain analyte concentrations. For best resolution, the titration should be performed around the expected *K*D of the interaction. Analysis is then performed on the entire sequence of injections ("global analysis," *see* [14]).

1.4 Fluorescent Labels for MST and SPR: Two Birds, One Stone

As described before, MST experiments rely on the fact the fluorescence of only one binding partner is observed during the assay. While MST instruments measuring the intrinsic fluorescence of proteins exist, their use is typically restricted for investigating binding events in which only one of the binding partners is a protein (e.g., protein-nucleic acid, protein-small molecule). Therefore, experiments involving most protein-protein interactions need a fluorescent label.

In contrast, SPR experiments are inherently label-free, but rely on fixing one of the binding partners (ligand) onto the sensor surface, usually by using coupling chemistries that may be detrimental to the activity of the ligand. Additionally, after each experimental cycle the ligand-analyte interaction interface must be regenerated; this means that the analyte must be removed quantitatively without disturbing the activity of the ligand. Generally, this procedure involves injecting a buffer which sufficiently changes the environment of the interaction interface for complete dissociation to occur. Often, this results in partial loss of ligand activity [15]. A possible solution to this problem is the use of a capture system. In this work flow, a so-called capturing molecule (usually an antibody)

Fig. 2 (continued) By convention, flow cell 1 (FC1) is used as a reference flow cell, while flow cell 2 (FC2) is used as the sample flow cell. *Right*: Simplified view of a single flow cell. The ligand (*white spheres*) is attached to flexible dextran chains (*black sticks*), which are in turn covalently fused to the gold surface (*shaded area*). (**b**) Compartment representations and sensorgrams for analyte injections. *Top*: If no interaction takes place between ligand (*white spheres*) and analyte (*black shapes*), the signal will rise temporarily during injection but fall back to baseline levels after injection has finished. Bottom: If an interaction takes place between ligand and analyte, the signal decrease after injection is more gradual and depends on the rate of dissociation between ligand and analyte. Note: axis labels are shown on first sensorgram only. (**c**) Exemplary SPR interaction kinetics analysis. *Left*: Data for a single analyte injection (*gray curve*, noisy) is being fitted to a mathematical model describing the interaction (*black curve*, smooth). *Right*: Kinetic titration experiment. Multiple injections of increasing concentration (c_1, c_2, c_3, c_4) have been performed during the same experimental run. The kinetic fit has been performed over the time frame of the whole experiment. Note: only the fit curve is shown for clarity

is covalently fixed onto the sensor surface using conventional coupling chemistry and the ligand is captured reversibly by using a high-affinity antibody-antigen interaction. This procedure is not without drawbacks; the interaction between capturing molecule and ligand has to be very stable to minimize ligand dissociation during the experimental cycle, and regeneration after each experiment involves dissociation of the ligand-analyte complex from the capturing molecule, increasing the quantity of ligand needed for repeated experiments. As reusing a sensor chip desirable, the capturing molecule should be stable over extended periods of time and be able to be regenerated without major activity loss. Finally, the capturing molecule should not sterically block the ligand-analyte interaction interface. A practical solution is the use of camelid single-domain antibodies (sdAbs, also called "nanobodies"), which are small, robust, bind epitopes with very high affinity and are readily available commercially [16].

By using a sdAb recognizing a fluorescent label (such as a GFP derivative) as capturing molecule in an SPR capture system, the same label can be used for both MST and SPR assays. This allows quantitative description of biomolecular interaction using two independent experimental procedures on the same types of samples, allowing streamlined preparatory work flows and higher fidelity in determining the binding affinity of a given biomolecular interaction.

2 Materials

2.1 Buffers, Reagents, and Solutions

Buffers are specified as working concentration. Preparation of sterile 10× stock solutions is convenient. Working concentration buffers should be either prepared freshly before the experiment or diluted freshly from sterile stocks; working concentration buffers do not need to be kept sterile, but should be filtered to ensure homogeneity.

2.1.1 Common Buffers

The composition of the experimental buffers may vary according to the specific solubility, stability, and activity needs of the biomolecules that will be studied. A basic buffer recipe that should be compatible with many cytoplasmic protein-protein interactions is given below. Care must be taken to ensure that buffers used within one experiment are prepared as a single aliquot to prevent buffer mismatch artifacts during experiments.

1. **Protein storage buffer**: 50 mM Tris pH 7.6 (adjusted with HCl), 150 mM NaCl.
2. **Assay buffer**: 50 mM Tris pH 7.6 (adjusted with HCl), 150 mM NaCl, 0.05% (v/v) Tween-20.

Ideally, this buffer should be identical to protein storage buffer except for detergent (also *see* **Note 1**). This buffer should be degassed (e.g., by extended vacuum filtration) for use in SPR assays. Detergent should be added after degassing.

2.1.2 MST Specific Buffers

Use standard assay buffer as given in Subheading 2.1.1 unless MST pretesting (*see* Subheading 3.2.1) indicates specific needs.

2.1.3 SPR Specific Buffers and Solutions

1. **Immobilization buffer**: 10 mM Acetate pH 5.5.
2. **Regeneration buffer**: 10 mM Glycine pH 2.0.
3. **Desorp solution 1**: 0.5% (w/v) sodium dodecyl sulfate (SDS).
4. **Desorp solution 2**: 50 mM Glycine-NaOH pH 9.5.
5. **20× sanitize solution**: Sodium hypochlorite (NaOCl), 8–15% active Chlorine. Dilute 1:20 immediately before use. Keep proper precautions when handling the concentrated solution.
6. **Immobilization reagents**: contained in the amine coupling kit (*see* Subheading 2.3.3).

2.2 Instruments and Software

Equipment manufacturers offer ranges of instruments for both MST and SPR experiments with different capabilities that cover specific needs. Selection of instruments according to some practical considerations for the type of experiments described in this chapter is discussed briefly here, but is not intended as an exhaustive guide for instrument choice. Due to the author's personal experience, this chapter works under the assumption of using specific instruments as listed below, but the procedures described in this chapter should be adaptable to other instrumental platforms with minor modifications.

2.2.1 Common Instruments for Sample Preparation

1. A **spectrophotometer** should be used to accurately quantify purified proteins using the specific extinction coefficient of the sample. Microvolume instruments (e.g., "NanoDrop," Thermo Scientific) are convenient to conserve sample.
2. A **centrifuge** (>20,000 × *g*) should be used to spin down protein samples before use to remove any particulate matter.

2.2.2 Instruments and Software for MST Assays

1. Main Instrument: **NT.115 platform** (Nanotemper; controlled via PC, bundled with instrument).
2. Software: **NT.115 evaluation software** (Nanotemper; bundled with instrument).
3. Accessories: **Capillary tray** (aluminum; Nanotemper).

MST technology was developed by Nanotemper (Munich, Germany) and at the time of writing Nanotemper was the only manufacturer offering instruments. MST instruments with the broadest spectrum of applications for protein-protein interactions (platform NT.115) are offered in multiple filtersets according to the excitation and emission wavelengths of protein labels. There are specific compatible combinations of instrument filter sets and labels; an overview is available on the manufacturer's web site.

2.2.3 Instruments and Software for SPR Assays

1. Main instrument: **Biacore ×100 platform** (controlled via PC; bundled with the instrument). Optional, but recommended: **Biacore ×100 "Plus package"** (contains in-line degasser and advanced controls).
2. Software: **Biacore Evaluation software** (bundled with the instrument).

SPR technology was developed by Biacore (Uppsala, Sweden), which is now a part of GE Healthcare. Instruments are now available from multiple manufacturers, but the classical Biacore platforms distributed by GE Healthcare remain the most popular by a wide margin; 89% of SPR-related publications in 2007 used Biacore systems [17]. The X100 platform that was used in the scope of this chapter is an entry level model with full support for kinetic analysis, but without the high-throughput screening capabilities of more advanced models.

2.3 Consumables

2.3.1 Common Consumables

Preparation of protein samples can generally be performed using standard issue laboratory plasticware. Some proteins may require special "low binding" vessels and pipette tips to minimize unspecific absorption and avoid quantitative differences in protein concentration after transfer.

2.3.2 Consumables Specific for MST Assays

1. **Capillary** sets (standard, hydrophobic, premium coated; Nanotemper). High-quality glass capillaries with specific properties are essential for MST assays; the ideal type of capillary must be determined during pretesting (Subheading 3.2.1).
2. **Small-capacity microcentrifuge tubes** (e.g., "PCR tubes," 200 μL capacity), with lids. Use of small capacity vessels is important to minimize volumetric concentration errors due to unspecific liquid adsorption and sample evaporation, a critical factor for MST assays.

2.3.3 Consumables Specific for SPR Assays

1. **Plastic vials without lids** (outer diameter **11 mm**, GE Healthcare) and corresponding **rubber caps** (Type II, Kraton G, ventilated; GE Healthcare). The rubber caps act as a septum for the injection needle during sample injection in SPR assays. Care must be taken to pair compatible plastic vials with the corresponding rubber caps, as standard-issue plastic vials without lids may not fit the rubber caps with sufficient tightness, which will cause issues during sample injection.

2. **Plastic vials without lids** (outer diameter **15 mm**, GE Healthcare) as water reservoir for cleaning the injection needle.
3. **Sensor chips** (type CM5). CM5 is a general all-purpose chip. Other sensor chips are available for specialized immobilization procedures and experimental applications (*see* the manufacturer's consumable catalogue); type CM5 is used for the scope of this chapter.
4. **Amine Coupling Kit** (GE Healthcare). Used for immobilization of the capturing molecule on the sensor chip surface (*see* **Note 2** for handling).
5. **Capturing molecule** of choice, e.g., purified sdAbs.

3 Methods

3.1 Preparation of Protein Samples

3.1.1 Basic Assumptions Regarding Samples

This chapter assumes purified protein samples as starting point. The research project that formed the experimental basis for this chapter used common bacterial expression systems (Novagen; Invitrogen) for protein expression and fast protein liquid chromatography (FPLC) systems (ÄKTA family, GE Healthcare) for protein purification, using hexahistidine tags as affinity purification tags. Descriptions and protocols for these methods are beyond the scope of this chapter; novice readers are guided toward several excellent review articles [18, 19, 20]. As the last step in protein purification, samples should be dialyzed into protein storage buffer (*see* Subheading 2.1.1), flash-frozen in liquid nitrogen, and stored at −80 °C, unless previous experience with the sample type indicates otherwise.

Furthermore, this chapter assumes binary protein-protein interactions in which one putative interaction partner is tagged with a fluorescent label. The exact type of the label is not important, but care must be taken to choose a label that is compatible with both the instrument filter set to be used in MST assays and the capturing molecule to be used in SPR assays. The research project that formed the experimental basis for this chapter used an mOrange fluorescent fusion protein tag [21] including a flexible linker domain [22] on a bacterial effector (E) protein sample, and no fluorescent label on effector target (ET) protein samples. The capturing molecule for SPR assays was an RFP-specific sdAb ("RFP-Trap," Chromotek, Munich). To exclude interactions between free label and the unlabeled molecule, purified free label should be used in place of the labeled protein of interest in control experiments.

3.1.2 Common Sample Preparation for MST and SPR

1. Thaw stored protein samples quickly (e.g., warming in hand) and mix by vortexing. After no visible ice crystals remain within the sample, store samples on ice.

2. Spin down protein samples at maximum speed (>20,000 × *g*, 5 min, 4 °C).
3. Quantify molar protein concentration by spectrometry using an aliquot of the supernatant.
4. Dilute the labeled protein samples in assay buffer to twice the empirically determined concentration for MST assay setup. If doing the protocol for the first time, start with the highest concentration practically attainable from the frozen stocks and additionally prepare the following serial dilutions in assay buffer for initial testing: 10, 100, 1000 fold.
5. Dilute the unlabeled protein samples in assay buffer to twice the highest concentration that will be used in the final assay. If doing the protocol for the first time, use the highest concentration practically attainable from the frozen stocks as a starting point for initial testing (*see* **Note 3**).
6. Use protein samples at the final concentrations for setting up the interaction experiments as described in the following sections.

3.2 MST Analysis

Practical instrument handling: The instrument should be switched off after each experiment. No regular maintenance is needed.

3.2.1 Screening of Fluorescence Intensity of Labeled Sample

Time scale: approx. 30 min to 1 h.

Concentration of the fluorescent molecule in MST assays must be chosen so that the final concentration, which is kept the same in all capillaries, results in a fluorescence intensity that falls into the linear range of the instrument. Additionally, best results are usually obtained when the concentration of the labeled molecule in the assay is kept close to the *KD* of the interaction.

1. Transfer 20 μL each of the tenfold serial dilutions prepared in Subheading 3.1.2, **step 4** to small capacity plastic vials (maximum capacity ≤200 μL) and place the tubes into a suitable rack.
2. Put one glass capillary each of the predetermined type into each sample tube (simply drop it in vertically—also *see* **Note 4** for capillary handling). If doing the protocol for the first time and no specific capillary type was determined yet, use "standard" (uncoated) capillaries for initial screening. Let the liquid rise by capillary action (*see* **Note 5**).
3. Put the glass capillaries containing the serial dilutions of the labeled protein into the MST capillary tray and note the positions of each sample.
4. Start the instrument, the control PC, and the control software. Open the instrument door. Put the capillary tray into the sample loading bay, and close the instrument door.

5. Set a new project in the control software and specify a name for saving the file. Select the correct instrument color channel for your label. Set the following parameters in the control software: LED power: 50%; range of capillaries: the amount of capillaries loaded into the tray.
6. Initiate a capillary scan. The detector will move across each capillary (starting from the back of the tray) and graph the fluorescence. Wait for the scan to complete.
7. Examine the signal peaks for magnitude. Refer to Fig. 3a (left) for reference. If you used a 1000-fold dilution range as recommended above, you should see signals that reach the saturation of the detector (top of peak is cut off flatly; high concentrations of sample) and capillaries that fail to show a visible signal (low concentrations of sample). Ideally, you should see at least one sample concentration in which the top of the peak is of a rounded shape. Observe the vertical axis scale and look for signals that are in a range between 100 and 1500 units (ideal linear detection range). If signals of the dilution best fitting to these criteria are slightly above 1500 units or slightly below 100 units, adjust the concentration of the sample accordingly (also *see* **Note 6**).
8. Examine the signal peaks for regular shape in the top part of the peak. Refer to Fig. 3a (right) for reference. Use the software zoom tool to enlarge the relevant area. Look for any deviations from a regular "rounded" peak top shape. Extreme deviations from ideal shape usually show as double peaks; this is typically due to unspecific adsorption of labeled protein to the side walls of the capillary. If this problem occurs mildly, standard procedures to minimize unspecific protein adsorption to surfaces can be tested (addition of detergent, addition of "background" proteins such as BSA). If this problem occurs strongly, it is usually advisable to test a different type of capillary.
9. Wait for 10 min (leaving the capillaries in the instrument), then repeat the capillary scan. Any deviations from ideal peak shape should intensify during the incubation period.
10. If both magnitude and shape of the peaks are satisfactory after the 10 min incubation, proceed to Subheading 3.2.2.

3.2.2 MST Experiment

Time scale: approx. 1.5 h.

Once the ideal concentration of the fluorescently labeled molecule and a suitable capillary type has been determined, an initial MST experiment can be set up. As accurate and consistent pipetting is critical for the success of MST experiments, make sure to stereotypically use the procedure outlined below and refer to **Note 7** for general pipetting recommendations.

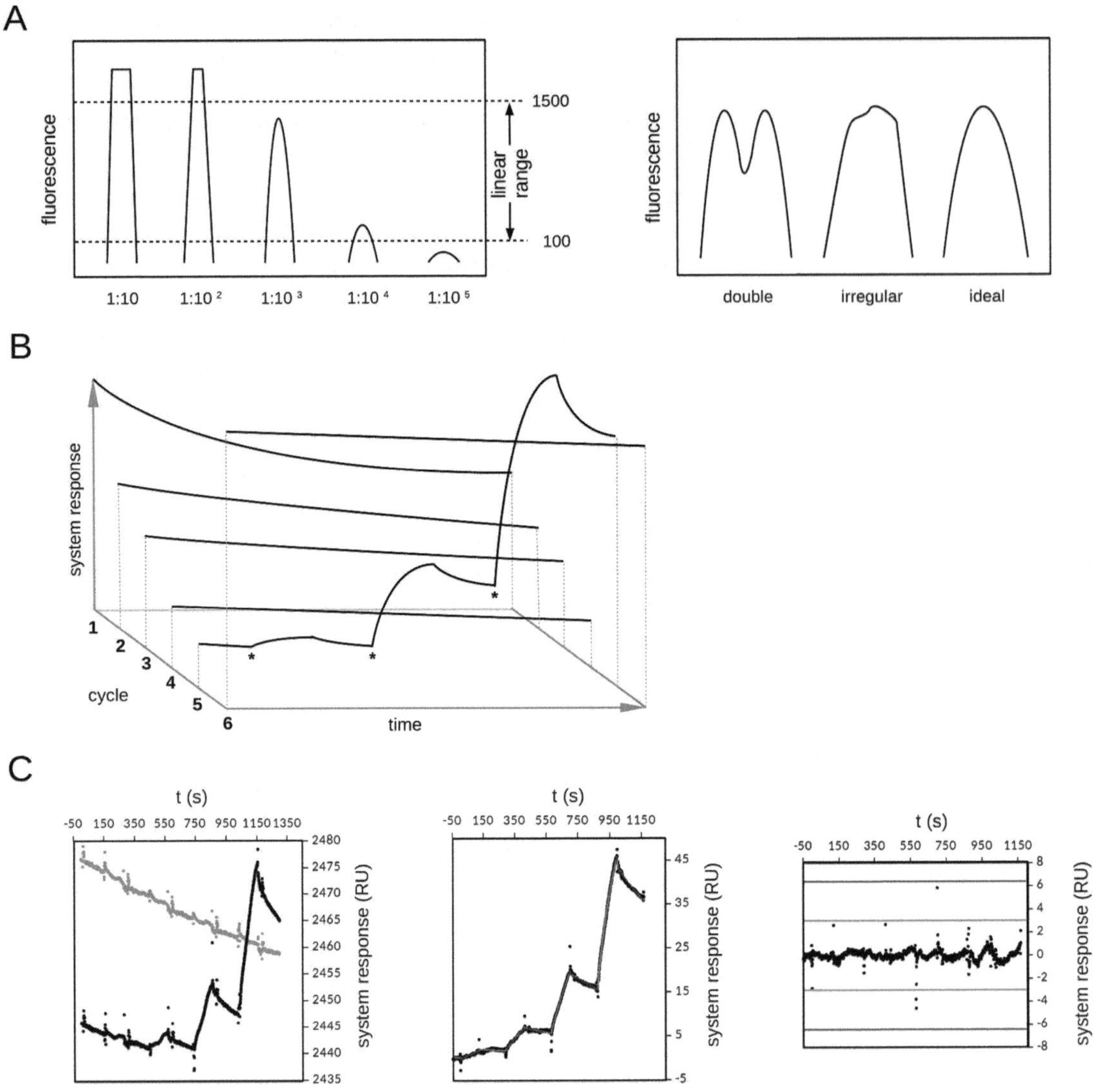

Fig. 3 Practical aspects of MST and SPR. (**a**) MST capillary scan quality control. *Left*: Quality control for signal magnitude. The signal should be in the detector's linear range (roughly 100–1500 fluorescence units). In this example, dilutions 1:10^3 and 1:10^4 would be usable. *Right*: quality control for peak shape (enlargement of peak tops). Peaks should be regularly rounded. Double peaks are indicative of sample absorption to the capillary. (**b**) Example of practical experimental cycle setup during a kinetic titration experiment using SPR. The cycle number is indicated on the z axis. Cycles 1–3 are buffer cycles and show that the baseline has not been stable before starting the experiment; the baseline curvature decreases during the buffer cycles and reaches the linear phase in the third cycle. Cycle 4 is the first blank cycle, providing a stable baseline. Cycle 5 is the main experimental cycle in which the kinetic titration of analyte takes place (three injections are shown, marked by *asterisks*). Cycle 6 is the second blank cycle, and was performed after the analyte injections to ensure the baseline is the same before and after the main experimental cycle. (**c**) Kinetic evaluation of SPR data. *Left*: A suitable blank cycle (*gray data points*) serves as a baseline to be subtracted from the experimental titration curve (*black data points*). Note that an extreme example of a sloping baseline has been chosen for illustrative purposes; the slope of the baseline should generally be less pronounced. *Center*: The blank-subtracted experimental titration curve (*black data points*) can be subjected to kinetic fitting (*orange line*). *Right*: The residuals plot shows deviations of experimental signal from the calculated fitting curve. Inner and outer deviation thresholds to evaluate the quality of the fit are indicated by the *green* and *red lines*, respectively

1. Place 16 small capacity plastic vials (maximum capacity ≤200 μL) into a suitable rack.
2. Aliquot 10 μL of assay buffer into all vials except the first (vials 2–16). Do not change the pipette tip during this procedure.
3. Aliquot 20 μL of the unlabeled protein at twice the highest concentration for the assay (*see* Subheading 3.1.2, **step 5**) into the first vial. Do not change the tip.
4. Transfer 10 μL from the first vial to the second vial. Mix carefully and thoroughly using the pipette. Do not change the tip. Transfer 10 μL from the second vial to the third vial. Repeat the procedure until reaching the last vial. Remove 10 μL from the last vial and discard the tip. This procedure will result in 16 vials of 10 μL serially diluted (1:1) unlabeled protein each.
5. Using a fresh tip, add 10 μL of labeled protein at twice the previously determined ideal concentration (*see* Subheading 3.2.1) to the first vial. Mix carefully and thoroughly using the pipette. Discard the tip.
6. Repeat the previous step for the rest of the vials until 10 μL of labeled protein was added to each vial (using a fresh pipette tip each time), resulting in 16 vials of 20 μL each.
7. Place the vials into a centrifuge and spin down for a few seconds at low speed to remove any droplets at the vessel sidewall or air bubbles that might have formed.
8. Place the vials into the rack again. Add one glass capillary of the previously determined suitable type (Subheading 3.2.1, **step 8**) to each vial. Fill the capillaries as described before (Subheading 3.2.1, **step 2**).
9. Start the instrument, the control PC, and the control software. Open the instrument door. Put the capillary tray into the sample loading bay, and close the instrument door.
10. Set a new project in the control software and specify a name for saving the file. Set the correct instrument color channel for the label being used. Set the following parameters in the control software. LED power: the value previously determined to yield a signal in the linear range (Subheading 3.2.1). Set three experimental runs with the following MST powers: 20, 40, 60%. Keep the default values for the rest of the experimental run parameters. Set the highest concentration (in nM) of the unlabeled molecule in the appropriate capillary field and click-draw (*see* **Note 8**) across the other capillary fields. The dilutions should be calculated automatically. For safety, double check the calculations for each capillary.
11. Perform a capillary scan (as in Subheading 3.2.1). Observe signal peaks for magnitude and shape (as in Subheading 3.2.1, **step 7/8**). The signal magnitude should vary less than 10%

among all capillaries; a single outlier may be acceptable (the data point may be discarded from evaluation). Do not proceed if the signal varies by more than 10% over more than two capillaries—this indicates either inaccurate pipetting (*see* **Note 7** for recommendations) or sample inhomogeneity. In this case, repeat all the previous steps. If the effect persists, sample quality should be increased by optimizing buffer conditions or selecting a different capillary type.

12. If the capillary scan passes quality control, proceed by starting the MST run(s). Each run will take approximately 20 min to complete (60 min total experiment time for 3 MST powers as recommended before).
13. After the last run has finished, proceed to evaluation (Subheading 3.2.3).

3.2.3 Evaluation of a Single MST Experiment

Time scale: approx. 10 min

1. Open the evaluation software. Load the result file (specified when setting up the experiment). Select a single MST power experiment (different experiments can be opened subsequently).
2. Navigate to the "Thermophoresis + T Jump" pane (*see* **Note 9**).
3. Initialize fitting values by pressing the corresponding button.
4. Specify the concentration of the labeled molecule in the corresponding field and check the "fix" box to let the fitting use this parameter.
5. Check the "*KD* fit" option. Press the fit button. The affinity will be calculated and displayed. Visually observe the deviation of the curve from the data points (*see* **Note 10**).

3.2.4 Evaluation of Multiple MST Experiments

Time scale: depends on number of experiments

1. Perform the evaluation as described in Subheading 3.2.3, then export the numerical values to a text or spreadsheet file by using the corresponding function of the evaluation software. Repeat for all replicate experiments.
2. Normalize the values using typical spreadsheet software (subtract the minimum value and divide by the amplitude).
3. Calculate mean and SD of the replicate normalized values. Graph using spreadsheet software. Normalized data should be displayed as "fraction bound" in graphs.

3.3 SPR Analysis

Practical instrument handling: The instrument should be left switched on and under constant flow when not used for short periods of time ("standby mode," engaged using the corresponding command in the control software). Standby mode using the experimental buffer may be performed for up to 24 h. Standby mode using deionized water may be performed for up to 7 days. If the instrument will not be used for more than 7 days, the flow system should be purged using the built-in methods and the instrument should be switched off in a dry state. Periodical cleaning procedures using desorp and sanitize solutions (Subheading 2.1.3) should be performed regularly as advised by the internal timer of the control software, or more often if multiple users perform experiments and buffer or sample conditions change regularly (also *see* **Note 11**). Always use the included "Maintenance chip" for all maintenance procedures, as experimental sensor chips will usually be damaged if exposed to cleaning solutions. For detailed handling instructions, refer to the instrument's user manual.

3.3.1 Immobilization of Capturing Molecule on the Sensor Chip

Time scale: approx. 2 h, not counting final equilibration

Before loading the ligand, the molecule that will be used to capture the ligand must be covalently coupled to the sensor chip surface ("immobilization"). This step must be performed once for each sensor chip. Once a sensor chip has been prepared, it can usually be reused multiple times for experiments depending on experimental conditions, chip age, and storage conditions between experiments.

This protocol assumes the use of camelid sdAbs directed against GFP derivates as capturing molecules. The capturing parameters (e.g., ideal concentration of the molecule) may vary slightly if a different molecule is used.

1. Start the instrument. Make sure the instrument was cleaned using the desorp procedure (*see* practical instrument handling, Subheading 3.3) before performing this step; the inlet should be connected to deionized water and a maintenance chip should be docked to the instrument. If it cannot be verified whether the instrument has been cleaned, clean the instrument using the desorp procedure.
2. Pause standby flow using the corresponding command in the control software. Connect immobilization buffer (Subheading 2.1.3) to the inlet. Resume standby flow.
3. Perform a "prime" command to flush the flow system with immobilization buffer.
4. Allow the system to remain in standby for 15 min for equilibration of the flow system. In the meantime, continue to the next step.
5. Take the sealed plastic pouch containing new sensor chip from storage (it should be stored at 4 °C). Do not open the pouch. Let the sensor chip equilibrate at room temperature for 10 min.
6. Undock the maintenance sensor chip using the corresponding command.
7. Open the pouch containing the new sensor chip. Using a permanent marker, label the new sensor chip clearly on the outer

plastic casing. Useful information would be the day of immobilization and the immobilized molecule.

8. Dock the new sensor chip using the corresponding command. Let the sensor chip equilibrate with immobilization buffer by leaving the system in standby for 15 min. In the meantime, prepare the immobilization reagents (next step).
9. Retrieve the immobilization reagents from storage. Thaw the frozen EDC and NHS aliquots. Make sure the solutions are thawed to homogeneity before continuing.
10. Dilute the capturing molecule to 50 µg/mL in immobilization buffer.
11. Start a predefined immobilization procedure work flow using the control software. Use the following parameters: Chip CM5, amine coupling chemistry, immobilize both flow cells, activation time 420 s, ligand contact time 420 s, deactivation time 420 s, flow rate 10 µL/min. As soon as prompted by the software, aliquot the needed amounts of reagents and solutions (*see* **Note 12**) into sample tubes, close them with rubber caps, and place the tubes into the appropriate positions predefined by the software.
12. Start the immobilization procedure. This is fully automated and will take approximately 1 h.
13. After immobilization is complete, check the immobilization level that is displayed by the software. Using the procedure described above with sdAbs should result in binding levels between 2000 and 3000 RU on both flow cells; the variation between both flow cells should be less than 20%.
14. Stop standby, and exchange the inlet buffer to the experimental buffer that will be used in the assay. Prime the system using the corresponding command and let the newly immobilized sensor chip equilibrate in this buffer for at least 1 h (ideally 2–3 h). After equilibration, either use the sensor chip directly for experiments (Subheading 3.3.2) or alternatively, proceed to the next step for storage.
15. If not using the chip for more than 24 h, undock the sensor chip. Remove the chip from the instrument and directly place it into a 50-mL centrifuge tube. Close the tube tightly, and store it at 4 °C. *See* **Note 13** for long-term storage.

3.3.2 Capturing Ligand on the Sensor Chip

Time scale: approximately 2 h.

As the first step in each interaction experiment, the ligand (which corresponds to the labeled molecule) must be reversibly captured on the sensor chip using the strong affinity between the capturing molecule and the ligand's label. The amount of ligand necessary to obtain useful signals from analyte injections in the interaction experiment varies according to various factors (relative size of ligand and analyte, active fraction of ligand, etc.) and should be determined empirically. Therefore, it may be necessary to repeat this procedure and subsequent interaction experiments several times until the assay is well established.

1. Start with a clean instrument equilibrated in assay buffer and with the appropriate sensor chip docked. Start a manual run, selecting both flow cells (check reference subtraction "2-1") and a flow rate of 10 μL/min.
2. Let the system run without further input for 10 min.
3. If not already preselected, choose the referenced signal curve by choosing the "2-1" graph (*see* **Note 14** for graphing modes). Observe the baseline signal in the "2-1" graph. The baseline should be a horizontal, straight line.
4. Observe the individual signal graphs for flow cells 1 and 2. They should deviate minimally in shape (any deviation will also be reflected in the "2-1" graph). Ideally, both flow cells should show horizontal base lines; small slopes may be acceptable as long as they are uniform in both flow cells (*see* **Note 15**).
5. Switch the flow path to flow cell 2 only.
6. Inject the lowest serial dilution of the labeled molecule that was prepared beforehand (e.g., 1:1000; *see* Subheading 3.1.2, **step 4**), using a manual injection command with a contact time of 30 s. Observe the signal for any binding (the view should be magnified until systematic noise is visible to detect weak signals). If binding is apparent, wait for the signal to stabilize after the end of injection and note signal stability over a period of at least 20 min. If no binding is apparent, proceed to next step; if there is binding, skip the next step.
7. Repeat the previous step using the next-highest serial dilution (e.g., 1:100) of the labeled molecule. If there is still no apparent binding, repeat the step using higher concentrations until binding is visible (if no visible binding is occurring at injected ligand concentrations of roughly 1 μM or above, there is probably a fundamental problem in the capturing system, e.g., the capturing molecule is inactive, the specificity does not match the label, or the label on the labeled molecule is not functional).
8. Once there is significant binding that is stable over longer durations of time (>20 min), observe the magnitude of binding signal in the "2-1" graph. A very rough estimate target to use for the first immobilization is around 100 RU of relative signal (*see* **Note 16**). If the signal is less than 100 RU, proceed with further injections until this signal level is reached (*see* **Note 17**).

9. To test regeneration, inject regeneration solution (Subheading 2.1.3). A single injection (contact time 45 s) should be sufficient to quantitatively remove the bound ligand. The signal should return to a baseline value identical to the value before ligand capturing.
10. Perform a second injection of regeneration solution. At this point, this should not result in any further changes in signal.
11. Proceed to reload ligand as described in the previous steps. The same capture levels should be reached if using the same parameters for injection (if this is not the case, the activity of the capturing molecule may be negatively affected by regeneration, limiting reusability of the chip).
12. Allow the system to equilibrate for at least 1 h. Observe the signal in the "2-1" graph. Typically, the signal should decrease quickly directly after ligand injection (washing of unbound or loosely bound ligand from the capturing molecule) and reach a stable linear phase of a slightly decreasing baseline after 30 min to 1 h. Reaching a "flat" baseline without any negative slope should not be expected; a certain amount of continuous dissociation is inherent to a non-covalent capturing system. This is not problematic as long as the slope of the decreasing baseline is small (<1 RU/min) and constant (no curvature in the signal trace; *see* **Note 18** for a practical way to check linearity).
13. If the baseline is stable as defined in the previous step, end the manual run and proceed with the interaction experiment as outlined in Subheading 3.3.3.

3.3.3 Kinetic Interaction Experiment

Time scale: approximately 2.5–5 h.

Once the ligand has been stably captured on the sensor surface, a kinetic interaction experiment using the analyte of interest can be performed. It is generally advisable to use the predefined methods ("application wizards") in the control software, as this allows automated curve fitting by the evaluation software after the experiments.

1. Decide on a sensible range of analyte concentrations to use during titration. For best results, concentrations of the analyte should be spaced around the expected *KD* of the interaction. Dilution factors between 2 and 5 for each serial dilution step are common. If no previous experience is available, start with a highest analyte concentration of 3 μM and use five consecutive threefold dilutions; this will cover the nanomolar affinity range common for many biomolecular interactions. Do not prepare the dilutions yet.
2. Select the "kinetics/affinity" application wizard in the control software.

3. Follow the wizard to set up the experiment. For the first prospective experiment, it is advisable to keep default values for most parameters. Analyte injections should be performed in both flow cells, with flow cell 1 being used for reference. When choosing the amount of experimental cycles, include one "buffer" cycle and two "blank" cycles flanking the experimental titration cycle. *See* **Note 19** for explanation of cycles. Fig. 3b shows an example of the practical arrangement of experimental cycles.
4. Continuing the assay setup wizard will reach a step where analyte types and concentrations can be specified (*see* **Note 20** on sample labeling). Use the concentrations decided on in **step 1**. The dilutions for each vial can be calculated automatically by specifying the highest concentration and the dilution factor.
5. Note the volumes of buffer and analyte samples required which will be displayed by the software. Proceed to prepare the corresponding analyte dilutions in sample vials, taking care to ensure accurate pipetting. As noted before, be sure to slightly exceed the specified volumes to include a margin of error.
6. Start the assay. All further steps will be automated. Total run time depends on the number of buffer and blank cycles chosen, and will usually be in the range of 2–4 h (also *see* **Note 21**).
7. After the experiment has finished, start a new manual run.
8. Inject regeneration solution (as described previously in Subheading 3.3.2, **step 9**) to both flow cells to regenerate the sensor surface. This should result in quantitative removal of ligand-analyte complex from the capturing molecule. Equilibrate the chip for 30 min after regeneration.
9. The chip is now ready for a new ligand loading or storage (as described in Subheading 3.3.1, **step 15**).

3.3.4 Evaluation of Kinetic Data

Time scale: approximately 10 min

1. Start the evaluation software and open the result file.
2. Visually screen the curves for any obvious artifacts such as large signal deviations or "wavy" patterns. Detailed description of determining the quality of data curves is beyond the scope of this article; there is a series of insightful review articles dedicated to this topic [17, 23]. As a rough guideline, all blank cycles should be stable baselines and the experimental titration cycle should show a dose-dependent increase in signal corresponding to the individual analyte dilutions used. Association and dissociation phases of analyte injection should resemble segments of simple exponential curves.
3. Select the corresponding button for kinetics analysis.
4. Refer to Fig. 3c for a visual example of the evaluation process. Select one or more (an average will be used in this case) blank

cycles to use for baseline subtraction (Fig. 3c, left). A good blank cycle should exhibit the same systematic injection artifacts (usually visible as very narrow signal spikes at the beginning and end of the injection), and the same baseline slope as the experimental titration cycle. As the slope of the experimental titration cycle may be hard to see, compare the slopes of the blank cycle immediately before and immediately after the experimental titration cycle; the slopes should be identical (indicating that it is reasonable to assume this baseline slope from the experimental cycle, too). Do not use any blank cycle exhibiting obvious, non-recurring artifacts.

5. Follow the evaluation wizard to subtract the blank cycle(s) from the experimental cycle. This should eliminate most systematic injection artifacts and the baseline slope (Fig. 3c, center).
6. Remove any remaining major injection artifacts manually by right-clicking and drawing selections that encompass artifact(s), then click the "remove" button. Minor artifacts will usually be successfully ignored by the fitting algorithm and do not have to be removed manually.
7. Proceed in the evaluation wizard to initiate the fitting procedure. Select a 1:1 binding model if you do not have any previous experimental indication that the putative interaction follows a different stoichiometry (*see* **Note 22**).
8. Visually observe the correspondence of the fitting curve to the data points (Fig. 3c, center). Also note the "residuals plot" in the corresponding tab (*see* **Note 23** and Fig. 3c, right).
9. Kinetic parameters and affinity calculated from the fitting procedure will be displayed in table format in the corresponding tab.

4 Notes

1. Minimizing the amount of handling time and buffer exchange steps is generally advisable to maintain protein stability. Ideally, the buffer used for protein extraction, protein purification, protein storage, and interaction assays should be as similar as possible. If the concentration of frozen protein stocks is high enough, any final small modifications in buffer composition (e.g., addition of detergent) can be realized by diluting the sample into the final buffer.
2. Upon arrival of the amine coupling kit, directly proceed to aliquot and store the reagents as described in the kit's manual. The EDC and NHS aliquots are single-use and should be stored at −20 °C; their efficiency will decrease over prolonged storage time, as reflected in lower attainable immobilization levels.

3. As a very rough estimate, good starting concentrations may typically be in the range of 1–100 μM, or at least 10 times the expected affinity of the interaction, if any previous information is available from biologically similar interactions.
4. Remove capillaries from the containing glass vessel by tilting the open vessel to expose the tips of a small number of capillaries and aid capillary separation. Capillaries may be touched with gloved or ungloved hands at one end, provided only the other end will be inserted into the sample liquid. Never touch the central part of the capillary, as the instrument reading will be taken at this position (note the corresponding cutout on the capillary tray to see the extent of the central part). Any capillary that has been touched by accident in this part or that has been dropped should be discarded.
5. Filling by capillary action can be sped up by tilting the rack containing the vessels and the capillaries slightly sideways (at an angle of approx. 45–70 °). Shake the rack slightly or tilt back and forth to increase the effect, taking care not to let the capillaries drop out of the tube. The capillaries do not have to be filled completely; as long as the central portion (approx. 50%) is filled, one or both capillary ends may contain air. Sealing of capillary ends after sample loading is unnecessary and not recommended.
6. Beside the concentration of the labeled molecule, the signal intensity can be finely adjusted by using the excitation (LED) power setting in the control software. The linear range of the excitation LED level is between the 10 and 100% setting. Typically, excitation LED power has no additional effects on MST assays.
7. Although good pipetting technique is important in all routine laboratory work and should be common knowledge, its critical importance for the success of MST assays warrants the repetition of some general guidelines. To increase consistency, most pipetting steps in the preparation of MST samples have been adjusted to a pipetting volume of 10 μL. Use the most accurate pipette available for this volume; this should typically be a fixed volume pipette or a 2–20 μL pipette. Empirically test the accuracy of the specific pipette that will be used for the assay (e.g., by pipetting water onto a precision balance with a measuring accuracy of at least 0.1 mg). Only use well-fitting tips. Always hold the pipette strictly vertically during sample transfer to minimize differential air pressure effects. Take care to submerge the pipette tip only as far as absolutely necessary into the solution to limit adsorption of sample liquid to the outside of the pipette tip. If applicable, always release volumes from the pipette by dipping the tip into the liquid already present in

the vessel (again, submerging only as far as absolutely necessary); if there is no liquid, release the volume against the vessel sidewall. Visually observe the sample volume in the pipette tip during any transfer step.

8. This action is performed just as in typical spreadsheet software (like Microsoft Excel). Values may also be entered manually.
9. MST evaluation software can be used to interpret responses obtained from distinct time points of each measurement trace (represented by the "cursor," the adjustable interval in which the response is measured). Evaluation of these different time points reflects different interaction-dependent molecular phenomena that occur on separate time scales after induction of a temperature gradient, and presets are given in the software for analysis of "thermophoresis," "T-jump," and "Thermophorsis + T-Jump." In-depth description about the different evaluation modes is beyond the scope of this chapter, and readers are guided toward the relevant literature [6]; however, a practical summary is given here. In brief, T-jump refers to the differential response of a fluorophore quickly (<1 s) after a temperature change (e.g., induction of temperature gradient), which may be dependent on the presence of a molecular complex, and thermophoresis refers to the actual movement of molecules within the temperature gradient. Interaction can be observed as effects on these phenomena, and evaluation can be performed on a single time scale (T jump only, thermophoresis only), on both time scales simultaneously (T-jump + thermophoresis), or on a freely user-adjustable timescale (manual). Not all interactions may be visible on all time scales. For practical purposes, the T-jump + thermophoresis analysis mode should be chosen as default.
10. Use your critical judgment when evaluating fits. The software does not perform quality control, nor does it judge whether the data is biologically reasonable. The fit should contain the inflection point of the sigmoidal fitting curve and at least either the baseline or the saturation zone. Ideally, both baseline and saturation should be reached. Adjust sample concentrations accordingly, if possible.
11. The amount of necessary cleaning and maintenance varies heavily according to instrument use. In general, SPR instruments are highly susceptible to experimental artifacts arising from improper maintenance and dirty flow system. As a practical rule of thumb, the desorp procedure should be performed liberally and prophylactically. If there is a significant chance of aggregations within the flow system, e.g., by using biological samples that are not purified, highly soluble proteins, the desorp procedure should be performed after each experiment to prevent carry over effects and clogging.

12. The predefined software work flows state necessary sample volumes. Although these contain small margins of error, it is recommended to use slightly higher volumes than recommend to safely eliminate the possibility of injecting air into the flow system. When loading samples into vessels, always make sure that there are no stray liquid drops on the side walls of the vessel (e.g., by centrifuging shortly) and that there is no air bubble trapped at the lower tip of the vessel, as the injection needle will draw liquid from this part. Smaller air bubbles at the surface of the liquid are not problematic.
13. For short to mid-term storage (up to a few months) of the sensor chip, the "dry storage" method (at 4 °C) as outlined in Subheading 3.3.1, **step 15** is adequate for inherently stable capturing molecules such as commercial sdAbs. For longer term storage or sensitive capturing molecules, it may be beneficial to remove the sensor chip from its cassette and place it in sterile buffer solution ("wet storage").
14. In SPR experiments using chips with two flow cells, three graphs are usually recorded: individual graphs for each flow cell ("1," "2") and a subtraction graph ("2-1," as flow cell 1 is used as reference by convention). Showing all three graphs in the same window usually obscures detail, as the axis will scale to display all graphs simultaneously. It is usually more informative to show only one graph, and switch between individual graphs to show more detail; the axis will scale automatically to the individual graph. The graphing window can be adjusted manually by drawing magnification rectangles with the mouse; it may be helpful to deselect the corresponding check box to disable automatic axis scaling.
15. Constantly decreasing or increasing baselines in both flow cells on a new chip may be indicative of incomplete equilibration. This should be remedied by extending equilibration time; however, a very slight, consistent slope (on the order of <1 RU/min) is usually unproblematic if it occurs similarly in both flow cells.
16. For kinetic experiments, a maximum response of 20–80 RU at the highest concentration of injected analyte is a reasonable aim. Assuming a 1:1 stoichiometry, equal molecular weight of ligand and analyte, and 100% ligand activity, 100 RU of captured ligand would yield 100 RU of maximum response at analyte saturation. In practice, 100% ligand activity is not possible. Calculating the theoretical maximum response (Rmax) is discussed in the "Biacore Assay Handbook" (GE Healthcare). Although a higher amount of immobilized ligand (and thus, higher potential analyte responses) leads to an increase in signal level, kinetic experiments benefit from low ligand densities as a measure to prevent mass transport limitations (quantitative

diffusion of analyte to ligand as rate-limiting step in complex formation; *see* [10]).

17. Fundamentally, it does not matter whether a given signal level for ligand capture is reached by several, consecutive injections of a dilute ligand solution, prolonged contact times, or a single injection using a more concentrated ligand solution. Using several consecutive, short injections usually allows more fine-grained control of final immobilization level; automated methods are available. However, it should be noted that the exact level of immobilization is not critical for the success of the assay.
18. A quick, practical way to observe curvature of the signal is to draw a magnification rectangle on a 10-min segment of the curve, aligning the top left corner of the rectangle with the highest signal (at relative time 0) and the bottom right corner of the rectangle with the lowest signal (at relative time 10 min). Any curvature in the signal should be easily recognizable by eyes in the resulting magnified view. If the signal is a straight line connecting the opposing corners of the rectangle, the signal decrease is linear for all practical purposes.
19. "Buffer" and "blank" cycles are essential to precondition the chip and serve as baselines for artifact subtraction. Both types of cycles are performed like the experimental analyte injection cycle, with the only difference being that the injected solution is running buffer. The fundamental difference between a "buffer" and "blank" cycle is the fact that the software will not allow using a "buffer" cycle for baseline subtraction. However, buffer cycles can be useful, e.g., by providing time before the actual experiment in which a freshly loaded ligand can equilibrate on the sensor surface. If the instrument has been equilibrated for extended periods after ligand loading, the buffer cycle step(s) may be skipped. Blank runs should never be skipped, and should be performed both before and after the main experimental cycle to allow evaluation of the stability of the baseline. Also *see* Fig. 3b for illustration.
20. When labeling the analytes used in each cycle, use exactly the same label for the main experimental cycle and its corresponding blank cycles. The software will recognize the connection between sample cycles and blank cycles by their identical name (i.e., be sure to label blank cycles corresponding to a sample cycle using molecule "ABC" as containing molecule "ABC" at a concentration of 0; do not label it something else like "buffer").
21. Because of the automation, it may be convenient to start a kinetic titration experiment late in the working day and let it run overnight. The system will automatically resume standby after finishing an experimental run (ensure that there is enough running buffer left to sustain extended standby). It is typically

unproblematic to leave the ligand/analyte complex bound to the capturing molecule for several hours and regenerate the chip on the following day.

22. The 1:1 binding model is the simplest model, and thus offers the least degrees of freedom. More complex binding models will allow the fitting algorithm to freely adjust more variables, which will always lead to better fitting. It is a tempting fallacy to be inclined to accept and use a complex model because it fits the data better than a simple model (this has infamously been dubbed "model surfing," [23]). The only correct way to determine the applicability of a complex binding model is to gain other experimental evidence that supports this model; thus, a 1:1 binding model should be assumed as default until other mechanistic data is available.
23. The residuals plot is generated automatically and shows the deviation between data points and model in response units, facilitating the visual evaluation of fitting quality. Acceptable ranges of deviations are automatically indicated by green lines (lower threshold) and red lines (upper threshold). A good fit is characterized by apparently random signal deviations within the lower threshold; obvious injection artifacts that have not been manually removed may show as a small number of data points outside the thresholds.

Acknowledgments

We thank Deutsche Forschungsgemeinschaft (SFB1101) for funding.

References

1. von Mering C et al (2002) Comparative assessment of large-scale data sets of protein-protein interactions. Nature 417:399–403
2. Aloy P, Russell RB (2004) Ten thousand interactions for the molecular biologist. Nat Biotechnol 22:1317–1321
3. Berggård T, Linse S, James P (2007) Methods for the detection and analysis of protein-protein interactions. Proteomics 7: 2833–2842
4. Van Der Merwe PA (2001) Surface plasmon resonance. In: Harding SE, Chowdhry BZ (eds) Protein–ligand interactions: hydrodynamics and calorimetry. Oxford University Press, Oxford, pp 137–170
5. Duhr S, Braun D (2006) Why molecules move along a temperature gradient. Proc Natl Acad Sci U S A 103:19678–19682
6. Jerabek-Willemsen M et al (2011) Molecular interaction studies using microscale thermophoresis. Assay Drug Dev Technol 9:342–353
7. Myszka DG (1997) Kinetic analysis of macromolecular interactions using surface plasmon resonance biosensors. Curr Opin Biotechnol 8:50–57
8. Stenberg E et al (1991) Quantitative determination of surface concentration of protein with surface plasmon resonance using radiolabeled proteins. J Colloid Interface Sci 143:513–526
9. Mayo CS, Hallock RB (1989) Immunoassay based on surface plasmon oscillations. J Immunol Methods 120:105–114
10. Myszka DG et al (1998) Extending the range of rate constants available from BIACORE: interpreting mass transport-influenced binding data. Biophys J 75:583–594

11. Johnsson B et al (1991) Immobilization of proteins to a carboxymethyldextran-modified gold surface for biospecific interaction analysis in surface plasmon resonance sensors. Anal Biochem 198:268–277
12. Karlsson R, Fält A (1997) Experimental design for kinetic analysis of protein-protein interactions with surface plasmon resonance biosensors. J Immunol Methods 200:121–133
13. Wofsy C, Goldstein B (2002) Effective rate models for receptors distributed in a layer above a surface: application to cells and biacore. Biophys J 82:1743–1755
14. Rich RL, Myszka DG (2000) Advances in surface plasmon resonance biosensor analysis. Curr Opin Biotechnol 11:54–61
15. Andersson K et al (1999) Identification and optimization of regeneration conditions for affinity-based biosensor assays. A multivariate cocktail approach. Anal Chem 71(13):2475–2481
16. Saerens D et al (2008) Single-domain antibodies as building blocks for novel therapeutics. Curr Opin Pharmacol 8:600–608
17. Rich RL, Myszka DG (2008) Survey of the year 2007 commercial optical biosensor literature. J Mol Recognit 21:355–400
18. Terpe K (2003) Overview of tag protein fusions: from molecular and biochemical fundamentals to commercial systems. Appl Microbiol Biotechnol 60:523–533
19. Sahdev et al (2008) Production of active eukaryotic proteins through bacterial expression systems: a review of the existing biotechnology strategies. Mol Cell Biochem 307:249–264
20. Young CL et al (2012) Recombinant protein expression and purification: A comprehensive review of affinity tags and microbial applications. Biotechnol J 7:620–634
21. Shaner NC et al (2004) Improved monomeric red, orange and yellow fluorescent proteins derived from Discosoma sp. red fluorescent protein. Nat Biotechnol 22:1567–1572
22. Prescott M et al (1999) The length of polypeptide linker affects the stability of green fluorescent protein fusion proteins. Anal Biochem 273:305–307
23. Rich RL, Myszka DG (2010) Grading the commercial optical biosensor literature-Class of 2008: 'The Mighty Binders'. J Mol Recognit 23:1–64

Chapter 8

In Vitro Ubiquitination Activity Assays in Plant Immune Responses

Giulia Furlan and Marco Trujillo

Abstract

Ubiquitination is a central posttranslational modification that impinges on the fate of proteins. While attachment of K48-linked chains onto soluble proteins marks them for proteolysis via the 26S proteasome, mono-ubiquitination or K63-linked chains result in the endocytosis and sorting through the endomembrane system of integral membrane proteins, such as pattern recognition receptors. In vitro ubiquitination assays allow the biochemical analysis of all individual components of the ubiquitination machinery and its potential substrates. Here, we describe how to reconstitute the ubiquitination cascade in vitro and detail different variations of the assay, the required controls and how to interpret the obtained results.

Key words Ubiquitination, Posttranslational modification, Degradation

1 Introduction

Ubiquitination is the attachment process of one or more ubiquitin molecules to a substrate protein. Ubiquitin is a 76 amino acid protein that is highly conserved in all eukaryotes; the plant ubiquitin differing in only three positions to the sequence of its human counterpart. First characterized for its function in protein degradation, roles of ubiquitin attachment have expanded to include endocytosis of receptors, DNA repair, and other non-proteolytic functions. Nevertheless, ubiquitin is the common denominator for all major degradatory pathways, including endocytic vacuolar/lysosomal, autophagy, and the 26S proteasome. As ubiquitination is involved in most, if not all, cellular processes, it plays a fundamental role in the regulation of the plant immune response [1].

The process of ubiquitination proceeds by the sequential action of three enzymes mediating three different steps:

1. First, ubiquitin activation takes place in a two-step reaction mediated by a ubiquitin-activating enzyme (E1) in an ATP-dependent manner. The first step consists in the generation of

Libo Shan and Ping He (eds.), *Plant Pattern Recognition Receptors: Methods and Protocols*, Methods in Molecular Biology, vol. 1578, DOI 10.1007/978-1-4939-6859-6_8, © Springer Science+Business Media LLC 2017

a ubiquitin-adenylate intermediate. ATP and ubiquitin are bound simultaneously to catalyze the acyl-adenylation of the C-terminal glycine of the ubiquitin moiety. In the second step, ubiquitin is transferred to the active site cysteine residue releasing AMP. The result is a thioester linkage between the C-terminal carboxyl group of ubiquitin and the E1 cysteine sulfhydryl group. The *Arabidopsis thaliana* (from here on referred to as *Arabidopsis*) genome is predicted to encode two E1 enzymes.

2. The ubiquitin-conjugating enzymes (E2s) are responsible for catalyzing the transfer of ubiquitin from the E1 to their own active site cysteine via trans(thio)esterification. To carry out this reaction, the E2 simultaneously binds to the activated ubiquitin and the E1. E2s are key mediators of chain assembly having an influence on the linkage between the ubiquitin moieties, chain length, and processivity. They are characterized by a core ubiquitin-conjugating (UBC) domain that is highly conserved. While only two E1s are found in *Arabidopsis*, 38 E2s are predicted.
3. The E3 ubiquitin ligases catalyze the final step of the ubiquitination cascade. They carry out dual tasks as both catalyst and in substrate selection, acting as a scaffold bringing together the right E2 with a specific target protein and enhancing the transfer rate of ubiquitin. The result is an isopeptide bond between a ε-amino group of a lysine in the target protein and the C-terminal glycine of ubiquitin. The number of genes encoding E3 ligases in plants has greatly expanded in comparison to other organisms and includes approximately 1500 genes. These can be broadly subdivided into four classes.

Homologous to the E6-AP Carboxyl Terminus (HECT): seven HECT E3 ligases are predicted in *Arabidopsis.* They are approximately 3700 amino acids large and are characterized by a HECT domain. HECT ligases link ubiquitin via a thioester bond to a catalytic cysteine before transferring it to the substrate. HECT domains have been proven active in ubiquitination assays with UBC8 [2, 3].

Really Interesting New Gene (RING): 469 RING containing E3 ligases are predicted in *Arabidopsis* [4]. RING ligases catalyze the transfer of ubiquitin from the E2 enzyme directly to the substrate. The RING domain contains a $Cys_3HisCys_4$ motif that chelates two zinc cations and forms a type of RING finger domain. Stone et al. in 2005 [4] investigated the functionality of the *Arabidopsis* RING family employing in vitro ubiquitination assays with UBC8 (*see* **Note 1**).

Plant U-box proteins (PUBs): 64 *Arabidopsis* genes are predicted to encode PUBs [5]. The U-box is similar to the RING domain.

However, the conserved zinc binding sites in the RING domain are replaced by more dynamic hydrogen bond networks in U-boxes [6]. Mudgil et al. [5] and Wiborg et al. [7] characterized the in vitro ligase activity of selected PUB proteins.

Cullin-RING-Ligases (CRLs): CRLs are multi-subunit E3s that contain a cullin, RING-box 1 (RBX1; which binds to the E2), and a variable target recognition module (approx. 850). The complexity of CRLs has precluded their analysis by using in vitro ubiquitination assays in plants.

The reconstitution of the ubiquitination cascade in vitro is a powerful tool to assess the activity of a putative E3 ligase. For this purpose, the assay makes use of the self-catalyzed ubiquitination (autoubiquitination), a puzzling feature displayed by the majority of E3 ligases. In the presence of all required components, namely, ATP, ubiquitin, E1, E2, and the E3 ligase, it catalyzes the attachment of ubiquitin onto itself. This modification can easily be visualized by immunoblot, characterized by a stepwise increase in size of the E3 as a result of the addition of ubiquitin moieties.

A critical aspect that needs to be considered independently for each E3 ligase is the pairing with an appropriate E2 enzyme [8]. The E2 is responsible for attaching ubiquitin to the E3 (or substrate), and in some instances, E3s may require a specific E2 to be active. E2s are also, to a certain extent, responsible for the type of ubiquitin chain that is synthesized. Ubiquitin molecules can be linked to each other through one of their seven lysine residues (K6, K11, K27, K29, K33, K48, K63). However, most ligases are active in the presence of a "generic" E2 enzyme such as UBC8 (At5g41700) that is highly promiscuous and possessive. Combining a specific E2–E3 pair with ubiquitin mutants in the ubiquitination assay, e.g., ubiquitin that lacks specific lysine residues (e.g., ubiquitin K48R) or back mutants that possess a single lysine (e.g., R48K) may give indications about the type of synthesized ubiquitin chains.

Additionally, in vitro ubiquitination assays can be used to test whether a known E3-interacting protein is also recognized as a substrate. Of note, interaction with an E3 ligase does not necessarily result in protein ubiquitination, as the interactor may bind in a way that it is inaccessible to the E2-Ub conjugate. These analyses, in combination with additional approaches, are a valuable tool to identify a *bona fide* substrate of an E3. An example from the field of plant immunity is the flagellin-sensitive 2 (FLS2). Using in vitro ubiquitination assays, the purified cytosolic domain of FLS2 was shown to be ubiquitinated by both the *Pseudomonas syringae* effector AvrPtoB [9] and the plant E3 ligases PUB12 and PUB13 [10], which results in its degradation in vivo. Similar approaches have been used to study the ubiquitination of chitin elicitor-receptor kinase 1 (CERK1), a LysM receptor kinase by AvrPtoB [11].

2 Materials

2.1 Bacteria and Plasmids for Protein Expression

1. BL21 (DE3) *Escherichia coli* competent cells.
2. Expression construct for one of the *Arabidopsis* E1s: *pDEST17-UBA1* (At2g30110).
3. Expression constructs for *Arabidopsis* E2s: for autoubiquitination activity, start with a "generic" E2 such as *UBC8* (*pDEST17-UBC8*; At5g41700) (*see* **Note 2**).
4. Expression construct for the E3 enzyme and interacting protein: the best tag and tag orientation for a specific E3 and the interacting protein need to be determined experimentally, e.g., *pMAL-c2X-PUB22* (MBP-PUB22) (*see* **Note 3**).

2.2 Protein Expression and Purification

1. LB medium: 10 g/L tryptone, 5 g/L yeast extract, 10 g/L sodium chloride. Autoclave and store at room temperature.
2. IPTG: prepare 1 M stock solution in water, aliquot in 1 mL, and store at −20 °C.
3. 4-(2-Aminoethyl)benzenesulfonyl fluoride hydrochloride (AEBSF): prepare 100 mM stock solution in water, aliquot in 1 mL, and store at −20 °C.
4. 1,4-Dithiothreitol (DTT): prepare 1 M stock solution in water, aliquot in 1 mL, and store at −20 °C.
5. Lysozyme.
6. Syringe filters with cellulose acetate membrane with 0.45 μm pore size.
7. Lysis-Elution-Washing (LEW) buffer: 50 mM NaH_2PO_4, 300 mM NaCl, adjust pH to 8.0 using NaOH and store at 4 °C.
8. Elution buffer: LEW supplemented with 200 mM imidazole.
9. Beads for 6× His fusion purification: Protino® Ni-Ted resin from Macherey Nagel.
10. Amicon® Ultra 4 mL Centrifugal Filters of 10K and 100K sizes from Merck Millipore.

2.3 In Vitro Ubiquitination Assay

1. Buffer solutions: prepare 1 M Tris–HCl pH 7.5, 0.1 M $MgCl_2$, 0.5 M KCl, 0.1 M DTT individual stock solutions in water. Aliquot in 1 mL and store at −20 °C.
2. ATP: prepare 0.1 M ATP stock solution in water. Aliquot in 100 μL and store at −20 °C.
3. Ubiquitin: WT untagged ubiquitin is commercially available, e.g., ubiquitin from bovine erythrocytes (e.g., Sigma Aldrich, catalogue no.U6253). Ubiquitin exclusively supports small N-terminal fusions and is available also as HA- and His-tagged recombinant protein (e.g., BostonBiochem, catalogue no.U-

110, His-ubiquitin) (*see* **Note 4**). Ubiquitin mutant variants can be employed to determine the linkage specificity of polyubiquitin chains synthesized by an E2–E3 pair (*see* **Note 5**). Dissolve the ubiquitin of choice at 5 mg/mL concentration and store in 50 μL aliquots at −20 °C. Ubiquitin stocks are stable after several freeze-thaw cycles.

4. Purified E1 ubiquitin-activating enzyme (UBA): UBA proteins are commercially available, including recombinant E1 enzymes UBA1 from *Arabidopsis* and *Ta*UBA1 from wheat (Cusabio) (*see* **Note 6**). *Sc*UBE1 GST-tagged (e.g., BostonBiochem, catalogue no. E-300) was also successfully used in ubiquitination assays with plant proteins. Alternatively, several *Arabidopsis* His-UBA1 fusion constructs have been used to express and purify the E1 in the lab (*see* **Note 7**).
5. Purified E2 ubiquitin conjugating enzyme (UBC): A commonly used E2 enzyme is the human UBE2D1 (e.g., UbcH5a BostonBiochem catalogue no. E2-616), which belongs to the subfamily VI, and is a "generic" E2 composed of mainly the core UBC domain and is most closely related to the *Arabidopsis* UBC8. Both enzymes display a high grade of promiscuity and processivity, making them ideal for in vitro ubiquitination assays (*see* **Note 8**).
6. Purified E3 ligase enzyme: around 1500 E3s have been identified in *Arabidopsis* which can be subdivided as mentioned above. Requirements for the purification of an active E3 ligase need to be determined individually (*see* **Note 9**).
7. Loading buffer: prepare a 4× loading sample buffer mixing 240 mM Tris–HCl pH 6.8, 0.5 M DTT, 10% SDS, 50% glycerol, 0.005% bromphenol blue. Aliquot in 1 mL and store at −20 °C.

2.4 Equipment

1. Shaker.
2. Sonicator.
3. Centrifuge.
4. Standard bench top incubator.
5. Protein electrophoresis apparatus.
6. Blotting apparatus.

3 Methods

3.1 AtUBA1 and AtUBC8 Protein Expression and Purification

1. Transform the *pDest17* plasmid carrying the fusion constructs for *His-UBA1* and *His-UBC8* into *E.coli* BL21(DE3) expression strain.
2. Use one colony obtained after transformation for expression of each protein to inoculate 5 mL of LB supplemented with the

appropriate antibiotics. Grow overnight in an incubator at 37 °C with shaking at 150 rpm.

3. Inoculate 500 mL of LB medium supplemented with the appropriate antibiotics with 5 mL overnight culture per protein. Incubate at 37 °C with shaking for approximately 3 h or until OD_{600} absorbance reaches 0.5.
4. Induce each culture with 0.5 mM IPTG and express UBA1 for 2 h at 28 °C and UBC8 for 4 h at 22 °C.
5. Harvest the cells by centrifugation for 10 min at 6000 × *g* at 4 °C (*see* **Note 10**).
6. Resuspend the pellets in 25 mL of LEW buffer freshly supplemented with 1 mM DTT and 1 mM AEBSF by vortexing.
7. Add 1 mg/mL of predissolved lysozyme, vortex, and incubate for 30 min on ice.
8. Lyse the cells by sonication (e.g., using Bandelin SONOPULS homogenizer probe VS70T) using 15 s pulses and 20 s pause cycles at 60% output six times on ice.
9. Centrifuge for 30 min at 23,000 × *g* at 4 °C.
10. Filtrate the supernatant through 0.45 μm cellulose acetate filters.
11. Transfer the supernatant for binding to 500 mg of Protino Ni-Ted resin under slow rotation for 1 h at room temperature (batch purification) or in columns (gravity flow purification) pre-equilibrated with LEW (*see* **Note 11**).
12. Wash resin with 15 mL LEW buffer (1 mM DTT and 1 mM AEBSF).
13. Elute three times with 3 mL of elution buffer.
14. Concentrate UBA1 using Amicon 100K centrifugal filter units and UBC8 using 10K centrifugal filter units following manufacturer's instructions.
15. Analyze protein quality and concentration by SDS-PAGE.
16. Store concentrated His-UBA1 and His-UBC8 at 1–10 μg/μL concentration in LEW buffer supplemented with 25% glycerol in 5 μL aliquots at −80 °C.

3.2 Autoubiquitination Assay

The simplest form of a ubiquitination activity test is the autoubiquitination assay which relies on the capacity of most E3s to mediate the attachment of ubiquitin to themselves.

1. Prepare:
   (a) Autoubiquitination reaction mix: mix stock solutions to the following final concentrations, 25 mM Tris–HCl pH 7.5, 5 mM $MgCl_2$, 50 mM KCl, 0.6 mM DTT to prepare the ubiquitination buffer, and add 2 mM ATP, 2 μg of

ubiquitin, 0.2 μg E1, 1.2 μg E2, and 2 μg E3 (e.g., for a total of 30 μL volume, mix 0.8 μL Tris–HCl pH 7.5, 1.5 μL $MgCl_2$, 3 μL KCl, 0.2 μL DTT, 0.6 μL ATP, 0.4 μL ubiquitin, and the relative amount of E1, E2, and E3 enzymes).

(b) Control reactions: except for the ubiquitination buffer components, prepare samples without one of the main components, namely (−) ATP control, (−) ubiquitin control, (−) E1 control, (−) E2 control, (−) E3 control (*see* **Note 12**).

Gently pipet up and down, do not vortex.

2. Incubate the samples for 1 h at 30 °C (*see* **Note 13**).
3. Stop the reaction by adding 10 μL of loading buffer and vortex.
4. Denature samples for 10 min at 68 °C.

3.3 Substrate ubiquitination assay

Proteins that interact with E3s, and are therefore potential substrates, can be tested in a substrate ubiquitination assay. In this version of the autoubiquitination assay, inclusion of the interactor leads to the formation of a complex with the E3, potentially resulting in its modification.

1. Prepare:

(a) Substrate ubiquitination reaction mix: mix all the components required for the complete reaction solution as described in Subheading 3.2, except for ubiquitin. Add 0.3 μg of the interacting protein and incubate for 1 h at 30 °C. Subsequently, add 2 μg of ubiquitin (*see* **Note 14**).

(b) Control reactions: except for the ubiquitination buffer components, prepare reactions without one of the main components, namely (−) ATP control, (−) ubiquitin control, (−) E1 control, (−) E2 control, (−) E3 control, and (−) substrate (*see* **Note 15**).

Proceed as in Subheading 3.2, **step 2**.

3.4 E2 Specificity

E3 ligases display specificity toward a set of E2s. The E2–E3 pairing may influence the capacity of an E3 to autoubiquitinate or to modify a substrate. Moreover, linkage specificity between ubiquitin moieties (i.e., K48 vs K63) is influenced by the E2. The specificity of autoubiquitination or substrate ubiquitination in the presence of different E2s can be tested as follows.

1. Prepare:

(a) Substrate or autoubiquitination reaction mix as described in Subheading 3.3 or 3.2 respectively, excluding the E2s.

(b) Add 1.2 μg of the purified E2s.

(c) Control reactions: include for each E2 enzyme, (−) E2 and (−) E3 controls. When testing an interactor, include for each E2 enzyme, (−) E2, (−) E3, and (−) substrate controls (*see* **Note 16**).

Proceed as in Subheading 3.2, **step 2**.

3.5 Ubiquitin Chain Linkage

Ubiquitin moieties are linked to one another via one of their seven lysine residues, giving rise to ubiquitin chains with different topologies. Ubiquitination assays showing ligase activity or substrate ubiquitination can be further refined to analyze the presence of a predominant, or in certain cases exclusive, type of linkage. This version of the assay is especially helpful after having determined specific E2–E3 pairs (*see* **Note 17**).

1. Prepare:

   (a) Substrate or autoubiquitination reaction mix as described in Subheading 3.3 or 3.2 respectively, without ubiquitin.

   (b) Add 2 μg of a selection of ubiquitin mutants (*see* **Note 5**) (e.g., using the Ub K48R mutant in combination with UBC8 usually results in reduced ubiquitination).

   (c) Control reactions: except for the ubiquitination buffer components, prepare reactions without one of the main components, as described above (Subheading 3.3).

   Proceed as in Subheading 3.2, **step 2** (*see* **Note 18**).

3.6 Analysis

1. Immunoblot: resolve samples by SDS-PAGE using a gel with the appropriate percentage (keep the stacking gel). Blot proteins onto nitrocellulose or PVDF membranes (including stacking gel). After blocking, probe the membranes with the antibody against the protein of interest (or tag) and with anti-ubiquitin antibodies (e.g., Santa Cruz Biotechnology, P4D1 anti-ubiquitin).

   (a) Analysis of the blot probed with antibodies against ubiquitin (lower panel): in the presence of all components, a typical ubiquitination smear signal should be detectable (*see* Fig. 1, lane 6). The signal extends throughout a wide range of sizes and is most intense around the size of the E3 as a result of the autoubiquitination activity (MBP-PUB22, arrowhead, 92 kDa). However, unspecific ubiquitination of co-purified impurities, or not fully translated E3, is also usually present. Ubiquitin-conjugated species of UBC8 are also generated during the assay (bottom) (*see* **Note 19**). The control samples in lane 1–5 show no ubiquitin signal, except for autoubiquitinated UBC8 in lane 5 (*see* **Note 19**).

   (b) Analysis of the blot probed with antibodies against the protein of interest (upper panel): reconstitution of the ubiquitination reaction leads to the modification of the

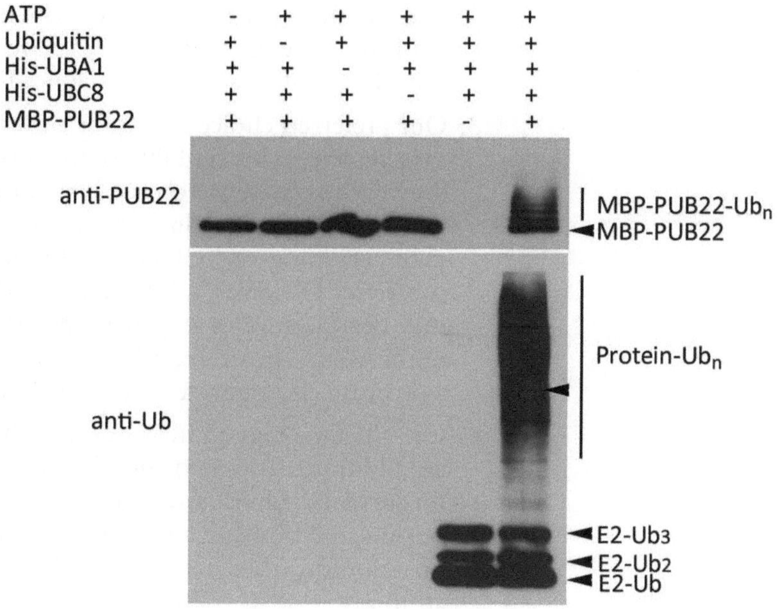

Fig. 1 MBP-PUB22 autoubiquitination assay: His-UBA1, His-UBC8 were purified as described above. MBP-PUB22 was purified using amylose beads, eluted and used within the same day for autoubiquitination assay. Complete reaction sample for testing autoubiquitination was prepared as described in Subheading 3.2 and loaded in lane 6. Lanes 1–5 show the following controls respectively (−) ATP, (−) ubiquitin, (−) E1, (−) E2, (−) E3

E3, which can be inferred by the typical laddering pattern (Fig.1, lane 6, *see* **Note 20**). No additional bands should be detectable in the control samples.

2. Coomassie: the autoubiquitination activity can also be visualized by simple coomassie staining, given that the purity and the concentration of the proteins is high enough. Alternatively, the modified protein can be put onto a second round of purification after the ubiquitination assay to improve results (after Subheading 3.2, **step 2**).
3. Liquid-chromatography tandem mass-spectrometric (LC-MS/MS) analysis: if available the mass-spectrometric analysis of in vitro ubiquitinated proteins can be employed to identify ubiquitination sites (*see* **Note 21**) and the ubiquitin chain linkage (*see* **Note 17**). Please refer to proteomic analysis of enriched samples by LC-tandem MS (MS/MS) detailed in Udeshi et al. [12].

4 Notes

1. The RING domain is stabilized by zinc chelation. However, the availability of different metal cations in the culture media may affect the selectivity of the recruitment of ions. During expression, do not include metal chelators.

2. Full length or partial coding sequences of 35 UBCs in pDONR (with the exception of UBC23) have been kindly deposited by Prof. Judy Callis [8] at TAIR (clones DQ027016–DQ027050).
3. Our preferred choice of tag for an uncharacterized E3 is MBP, as it improves the solubility of the fusion protein. However, the drawback is its relatively large size of 42.5 kDa. Position and size of the tag may influence the activity of the E3, by impairing correct folding or by blocking the interaction with the E2 or substrate. However, a large tag can serve as a substrate, which may be a positive or negative side effect depending on the aim of the assay. Therefore, using an expression vector that includes a protease cleavage site could prove useful.
4. Employing a tagged ubiquitin has the advantage that it allows an additional round of purification of ubiquitinated proteins for further analysis, if required. In addition, ubiquitination can be detected by using antibodies directed against the tag instead of ubiquitin, thus avoiding having to purchase an additional antibody.
5. Mutation of a single lysine into arginine (e.g., Ub K63R) renders a specific lysine unavailable during ubiquitin chain formation. Zero lysine mutants (Ub 0K) cannot form chains and allow exclusively mono- or multimono-ubiquitination. By contrast, back mutation into lysine from the Ub 0K variant (e.g., Ub R63K) provides a single lysine for the assembly of chains.
6. *Arabidopsis* possesses two E1 enzymes, *At*UBA1 and *At*UBA2, which share 81% amino acid sequence identity and do not display different enzymatic properties in vitro [13].
7. Although the *At*UBA1 is fairly large in size, its expression in *E. coli* is straightforward and long-term storage at −80 °C (25% glycerol) is unproblematic.
8. UBC8 and related E2s are well-behaved proteins that can easily be expressed and purified in house. Kraft and colleagues [8] published a comprehensive study that summarizes the solubility properties of *Arabidopsis* E2s and shows that 24 of the 30 E2s tested in vitro display E2 activity. Note that in the case of UBC35 and UBC36, ubiquitin conjugating enzyme variants (UEVs) are required for the activity, and to mediate the formation of free K63-linked chains in an E3-independent manner [14]. Additional E2 enzymes have been shown to possess activity in vitro in the absence of an E3, and to form free or self-conjugated ubiquitin chains [15].
9. Ligase activity of certain E3s may be drastically reduced by freezing. For example, the E3 ligase PUB22 loses approximately 75% of its activity after storage at −80 °C. Therefore, E3 enzymes storage conditions have to be determined individually.

10. Pellets can be frozen and stored at −20 °C for two weeks.
11. When using columns, pass the lysate through the column a minimum of three times.
12. In addition to the above-mentioned controls, it is advisable to include an inactive E3 mutant. Conserved cysteines and histidines can be mutated into alanines to inactivate the RING zinc finger domain, the U-box hydrogen bonds, or the HECT domain, thus inhibiting E2 interaction and E3 ligase activity.
13. Modified proteins can be detected in samples incubated for as short as 2 min in a ubiquitination assay. Time course analyses can help to determine the desired level of ubiquitination. In our experience, increased incubation times (e.g., overnight) may be employed without compromising the specificity of the ubiquitination reaction.
14. Preincubation in the absence of ubiquitin allows the E3 to interact with the potential substrate, previous to the initiation of the ubiquitination reaction. This point is key, because post-translational modifications, particularly phosphorylation, which regulate ubiquitination are missing. Phosphorylation of a substrate can result in its increased affinity to the E3. Thus, autoubiquitination can be favored to ubiquitination of a low-affinity-binding substrate. Furthermore, autoubiquitination can interfere with substrate interaction through steric hindrance or by modification of the interaction surface.
15. Substrate ubiquitination can be detected only by using antibodies against the substrate itself or its tag. The E3 ligase autoubiquitination, although it may be reduced in the presence of the substrate, is undistinguishable from the substrate ubiquitination, when using anti-ubiquitin antibodies. An additional purification of the substrate can be included after the ubiquitination reaction, to separate the two ubiquitination events.
16. Certain E2 enzymes can generate free ubiquitin chains in vitro in the absence of an E3 ligase. The comparison between the (−) E3 controls and the complete reaction samples will show whether the presence of the specific E3 is required, enhances the free chain formation, or results in E3 autoubiquitination or substrate modification.
17. Results regarding chain linkage specificity obtained from in vitro ubiquitination assays need to be considered with care as they may not necessarily represent the "physiological" activity of the E2–E3 pair, mainly because of the high concentrations of the components present in the reaction [16].
18. Ubiquitin chain linkage can be determined by employing specific antibodies (e.g., Cell Signaling, D7A11 anti-K63 polyubiquitin).

19. Sample preparation under nonreducing conditions preserves the E2~Ub thioester conjugates and can be detected as modified species of UBC8. Samples in Fig. 1 were analyzed under reducing conditions in which thioester bonds should be lost; however, ubiquitin conjugates of UBC8 are detected. This indicates that the E2 is also capable of autoubiquitination under the used conditions.
20. Most of the E3 ligases can autoubiquitinate in vitro. In order to confirm *bona fide* autoubiquitination activity, it is necessary to test an untagged version of the E3.
21. It must be kept in mind that in difference to many enzymes, most E3 ubiquitin ligases display an inherent low specificity toward specific lysines. Mutations of single lysine residues rarely result in the inhibition of ubiquitination. Most commonly, it is necessary to introduce multiple mutations, encompassing a large area of the protein, to inhibit ubiquitination.

Acknowledgments

The author would like to acknowledge the funding from the Leibniz Association and the state of Saxony-Anhalt.

References

1. Furlan G, Klinkenberg J, Trujillo M (2012) Regulation of plant immune receptors by ubiquitination. Front Plant Sci. 3:238. doi:10.3389/fpls.2012.00238
2. Bates PW, Vierstra RD (1999) UPL1 and 2, two 405 kDa ubiquitin-protein ligases from *Arabidopsis thaliana* related to the HECT-domain protein family. Plant J 20:183–195. doi:10.1046/j.1365-313x.1999.00590.x
3. Miao Y, Zentgraf U (2010) A HECT E3 ubiquitin ligase negatively regulates *Arabidopsis* leaf senescence through degradation of the transcription factor WRKY53. Plant J *63*:179–188. doi:10.1111/j.1365-313X.2010.04233.x
4. Stone SL, Hauksdóttir H, Troy A, Herschleb J, Kraft E, Callis J (2005) Functional analysis of the RING-type ubiquitin ligase of Arabidopsis. Plant Physiol 137:13–30. doi:10.1104/pp.104.052423
5. Mudgil Y, Shiu S, Stone SL, Salt JN, Goring DR (2004) A large complement of the predicted *Arabidopsis* ARM repeat proteins are members of the U-Box E3 ubiquitin ligase family. Plant Physiol 134:59–66. doi:10.1104/pp.103.029553
6. Ohi MD, Vander Kooi CW, Rosenberg JA, Chazin WJ, Gould KL (2003) Structural insights into the U-box, a domain associated with multi-ubiquitination. Nat Struct Biol 10:250–255. doi:10.1038/nsb906
7. Wiborg J, O'Shea C, Skriver K (2008) Biochemical function of typical and variant *Arabidopsis thaliana* U-box E3 ubiquitin-protein ligases. Biochem J 413:447–457. doi:10.1042/BJ20071568
8. Kraft E, Stone SL, Ma L, Su N, Gao Y, Lau O, Deng X, Callis J (2005) Genome analysis and functional characterization of the E2 and RING-type E3 ligase ubiquitination enzymes of *Arabidopsis*. Plant Physiol 139:1597–1611. doi:10.1104/pp.105.067983
9. Göhre V, Spallek T, Häweker H, Mersmann S, Mentzel T, Boller T, de Torres M, Mansfield JW, Robatzek S (2008) Plant pattern-recognition receptor FLS2 is directed for degradation by the bacterial ubiquitin ligase AvrPtoB. Curr Biol 18:1824–1832. doi:10.1016/j.cub.2008.10.063
10. Lu D, Lin W, Gao X, Wu S, Cheng C, Avila J, Heese A, Devarenne TP, He P, Shan L (2011) Direct ubiquitination of pattern recognition receptor FLS2 attenuates plant innate immunity. Science 332:1439–1442. doi:10.1126/science.1204903

11. Gimenez-Ibanez S, Hann DR, Ntoukakis V, Petutschnig E, Lipka V, Rathjen JP (2009) AvrPtoB targets the LysM receptor kinase CERK1 to promote bacterial virulence on plants. Curr Biol 19:423–429. doi:10.1016/j.cub.2009.01.054
12. Udeshi ND, Mertins P, Svinkina T, Carr SA (2013) Large-scale identification of ubiquitination sites by mass spectrometry. Nat Protoc 8:1950–1960. doi:10.1038/nprot.2013.120
13. Hatfield PM, Gosink MM, Carpenter TB, Vierstra RD (1997) The ubiquitin-activating enzyme (E1) gene family in *Arabidopsis thaliana*. Plant J 11:213–226. doi:10.1046/j.1365-313X.1997.11020213.x
14. Wen R, Torres-Acosta JA, Pastushok L, Lai X, Pelzer L, Wang H, Xiao W (2008) *Arabidopsis* UEV1D promotes lysine-63-linked polyubiquitination and is involved in DNA damage response. Plant Cell 20:213–227. doi:10.1105/tpc.107.051862
15. Callis J (2014) The ubiquitination machinery of the ubiquitin system. Arabidopsis Book. doi:10.1199/tab.0174
16. Kim HT, Kim KP, Lledias F, Kisselev AF, Scaglione KM, Skowyra D, Gygi SP, Goldberg AL (2007) Certain pairs of ubiquitin-conjugating enzymes (E2s) and ubiquitin-protein ligases (E3s) synthesize nondegradable forked ubiquitin chains containing all possible isopeptide linkages. J Biol Chem 282:17375–17386. doi:10.1074/jbc.M609659200

Chapter 9

Bioinformatics Analysis of the Receptor-Like Kinase (RLK) Superfamily

Otávio J.B. Brustolini, José Cleydson F. Silva, Tetsu Sakamoto, and Elizabeth P.B. Fontes

Abstract

Receptor-like kinases (RLKs) play key roles during development and in responses to the environment. In plant immunity, some members of RLKs function as pattern recognition receptors (PRRs), which, upon recognition of pathogen-associated molecular patterns (PAMP), are recruited into active complexes to induce pathogen-triggered immunity (PTI). In this chapter, we describe the bioinformatics tools and procedures for the identification and phylogenetic classification of RLKs from different plant species as a framework for understanding RLK function in signal transduction and immunity.

Key words Receptor-like kinase (RLK), Leucine-rich repeat (LRR), Ectodomain, Serine/threonine kinase domain, Pattern recognition receptors (PRR)

1 Introduction

Receptor-like protein kinases (RLKs) are a diverse group of plant cell surface receptors that have been implicated in a wide range of signal transduction pathways, in growth, development and defense. These transmembrane receptors are structurally organized into a divergent extracellular domain at the N-terminal portion (ectodomain) that confers ligand-binding specificity followed by an internal transmembrane segment and a predicted cytoplasmic signal-transducing domain, which includes a juxtamembrane domain, a conserved serine/threonine kinase domain, and a C-terminal region [1]. On the basis of their extracellular domain, these receptors have been organized into different structural classes [1]. The largest class comprises the extracytoplasmic leucine-rich repeats (LRR)-containing RLKs that can be further divided into 13 subfamilies (LRR I–XIII) on the basis of the LRR organization, ranging from three to 26 LRRs [2, 3]. The LRR-RLKs are often associated with signal transduction and immunity. In immunity, they function as PRRs, which, in response to ligand (PAMPs) binding to the LRR

Libo Shan and Ping He (eds.), *Plant Pattern Recognition Receptors: Methods and Protocols*, Methods in Molecular Biology, vol. 1578, DOI 10.1007/978-1-4939-6859-6_9, © Springer Science+Business Media LLC 2017

ectodomain, form active immune complexes with co-receptors to induce PTI (for review, *see* ref. 4). Although to a lesser extent, lysine motifs (LysM) and lectin motifs ectodomain-containing RLKs have also been identified as relevant PRRs [4].

To detect these transmembrane receptors by bioinformatics tools, two types of analyses can be done, (1) the homologous detection driven by alignment and phylogenetic (evolutionary) approach, and (2) function site recognition uncovered by machine learning algorithms [5]. Although methods based on homology relationship among proteins may allow low or no similarity to known proteins for comparison, the machine learning techniques can predict protein function by using sequence-derived properties absolutely independent of sequence similarity [6]. The combination of these two approaches reveals a powerful detection method that can be applied to help the annotation of new RLKs. Associated with algorithms and bundled as a suite of programs, a portal like Pfam [7] uses a set of methods and a clustered data to detect protein regions that share a significant degree of sequence similarity. It applies the algorithm hidden Markov model (HMM) to predict conserved regions such as the LRR and kinase domains. Furthermore, to enhance the reliability of the RLK detection as well as to achieve the subfamily structure, the methods based on the reconciliation of a gene tree or species tree should be indicated [2]. The aim of this chapter is to highlight and summarize how bioinformatics tools can be used to improve the knowledge of the evolutionary history of the RLK superfamily. It also helps to bring about the concepts well studied in plants evolution with the pattern recognition granted by the machine learning.

2 Materials

Here, we will describe how to identify and classify RLKs from plant species that have their genomes sequenced and the proteins identified. For this, we will take advantage of the RLK annotation of the reference plant organism *Arabidopsis thaliana*. In some steps, we will execute programs in the command line and will use some basic programming skills. If you are not familiar with any programming language (e.g., C, C++, Java, Perl, Python, etc.) or in executing programs in the command line, look for help from people who have bioinformatics skills. All local software cited in this protocol can be run either in a UNIX or in a Windows machine.

2.1 Sequences Databases

To start with, you have to seek the protein sequences in an online public database. Many sequences repositories are publically available online. For green plants, some of them, such as Phytozome (http://phytozome.jgi.doe.gov) and Ensembl Plants (http://plants.ensembl.org), gather the sequence data and annotation in the same place to simplify the analysis for the users. The database Phytozome (version 10.3) is a portal for comparative genomics among green plants [8]. This database has several tools that help the advanced search for DNA sequences, amino acids, and their functional annotations. Currently, the Phytozome hosts bulk data from approximately 63 plant species. The online system has a user-friendly interface, which allows a fast access to the data.

2.2 RLK Identification

1. Proteome of a plant organism of interest in FASTA format. Do not include protein isoforms (*see* **Note 1**).
2. RLK set of sequences from a reference plant organism (Arabidopsis) in FASTA format (*see* refs. 1 and 2).
3. No-RLK sequences to be used as outgroups in FASTA format (*see* refs. 1 and 2).
4. Internet access to Pfam [7] website (*see* **Note 2**).
5. Standalone BLAST+ tool [9].
6. Global alignment tool that can handle a large number of sequences (greater than a thousand sequences), e.g., MUSCLE [10], ClustalOmega [11], or MAFFT [12].
7. Phylogenetic tree reconstruction tool that can handle a large number of sequences (greater than a thousand sequences) like FastTree [13].
8. Tree visualization software, e.g., FigTree (http://tree.bio.ed.ac.uk/software/figtree/).

2.3 Inference on Duplication/Deletion Events

1. Amino acid sequence set of a RLK subfamily from the Arabidopsis and from the organism of your interest.
2. Species tree containing the organisms selected in **step 1** (*see* **Note 3**).
3. Global alignment tool.
4. Phylogenetic tree reconstruction tool.
5. Phylogenetic tool with reconciliation method implemented in it, e.g., Notung [14].

3 Methods

3.1 Retrieve Proteome Sequences in FASTA Format

3.1.1 Phytozome

1. Access the upper menu on the left side of the site to retrieve the sequences. Click the option species and choose the desired species.
2. At the center of the screen, five options will appear: keyword search, BLAST search, BLAT search, Jbrowser, and bulk data.
3. Click on the tag "keyword search" for retrieving sequences based on annotation terms. For example, you can write RLK

or receptor-like kinase and click GO button. The next page shows the genes list in a table format. The column "View" has two options: the letter "G" for visualizing gene information and "B" to access the Jbrowse tool for chromosome position visualization. Click on the option G.

4. The next page shows the information about functional annotations, genomics, sequences, and protein homologs. Click on the option "Sequences" and choose one of the following alternatives: genome sequence, transcript sequence, CDS, or peptide sequence.
5. The sequences are shown in FASTA format. Copy and paste the chosen sequence in a text editor (Gedit if you use Unix-like, TextWrangler for Mac OS X, or notepad for Windows).
6. For sequences of all chromosomes, genomic regions or amino acids, click on the option of download or bulk data of the desired specie.

3.1.2 Ensembl Plants

1. In "Search" box choose the desired species. Write a gene name or description. For example, write RLK and click on GO button. The result page displays a summary of the genes related to the RLK term. Click on the word after the Gene ID. The next screen will display a table with eight columns: Name, Transcript ID, bp, Protein, Translation ID, biotype, UniProt, and Flags.
2. Click on the accession number below the UniProt column. You will be redirected to the Uniprot website (www.uniprot.org/uniprot).
3. Click on the Format button, choose the option FASTA (canonical). Copy and paste the sequence in a text editor (Gedit if you use Unix-like, TextWrangler for Mac OS X, or Notepad for Windows).
4. For the sequences of all chromosomes, genomic regions or amino acids, go to download options at the top of the main page of Ensembl plants.

3.2 Looking for the Kinase Domain in Arabidopsis RLKs

Figure 1 summarizes the steps for the procedures described from Subheadings 3.2–3.4

1. Access the Pfam (http://pfam.xfam.org/search) website.
2. In "Batch search" tag, upload the FASTA file containing the RLKs from the Arabidopsis. Include your email address and submit the search. You will receive a table containing all predicted domains and their location in each submitted protein.
3. Filter the table leaving only those rows containing information of kinase domain.
4. For each predicted kinase domain, map the sequence and the position of the kinase domain in the sequence and write it into

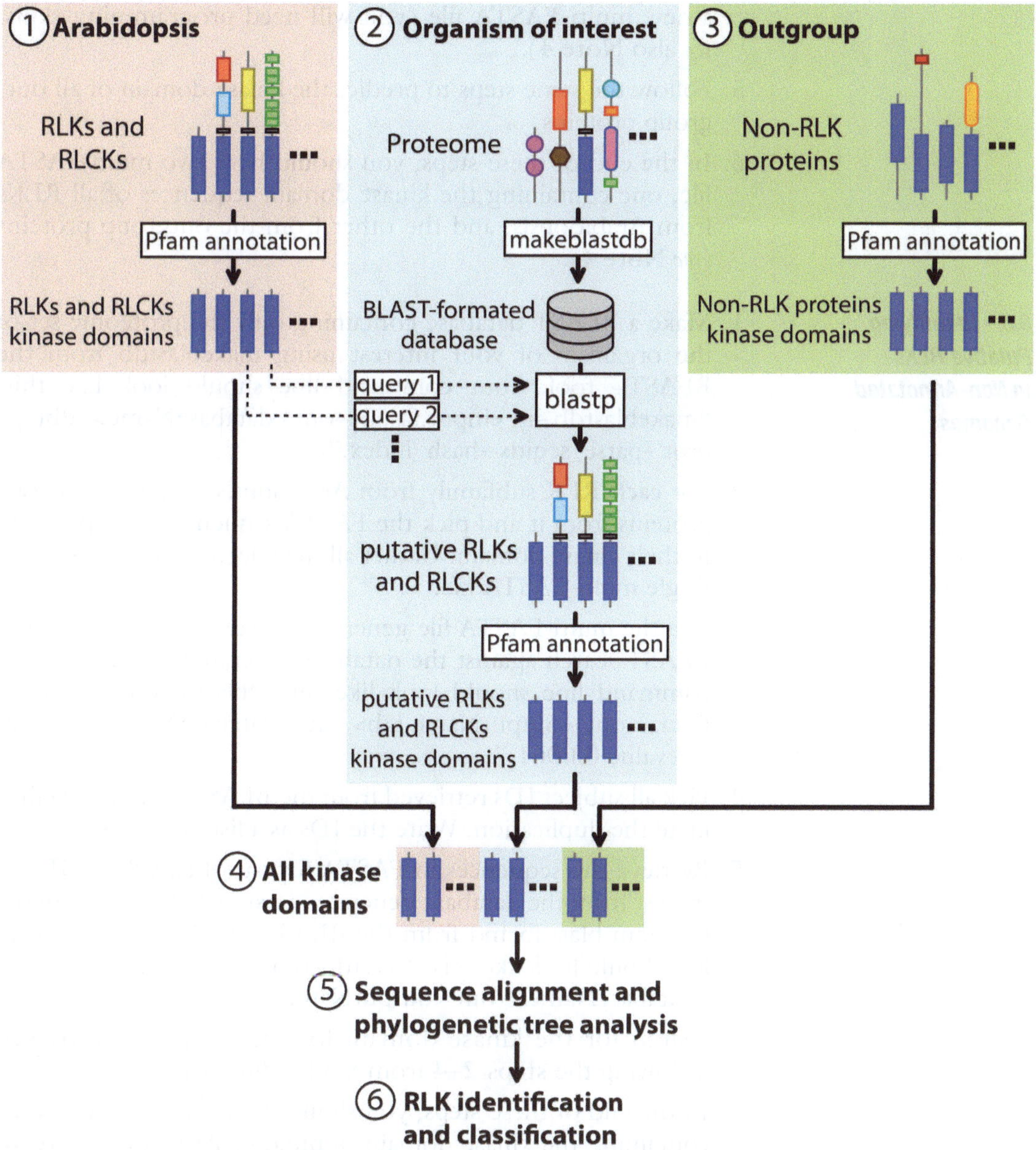

Fig. 1 Flowchart of the RLKs analysis using phylogenetics approach. The sequences from Arabidopsis RLKs are submitted to the Pfam website (http://pfam.xfam.org/search) to retrieve the kinase domain sequences [1]. Then, a BLAST-formatted database is made from query sequences. Using the Arabidopsis sequences from **step 1** as a prototype, a blastp search identifies putative RLK sequences from the query data set. The Pfam is applied to the query sequences to recognize and retrieve the kinase domains [2]. Parallely, the sequences from Arabidopsis non-RLKs are submitted to the Pfam website to retrieve the kinase domain sequences [3]. The retrieved kinase domains from the **steps 1**, **2**, and **3** are combined in a single multi-FASTA file [4]. Then, the sequences are aligned by ClustalOmega, MUSCLE or MAFFT and a phylogenetic tree is constructed by FastTree or any other tree reconstruction tool [5]. Finally, the query RLKs are identified, classified into subfamilies, which are structured in new query phylogenetic tree as compared to the Arabidopsis RLKs tree [6]

a new multi-FASTA file (you will need programming skills, *see* also **Note 4**).

5. Follow the same steps to predict the kinase domain of all outgroup proteins.
6. In the end of these steps, you should have two multi-FASTA file, one containing the kinase domain sequences of all RLKs from Arabidopsis, and the other from the outgroup proteins (*see* **Note 4**).

3.3 Identifying Putative RLKs in Non-Annotated Genomes

1. Make a BLAST database containing only the proteome set of the organism of your interest using makeblastdb from the BLAST+ tool. Your command line should look like this: "makeblastdb –in <input.fasta> –out <databaseName> -dbtype prot –parse_seqids –hash_index."
2. For each RLK subfamily from Arabidopsis, select one or two proteins from it and pick the FASTA sequence corresponding to their kinase domain. Gather all the selected sequences into a single multi-FASTA file.
3. Use the multi-FASTA file generated in **step 2** as query for the BLAST search against the database generated in **step 1**. Your command line should look like this: "blastp -query <input.fasta> -out <outputName.tab> -db <databaseName> -outfmt 6 -evalue 0.0001."
4. Pick all subject IDs retrieved from the BLAST search and eliminate the duplication. Write the IDs as a list into a file.
5. Retrieve the sequences in FASTA format of all subject IDs in the list from the database generated in **step 1**. For this, use the program blastdbcmd from the BLAST+ tool. Your command line should look like this: "blastdbcmd –entry <listFile.txt> -db <databaseName> -out <outputName.fasta>."
6. Search for the kinase domain for each sequence retrieved following the **steps 2–4** from Subheading 3.1.
7. In the end of these steps, you should have a multi-FASTA file containing the kinase domain of putative RLKs of the organism of your interest (*see* **Note 5**).

3.4 Identifying and Classifying RLKs

1. Merge all three multi-FASTA files in a single multi-FASTA file.
2. Align the sequences using ClustalOmega or a sequence alignment tool of your choice.
3. Generate a phylogenetic tree using FastTree or a tree reconstruction tool of your choice.

4. Visualize the phylogenetic tree with FigTree or any other tree-visualizing tool. Root the tree by the outgroup sequences. All proteins that do not cluster together with the outgroup proteins can be considered RLKs. Use the RLK subfamily classification from the Arabidopsis sequences to classify the other RLKs into a subfamily.

3.5 Inferring the Duplication/ Deletion Events in Each RLK Subfamilies

1. Select a RLK subfamily to procedure the duplication/deletion inference (*see* **Note 6**).
2. Retrieve the amino acid sequences of the kinase domain of the RLKs comprising the selected subfamily. Include also an outgroup sequence for tree rooting. It could be a kinase domain of another RLK subfamily.
3. Align the amino acid sequences and reconstruct a phylogenetic tree. You can choose tools that can be more time consuming, but not necessarily more accurate, for both procedures. E.g., PRANK [15] or TCOFFEE [16] for sequence alignment and tools based on maximum likelihood or Bayesian inference methods for phylogenetic reconstruction (*see* **Note** 7).
4. Open the Notung and load the phylogenetic tree generated in **step 3**. Also load the species tree generated in the NCBI Taxonomy Common Tree tool.
5. In your phylogenetic tree, go to the "Reconciliation" tag and click on "Reconcile/Rereconcile" button. If necessary, go to the "Rooting" tag and click on the branch that leads to your outgroup(s) for tree rooting.
6. Go to the "Rearrange" tag and click on "Perform Rearrangement" button. This will minimize the number of duplication and deletion events in tree by evaluating the node support value.
7. Now you can infer the number of protein members in the common ancestral of Arabidopsis and the organism of your interest by counting the number of speciation events on the tree. The program also shows how many genes were lost in each organism. Moreover, if there is a monophyletic clade (with more than one sequence) for a single organism, this implies the occurrence of a lineage-specific expansion after the speciation event of the two organisms. One could estimate the number of expansion events by counting the number of leaves that comprise these clades minus the number of monophyletic clades.

4 Notes

1. The FASTA header should contain some information to facilitate the further analysis. We recommend putting a species code in the beginning of the header and, for the Arabidopsis RLK, the RLK subfamily that the protein belongs to.
2. A suitable way to annotate a large number of proteins from Pfam or other platform (e.g., SMART) is by installing and downloading the tools and database and by performing a local search. If the web batch search presented here does not attend your demand, please refer to the FAQ page to see how to install them in your machine.
3. Species tree can be generated by NCBI Taxonomy Common Tree tool (http://www.ncbi.nlm.nih.gov/Taxonomy/CommonTree/wwwcmt.cgi). To obtain the tree, just type and add the species name of the organisms in the analysis (e.g., *Arabidopsis thaliana* and *Solanum lycopersicum*). Save the species tree as PHYLIP tree.
4. For scripts, examples, and tree files, access: http://www.inc-tipp.ufv.br/rlkchapter.
5. Some predicted kinase domains are fragmented and should be manually removed from the further analysis.
6. Figure 2 describes an example of Duplication/Deletion analysis using Notung (v2.6) of the LRRXV subfamily of RLK.
7. A combination of several alignment results from different tools has been demonstrated to increase the alignment accuracy [17].

Acknowledgment

This work was supported by the National Institute of Science and Technology in Plant-Pest Interactions, CNPq grants 573600/2008-2 and 447578/2014-6 and Fapemig grants APQ-00070-09 and CBB-APQ-01491-14.

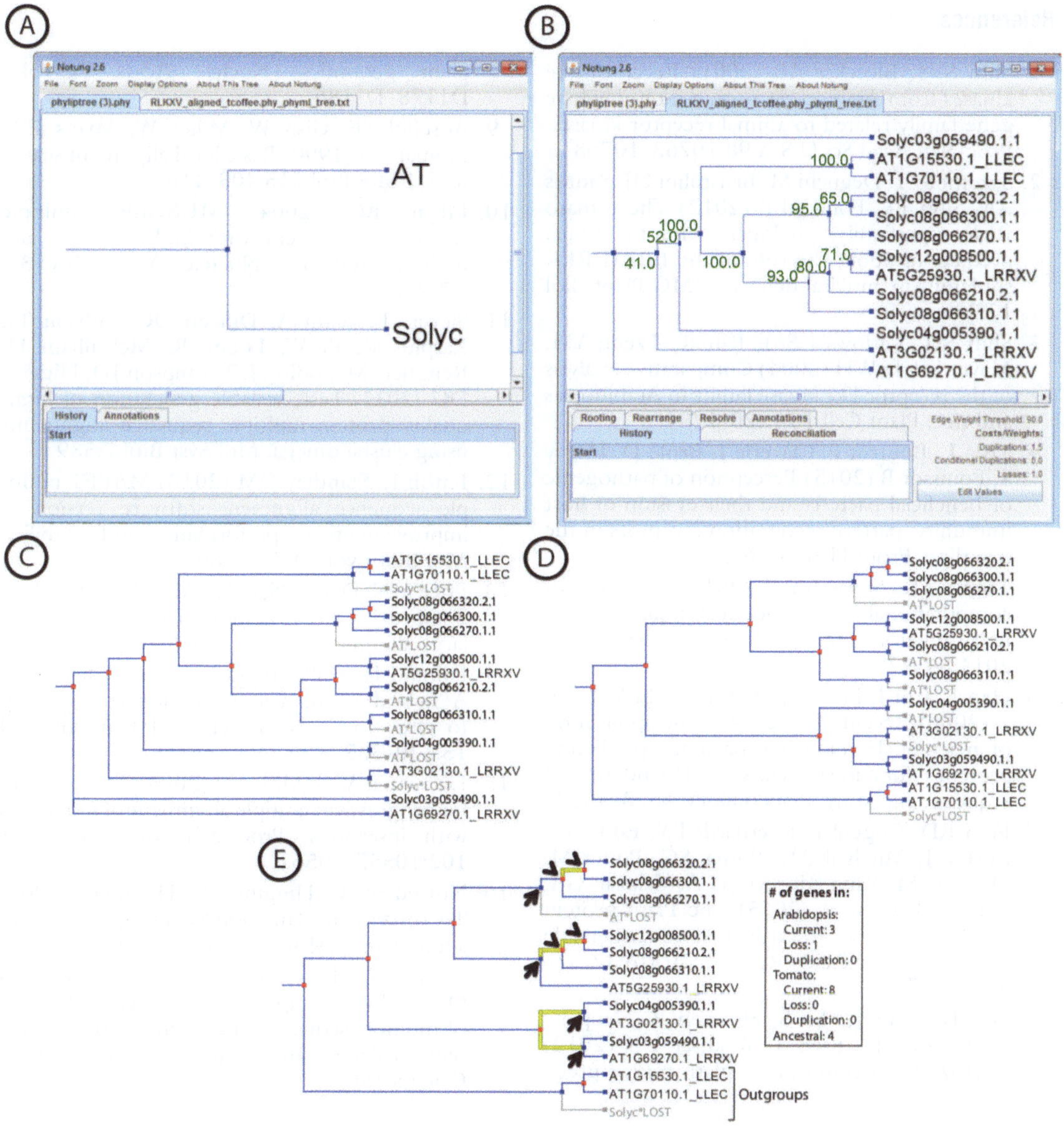

Fig. 2 Duplication/deletion analysis using Notung (v2.6) of the LRRXV subfamily of RLKs. The analysis starts by loading the species tree (**a**) and the gene tree (**b**) in the application. After reconciling the both tree (**c**), the tree internal nodes are marked as duplication (*red squares*) or speciation (*blue squares*) events. Also, gene loss events in each species are represented in *gray branches* that lead to a leaf with "*LOST" notation. In "Rooting" tab, we can select the correct branch for tree rooting. In our example (**d**), we added two Arabidopsis RLK representatives from subfamily L-lectin to be considered the outgroup. Finally, we use the "Rearrangement" tool to allow Notung to minimize the number of gene duplication and loss events. The final tree (**e**) can be used to estimate the duplication/deletion events that occurred in this RLK subfamily. The number of genes in the lowest common ancestor of Arabidopsis and tomato can be estimated by counting all internal nodes marked as speciation event (*arrow*). Gene expansion in each species can be estimated by counting the number of duplication events that occurred after a speciation event (*arrow head*). Finally, gene loss in each species can be estimated by counting the leaves with "*LOST" notation. Legend: AT: *A. thaliana*; Solyc: *S. lycopersicum*

References

1. Shiu S-H, Bleecker AB (2001) Receptor-like kinases from Arabidopsis form a monophyletic gene family related to animal receptor kinases. Proc Natl Acad Sci U S A 98:10763–10768
2. Sakamoto T, Deguchi M, Brustolini OJ, Santos AA, Silva FF, Fontes EP (2012) The tomato RLK superfamily: phylogeny and functional predictions about the role of the LRRII-RLK subfamily in antiviral defense. BMC Plant Biol 12:229
3. Shiu SH, Karlowski SM, Pan R, Tzeng YH, Mayer KF, Li WH (2004) Comparative analysis of the receptor-like kinase family in Arabidopsis and rice. Plant Cell 16:1220–1234
4. Trdá L, Boutrot F, Claverie J, Brulé D, Dorey S, Poinssot B (2015) Perception of pathogenic or beneficial bacteria and their evasion of host immunity: pattern recognition receptors in the frontline. Front Plant Sci 6:219
5. Tiwari AK, Srivastava R (2014) A survey of computational intelligence techniques in protein function prediction. Int J Proteomics 2014:845479
6. Han L, Cui J, Lin H, Ji Z, Cao Z, Li Y, Chen Y (2006) Recent progresses in the application of machine learning approach for predicting protein functional class independent of sequence similarity. Proteomics 6:4023–4037
7. Finn RD, Coggill P, Eberhardt RY, Eddy SR, Mistry J, Mitchell AL, Potter SC, Punta M, Qureshi M, Sangrador-Vegas A, Salazar GA, Tate J, Bateman A (2015) The Pfam protein families database: towards a more sustainable future. Nucleic Acids Res. doi:10.1093/nar/gkv1344
8. Goodstein DM, Shu S, Howson R, Neupane R, Hayes RD, Fazo J, Rokhsar DS (2012) Phytozome: a comparative platform for green plant genomics. Nucleic Acids Res 40: D1178–D1186
9. Altschul SF, Gish W, Miller W, Myers EW, Lipman DJ (1990) Basic local alignment search tool. J Mol Biol 215:403–410
10. Edgar RC (2004) MUSCLE: multiple sequence alignment with high accuracy and high throughput. Nucleic Acids Res 32: 1792–1797
11. Sievers F, Wilm A, Dineen DG, Gibson TJ, Karplus K, Li W, Lopez R, McWilliam H, Remmert M, Söding J, Thompson JD, Higgins DG (2011) Fast, scalable generation of high-quality protein multiple sequence alignments using clustal omega. Mol Syst Biol 7:539
12. Katoh K, Standley DM (2013) MAFFT multiple sequence alignment software version 7: improvements in performance and usability. Mol Biol Evol 30:772–780
13. Price MN, Dehal PS, Arkin AP (2010) FastTree 2-approximately maximum-likelihood trees for large alignments. PLoS One 5:e9490
14. Durand D, Halldorsson BV, Vernot B (2006) A hybrid micro-macroevolutionary approach to gene tree reconstruction. J Comput Biol 13:320–335
15. Löytynoja A, Goldman N (2005) An algorithm for progressive multiple alignment of sequences with insertions. Proc Natl Acad Sci USA 102:10557–10562
16. Notredame C, Higgins DG, Heringa J (2000) T-Coffee: a novel method for multiple sequence alignments. J Mol Biol 302:205–217
17. Thompson JD, Linard B, Lecompte O, Poch O (2011) A comprehensive benchmark study of multiple sequence alignment methods: current challenges and future perspectives. PLoS One 6:e18093

Chapter 10

Identification of MAPK Substrates Using Quantitative Phosphoproteomics

Tong Zhang, Jacqueline D. Schneider, Ning Zhu, and Sixue Chen

Abstract

Activation of mitogen-activated protein kinases (MAPKs) under diverse biotic and abiotic factors and identification of an array of downstream MAPK target proteins are hot topics in plant signal transduction. Through interactions with a plethora of substrate proteins, MAPK cascades regulate many physiological processes in the course of plant growth, development, and response to environmental factors. Identification and quantification of potential MAPK substrates are essential, but have been technically challenging. With the recent advancement in phosphoproteomics, here we describe a method that couples metal dioxide for phosphopeptide enrichment with tandem mass tags (TMT) mass spectrometry (MS) for large-scale MAPK substrate identification and quantification. We have applied this method to a transient expression system carrying a wild type (WT) and a constitutive active (CA) version of a MAPK. This combination of genetically engineered MAPKs and phosphoproteomics provides a high-throughput, unbiased analysis of MAPK-triggered phosphorylation changes on the proteome scale. Therefore, it is a robust method for identifying potential MAPK substrates and should be applicable in the study of other kinase cascades in plants as well as in other organisms.

Key words MAPK, MAPK substrate, Signal transduction, Phosphoproteomics, TMT

1 Introduction

Mitogen-activated protein kinase (MAPK) cascades are responsive to most environmental challenges faced by plants. The canonical MAPK kinase kinase (MAPKKK)—MAPK kinase (MAPKK or MKK)—MAPK pathways represent sequential phosphorylation and activation of the three players and subsequent phosphorylation of substrate proteins by MAPKs [1]. Pattern recognition receptors (PRRs) are typically proteins that span the cell membrane to perceive and transduce cellular signals [2, 3]. A growing body of evidence indicates that many of the MAPK cascades function downstream of PRRs such as receptor-like kinases (RLKs). For example, the MAPKKK4 (YDA)-MKK4/MKK5-MPK3/MPK6 cascade was demonstrated to be downstream of the RLK ERECTA to modulate cell proliferation and inflorescence architecture [4].

Libo Shan and Ping He (eds.), *Plant Pattern Recognition Receptors: Methods and Protocols*, Methods in Molecular Biology, vol. 1578, DOI 10.1007/978-1-4939-6859-6_10, © Springer Science+Business Media LLC 2017

Another PRR protein chitin elicitor receptor kinase 1(CERK1) recruits receptor-like cytoplasmic kinase 185 to activate the downstream MAPK cascade [5]. However, gaps still remain in the understanding of how MAPK cascades regulate cellular signaling and metabolic activities in the course of plant growth, development, and defense. Thus, identification of the MAPK substrates is a prerequisite for filling the knowledge gaps and furthering insight into plant signaling and metabolic networks.

Traditional methods to identify protein kinase substrates are usually performed at an individual protein level, relying heavily on time-consuming genetic analysis and biochemical assays of the protein-protein interactions [6–9]. Large-scale methods such as phosphoproteomics have been developed during the last decade as methodologies in mass spectrometry (MS)-based identification and quantification of phosphopeptides have improved [10–12]. In such a strategy, the global phosphorylation levels are compared between two groups of plants: either the wild-type versus kinase knock-out mutant plants [13], or plants expressing a constitutive active (CA) version of a kinase of interest versus wild type [14]. In both cases, proteins with significantly higher levels of phosphorylation in the former group of plants are considered candidate substrates.

Three major challenges should be kept in mind when utilizing phosphoproteomics for MAPK substrate identification. First, the MAPK of interest should be activated to phosphorylate the downstream substrates, as the activities of most MAPKs are tightly controlled and the basal MAPK activity is kept low [15, 16]. Second, in addition to the fact that many MAPK substrates (e.g., transcription factors) are low-abundance proteins, phosphorylated protein isoforms (proteoforms) are usually in low stoichiometry [17]. Thus, enrichment of phosphoproteins and/or phosphopeptides is critical for successful identification of phosphoproteins. Third, the changes in phosphorylation must be quantified accurately to achieve high confidence in determining putative kinase substrates.

Advances in genetics and molecular biology, biochemical characterization of MAPKs, quantitative proteomics, and enrichment of phosphopeptides have allowed the development of phosphoproteomics approaches for kinase substrate identification. For example, point mutations of amino acid residues within the conserved ATP binding pocket or the activation loop resulted in CA versions of MAPKs without affecting the substrate specificity [16]. In addition, transient expression of $MAPK^{CA}$, but not $MAPK^{WT}$, in infiltrated *Nicotiana benthamiana* leaves led to reactive oxygen species burst and cell death—a hallmark of MAPK cascade activation [18]. Thus, the phosphoproteomes between $MAPK^{CA}$ and $MAPK^{WT}$ infiltrated leaves can be compared to indicate potential MAPK substrates. To quantify $MAPK^{CA}$ triggered phosphorylation changes, tandem mass tags (TMT) have been the method of choice, as the tags were found to be effective in evaluating phosphorylation dynamics [19]. Furthermore, enrichment of phosphopeptides using metal dioxide has been proven to be one of the most successful and widely accepted methods in plant phosphoproteomics [17]. Here, we describe a workflow combining genetic engineering and transient expression of $MAPK^{CA}$ and $MAPK^{WT}$, protein extraction and digestion, TMT labeling of the resulting peptides, phosphopeptide enrichment, and liquid chromatography tandem mass spectrometry (LC-MS/MS) for kinase substrate identification (Fig. 1). The methods for creating $MAPK^{CA}$ point mutation and expressing the proteins in *N. benthamiana* have been reported previously [16, 18]. Therefore, the phosphoproteomics method will be the focus of this chapter.

2 Materials

2.1 Protein Extraction

1. Extraction buffer: 0.1 M Tris–HCl pH 8.8, 10 mM EDTA, 0.2 M DTT or 0.4% 2-mercaptoethanol, 0.9 M sucrose. Protease and phosphatase inhibitor tablet is added in the extraction buffer (e.g., Pierce tablet catalog number: 88668, add one tablet per 10 mL extraction buffer).
2. Tris pH 8.8 saturated phenol.
3. Precipitation buffer: 0.1 M ammonium acetate in 100% methanol, prechilled at −20 °C.
4. 80% acetone and 100% cold acetone, prechilled at −20 °C.
5. Protein Dissolution buffer: 8 M urea, 4% CHAPS, and 50 mM Tris–HCl.
6. Protein quantification reagents (e.g., EZQ quantification kit from Thermo Scientific, catalog number: R-33200).

2.2 Reduction, Alkylation and Trypsin Digestion of Protein Samples

1. 100 mM triethylammonium bicarbonate (TEAB).
2. 500 mM iodoacetamide (IAM) stock. Prepare immediately before use.
3. 500 mM tris(2-carboxyethyl)phosphine stock (e.g., Pierce, catalog number: PG82080).
4. Sequencing grade modified trypsin (e.g., Promega, catalog number: V5111).
5. C18 desalting column (e.g., MacroSpin Column Silica C18, 30–300 μg capacity, catalog number: SMM SS18V, The Nest Group).
6. Speedvac lyophilization system.

2.3 TMT-Labeling

1. TMT mass tagging kit (catalog number: 90061, Thermo Scientific, *see* **Note 1**).
2. 50% hydroxylamine solution (e.g., Sigma, catalog number: 438227). Dilute to 8% before use.

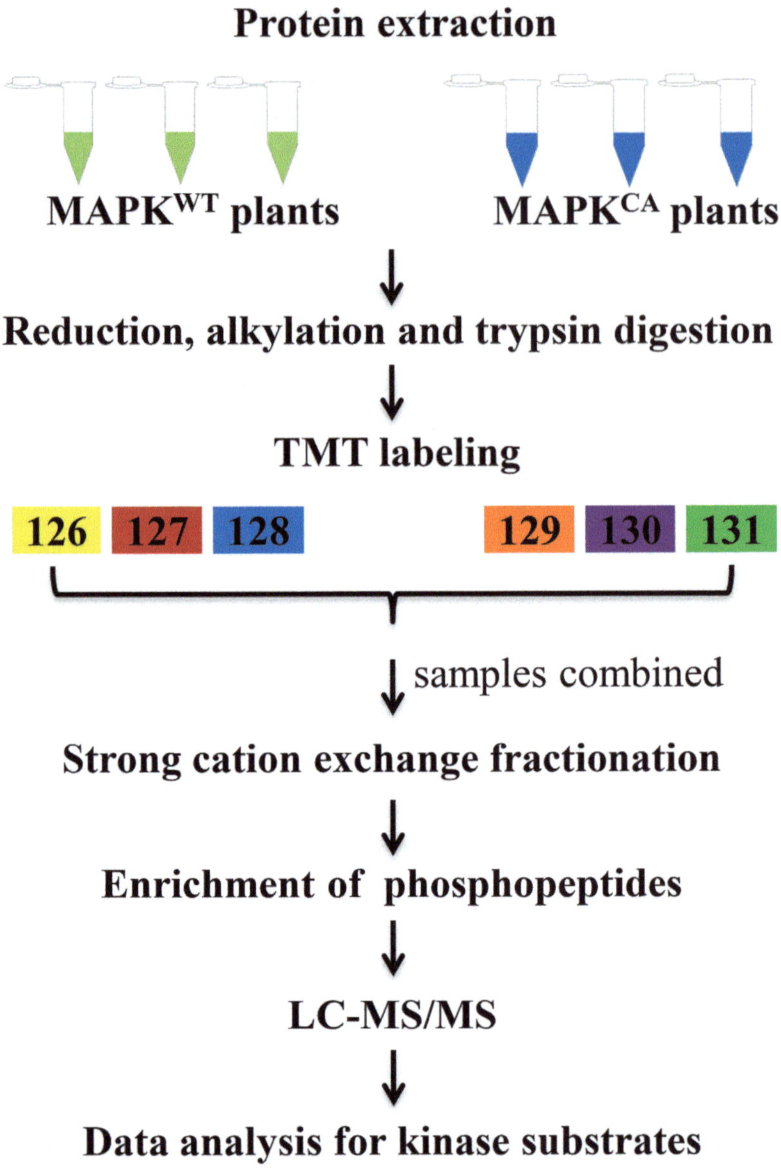

Fig. 1 Workflow of MAPK substrate identification using phosphoproteomics. MAPK^{WT} and MAPK^{CA} are transiently expressed in *N. benthamiana* leaves. Proteins are extracted from three biological replicates of MAPK^{WT} and MAPK^{CA} samples. Equal amount of total proteins from the two sets of samples is reduced by TCEP, alkylated by IAM and then digested by trypsin. The three MAPK^{WT} samples are then labeled with TMT tags 126, 127, and 128, while the three MAPK^{CA} samples are labeled with 129, 130, and 131. All the samples are combined after labeling and subjected to strong cation exchange LC fractionation. Phopshopeptides in different SCX fractions are enriched and analyzed by LC-MS/MS. The MS/MS data are analyzed using Proteome Discoverer for MAPK candidate substrate identification

2.4 Strong Cation Exchange (SCX)

1. SCX Buffer A: 25% acetonitrile (ACN), 10 mM ammonium formate, 0.1% formic acid, pH 2.8.
2. SCX Buffer B: 25% ACN, 500 mM ammonium formate, pH 6.8.
3. SCX column (e.g., PolySULFOETHYL A, 100 × 2.1 mm, 5 μM, 300 Å).

2.5 Phosphopeptide-Enrichment

1. Sampling buffer: 3% ACN, 0.1% acetic acid, 0.01% trifluoroacetic acid (TFA).
2. Binding buffer: 80% ACN, 5% TFA.
3. Washing buffer: 80% ACN, 1% TFA.
4. Elution buffer: 3% ammonium hydroxide.
5. Metal dioxide tips (e.g., NuTip TiO_2 + ZrO_2 from Glygen, catalog number: NT3TiZr.96, *see* **Note 2**).

2.6 Mass Spectrometry and Data Analysis

1. Nanoflow ultra performance liquid chromatography system.
2. HPLC Solvent A: 0.1% formic acid.
3. HPLC Solvent B: 0.1% formic acid, 99.9% ACN.
4. Q-Exactive Obitrap (Thermo Scientific).
5. Proteome Discoverer 1.4 (PD 1.4).
6. Microsoft Excel.

3 Methods

3.1 Protein Extraction

1. Grind 2 g leaves with mortar and pestle in liquid nitrogen. Do not allow the powder to thaw.
2. Add 4 mL Tris pH 8.8 saturated phenol and 4 mL extraction buffer to the mortar, and keep grinding until a slurry forms (*see* **Note 3**).
3. Transfer the extract to an Oakridge tube and agitate gently at room temperature for 30 min.
4. Centrifuge at 5000 × *g* for 10 min at 15 °C.
5. Transfer the supernatant (the phenol phase) to a new Oakridge tube. Measure the volume (*see* **Note 4**).
6. Add five volumes of the precipitation buffer, and invert the tubes 10 times. Precipitate for at least 1 h to overnight at −20 °C.
7. Centrifuge at 20,000 × *g* for 20 min at 15 °C. Discard the supernatant.
8. Wash the pellet twice with the precipitation buffer (*see* **Note 5**).
9. Wash the pellet twice with 80% acetone.

10. Wash the pellet one time with 100% acetone, and dry briefly on ice in a clean chemical hood (*see* **Note 6**).
11. Dissolve the pellet with 100 μL protein dissolution buffer (*see* **Note 7**).
12. Determine protein concentration (*see* **Note 8**) of the samples and aliquot out 1 mg.

3.2 Reduction, Alkylation, and Trypsin Digestion of Protein Samples

1. Dilute the protein samples with 100 mM TEAB to a concentration of urea less than 1 M (*see* **Note 9**).
2. Add TCEP to a final concentration of 10 mM. Incubate at 37 °C for 1 h (*see* **Note 10**).
3. Add IAM to a final concentration of 20 mM. Incubate at room temperature in the dark for 30 min.
4. Dissolve a vial of trypsin (20 μg) in 100 μL ddH_2O (*see* **Note 11**). Add 50 μL trypsin solution to the protein sample to reach a trypsin: sample ratio of 1:100. Incubate at 37 °C overnight.
5. Add 10 μL formic acid to terminate the digestion. Dry the peptides by speed vac.
6. Remove salts in the sample using mini-C18 columns (*see* **Note 12**). Dry the desalted samples by speed vac.

3.3 TMT-Labeling

1. Resuspend the peptides in 100 μL 100 mM TEAB.
2. Bring the TMT tags to room temperature, and add 41 μL acetonitrile to each tube (*see* **Note 13**). Vortex briefly.
3. Transfer each of the peptide samples to the correspondent reconstituted TMT tags. Vortex briefly and incubate at room temperature for 2 h (*see* **Note 14**).
4. Add 8 μL 5% hydroxylamine to each sample and incubate at room temperature for 15 min.
5. Combine all the samples into one tube and dry it by speed vac.
6. Desalt the combined sample using mini C18 columns, and dry the resulting sample using speed vac.

3.4 Strong Cation Exchange

1. Dissolve the peptides in 103 μL SCX Buffer A.
2. Set the flow rate of the HPLC to be 0.2 mL/min. Set the gradient as follows: 0–10 min: 100% A; 10–90 min: 80% B and 20% B; 90–95 min: 100% B. Use time-based method to collect 28 fractions (*see* **Note 15**).
3. Inject 100 μL sample using the autosampler. After fractionation, pool the 28 fractions into approximately ten tubes according to the peak intensity of the individual fractions (*see* **Note 16**).

4. Dry the combined ten fractions using speed vac.
5. Desalt the combined fractions using C18 columns, and dry using speed vac.

3.5 Phosphopeptide Enrichment

1. Dissolve each of the fractions in 40 μL sampling buffer and 40 μL binding buffer.
2. Equilibrate the enrichment tip by washing the resin 10× with 40 μL binding buffer.
3. Load 40 μL peptide sample into the tip. Pipette the sample up and down slowly for 10 min. Repeat loading and binding for the remaining 40 μL sample.
4. Wash the tip with 40 μL washing buffer. Repeat 25 times.
5. Elute the peptide with 40 μL washing buffer by aspiring and expelling for 1 min. Repeat for a second time (*see* **Note 17**).
6. Dry the eluted samples in speed vac.

3.6 Mass Spectrometry and Data Analysis

1. Dissolve the enriched samples in 13 μL of 0.1% formic acid, and transfer 12 μL into HPLC autosampler vials.
2. Set the flow rate of the HPLC to be 350 nL/min. Set the gradient as follows: 0 min: 98% A and 2% B; 100 min: 70% A and 30% B; 110 min: 2% A and 98% B.
3. Set the parameters of the full MS scan in Q Exactive method as follows: Resolution: 70,000; AGC target: 1 × E6; Maximum IT: 100 ms; and Scan range: 400–2000 m/z. Set the data-dependent MS2 as follows: Resolution: 17,500; AGC target: 1 × E6; Maximum IT: 60 ms; Loop count: 10; Isolation window: 2.0 m/z; Fixed first mass: 105.0 m/z; and NCE: 28.
4. Inject 10 μL samples into the HPLC, and then acquire MS1 and MS2 spectra.
5. Search the raw data against the NCBI green plant nonredundant database using Proteome Discoverer 1.4 (PD). In the SEQUEST algorithm, set the maximum missing of tryptic cleavage sites as 2, and set carbamidomethylation (+57.021) on cysteine residues as fixed modification. For the dynamic modification, choose oxidation of methionine (+15.996), TMT-six plex labeling (+229.143) on N-terminus and lysine residues, phosphorylation (79.966) on tyrosine, serine, and threonine residues.
6. When the searching is completed, filter the results using the following criteria: peptide confidence level as 1%; minimum peptide per protein as two for protein identification.
7. Extract the identified peptides with the ion intensity of the six tags from PD, and then export the raw data into Excel.
8. Perform *t*-test in Excel to see the differential phosphorylation in the two groups of samples.

4 Notes

1. Other isobaric tags such as iTRAQ have also been used to quantify phosphopeptides. However, iTRAQ-labeled peptides are not easy to enrich using the metal dioxide method.
2. Peptides can have one or more phosphorylation sites. TiO_2 and ZrO_2 are good for capturing singly and multiply phosphorylated peptides, respectively. Thus, an enrichment matrix that contains both materials is preferred for high coverage.
3. It is important to keep the concentration of sucrose in the water phase at 0.9 M to have a separation of three layers after centrifuge. Therefore, grind the samples fast since prolonged grinding will result in absorption of water from the atmosphere into the samples, thus diluting the samples.
4. Do not disturb the middle layer during the transfer of the supernatant. The bottom phase (aqueous phase with sucrose) can be extracted for a second time with another 1 mL Tris–phenol. Repeat **steps 2** through **4** and combine the supernatant with the previous one.
5. For **steps 8** and **9**, break the pellet thoroughly by vortexing or using a glass transfer pipet tip. Allow precipitation for at least 30 min before centrifuge. Conditions for centrifugation are listed in **step** 7.
6. Samples should be placed on ice to prevent protein degradation. Do not over-dry the samples, which will make re-solubilization difficult. The drying process can be stopped when the surface of the pellet no longer appears wet.
7. Acetone-precipitated protein pellet is not easy to dissolve. Urea at high concentration is very efficient for solubilizing and denaturing the proteins. However, thiourea should be avoided, since it reacts with IAM and decreases the alkylation efficiency.
8. Accurate protein quantification is a prerequisite for comparison of protein phosphorylation changes using TMT tags. Protein quantification methods that are sensitive to urea should be avoided. The EZQ protein quantification assay is recommended.
9. Typically, the protein concentration is approximately 10 μg/μL, and 100 μL will amount to 1 mg. In such cases, all samples can be diluted to 1 mL to make sure the urea is less than 1 M to be compatible with the following trypsin digestion step.
10. The reduction reaction can be performed at a high temperature such as the 55 °C suggested in the TMT manufacturer's instructions. However, high temperature should be avoided when urea is present in the samples as urea may decompose and introduce artificial carbamylation modification to the proteins.

11. Alternatively, dissolve the trypsin in 100 mM TEAB or 50 mM ammonium bicarbonate, since trypsin works best at pH 8.5. Dissolving the enzyme with the suspension buffer (50 mM acetic acid) that comes along with commercial available trypsin products is recommended if the left-over will be stored for later use.
12. It is very important to remove all the salts before the labeling. Amine groups in the protein dissolution buffer such as Tris–HCl and urea react readily with the TMT tags, which significantly reduces the labeling efficiency of the tags.
13. Accurate measurement of ACN is important to ensure the concentration of ACN in every sample is the same.
14. The labeling time is extended to 2 h, a modification to the manufacturer's instruction.
15. The gradient and the number of fractions collected depend on the complexity of the sample, as well as on the efficiency of the SCX column.
16. Depending on the complexity of the sample, a variable number of fractions can be pooled. However, the first four or five of the initial 28 fractions should not be combined with other fractions. They are rich in phosphopeptides, which elute early in SCX chromatography.
17. Formic acid can be added if the sample cannot be analyzed by mass spectrometry immediately, since the phosphate group tends to be lost in high pH.

Acknowledgments

This research was supported by grants from the National Science Foundation (MCB 0818051 and MCB 1412547) to S. Chen.

References

1. Pitzschke A (2015) Modes of MAPK substrate recognition and control. Trends Plant Sci 20:49–55
2. Chinchilla D, Zipfel C, Robatzek S, Kemmerling B, Nurnberger T, Jones JDG et al (2007) A flagellin-induced complex of the receptor FLS2 and BAK1 initiates plant defence. Nature 448:497–500
3. Lu DP, Lin WW, Gao XQ, Wu SJ, Cheng C, Avila J et al (2011) Direct ubiquitination of pattern recognition receptor FLS2 attenuates plant innate immunity. Science 332:1439–1442
4. Meng XZ, Wang HC, He YX, Liu YD, Walker JC, Torii KU et al (2012) A MAPK cascade downstream of ERECTA receptor-like protein kinase regulates Arabidopsis inflorescence architecture by promoting localized cell proliferation. Plant Cell 24:4948–4960
5. Yamaguchi K, Yamada K, Ishikawa K, Yoshimura S, Hayashi N, Uchihashi K et al (2013) A receptor-like cytoplasmic kinase targeted by a plant pathogen effector Is directly phosphorylated by the chitin receptor and mediates rice immunity. Cell Host Microbe 13:347–357
6. Mao GH, Meng XZ, Liu YD, Zheng ZY, Chen ZX, Zhang SQ (2011) Phosphorylation of a WRKY transcription factor by two pathogen-responsive MAPKs drives phytoalexin biosynthesis in Arabidopsis. Plant Cell 23: 1639–1653

7. Andreasson E, Jenkins T, Brodersen P, Thorgrimsen S, Petersen NHT, Zhu SJ et al (2005) The MAP kinase substrate MKS1 is a regulator of plant defense responses. EMBO J 24:2579–2589
8. Xu J, Chua NH (2012) Dehydration stress activates Arabidopsis MPK6 to signal DCP1 phosphorylation. EMBO J 31:1975–1984
9. Feilner T, Hultschig C, Lee J, Meyer S, Immink RGH, Koenig A et al (2005) High throughput identification of potential Arabidopsis mitogen-activated protein kinases substrates. Mol Cell Proteomics 4:1558–1568
10. Minkoff BB, Stecker KE, Sussman MR (2015) Rapid phosphoproteomic effects of ABA on wildtype and ABA receptor-deficient *A. thaliana* mutants. Mol Cell Proteomics 14:1169–1182
11. Wang P, Xue L, Batelli G, Lee S, Hou YJ, Van Oosten MJ et al (2013) Quantitative phosphoproteomics identifies SnRK2 protein kinase substrates and reveals the effectors of abscisic acid action. Proc Natl Acad Sci USA 110:11205–11210
12. Kline-Jonakin KG, Barrett-Wilt GA, Sussman MR (2011) Quantitative plant phosphoproteomics. Curr Opin Plant Biol 14:507–511
13. Reiland S, Finazzi G, Endler A, Willig A, Baerenfaller K, Grossmann J et al (2011) Comparative phosphoproteome profiling reveals a function of the STN8 kinase in fine-tuning of cyclic electron flow (CEF). Proc Natl Acad Sci USA 108:12955–12960
14. Hoehenwarter W, Thomas M, Nukarinen E, Egelhofer V, Rohrig H, Weckwerth W et al (2013) Identification of novel in vivo MAP kinase substrates in *Arabidopsis thaliana* through use of tandem metal oxide affinity chromatography. Mol Cell Proteomics 12:369–380
15. Cheng Z, Li JF, Niu Y, Zhang XC, Woody OZ, Xiong Y et al (2015) Pathogen-secreted proteases activate a novel plant immune pathway. Nature 521:213–216
16. Berriri S, Garcia AV, Frey NFD, Rozhon W, Pateyron S, Leonhardt N et al (2012) Constitutively active mitogen-activated protein kinase versions reveal functions of Arabidopsis MPK4 in pathogen defense Signaling. Plant Cell 24:4281–4293
17. Li JN, Silva-Sanchez C, Zhang T, Chen SX, Li HY (2015) Phosphoproteomics technologies and applications in plant biology research. Front Plant Sci 6:430
18. Zhang T, Zhu MM, Song WY, Harmon AC, Chen SX (2015) Oxidation and phosphorylation of MAP kinase 4 cause protein aggregation. Biochim Biophys Acta (BBA-Proteins Proteom) 1854:156–165
19. Chen Y, Britton DJ, Paraiso K, Fedorenko I, Wood ER, Magliocco A et al (2014) Quantification of biomarker expression, phosphorylation, and mutation in cancer using TMT labeling prior to liquid chromatography-multiple reaction monitoring mass spectrometry. Cancer Res 74:2491

Chapter 11

Analysis of PAMP-Triggered ROS Burst in Plant Immunity

Yuying Sang and Alberto P. Macho

Abstract

The plant perception of pathogen-associated molecular patterns triggers a plethora of cellular immune responses. One of these responses is a rapid and transient burst of reactive oxygen species (ROS) mediated by plasma membrane-localized NADPH oxidases. The ROS burst requires a functional receptor complex and the contribution of several additional regulatory components. In laboratory conditions, the ROS burst can be detected a few minutes after the treatment with an immunogenic microbial elicitor. For these reasons, the elicitor-triggered ROS burst has been often exploited as readout to probe the contribution of plant components to early immune responses. Here, we describe a detailed protocol for the measurement of elicitor-triggered ROS burst in a simple, fast, and easy manner.

Key words Reactive oxygen species, ROS, flg22, PAMP, Immunity

1 Introduction

Plants can perceive specific molecular patterns that act as danger signals. When these molecules are non-self, they are usually termed pathogen-associated molecular patterns (PAMPs). Their perception leads to the activation of pattern-recognition receptors (PRRs) and subsequent immune responses [1]. The regulation of PRR activation has received increasing attention over the past years, leading to the development of several assays to test immune responses.

Reactive oxygen species (ROS) play important roles in several biological processes in plants, including development and the response to environmental signals [2–4]. PRR activation at the plasma membrane is often accompanied by a rapid and transient ROS burst, detectable a few minutes after elicitor treatment in laboratory conditions. Although ROS are associated to other immune and nonimmune responses, this article focuses on the rapid PRR-mediated ROS burst. This ROS production is mediated by NADPH oxidases at the plasma membrane, and, in *Arabidopsis thaliana* (hereafter, Arabidopsis), RBOHD plays a major role (reviewed in [5]). The activation of RBOHD triggered by micro-

Libo Shan and Ping He (eds.), *Plant Pattern Recognition Receptors: Methods and Protocols*, Methods in Molecular Biology, vol. 1578, DOI 10.1007/978-1-4939-6859-6_11, © Springer Science+Business Media LLC 2017

bial elicitors is regulated at multiple levels, and requires appropriate activation of PRR complexes (reviewed in [6]). Therefore, production of RBOHD-dependent ROS is susceptible to alterations on RBOHD, PRR complexes, and their respective regulators [7, 8]. For these reasons, the elicitor-triggered RBOHD-dependent ROS burst has often been exploited as readout to probe the contribution of potential regulators of early immune responses. The measurement of ROS burst can also be used to identify or probe additional plant processes that contribute to early immune responses upstream or independent of PRR activation, such as accumulation of immune components [9, 10] or plant physiological functions [11].

A straightforward manner to detect the PRR-mediated ROS is a luminol-based assay, where the oxidation of luminol by ROS, catalyzed by a peroxidase, produces bioluminescence, which, in the appropriate experimental conditions, is directly proportional to the concentration of the produced ROS. This assay can be conveniently performed in 96-well plates using leaf discs, allowing the rapid and simultaneous detection of ROS burst in several samples [12]. The major requirement is a plate luminometer and the corresponding elicitor, but all other materials can be found in most laboratories. This technique can be applied to the detection of ROS generated after treatment with several different elicitors that represent danger signals, either from the plant or from other organisms (e.g., bacterial flg22, bacterial elf18, fungal chitin, PEPs, oligogalacturonides, etc.).

The assay described here can be used to detect quantitative differences on ROS production in mutants, stable transgenic lines, plant tissues transiently expressing genes, and/or after chemical treatments. Given the versatility of this assay, it can be adapted for high-throughput screening purposes using multiwell plates [9] or other technical layouts [13]. Elicitor treatment can also be substituted for bacterial extracts or living bacteria [14].

Here, we describe a detailed protocol for the measurement of elicitor-triggered ROS burst in a simple, fast, and easy manner. To allow a high level of detail, we focus on the well-established flg22-triggered ROS burst in the model plants Arabidopsis and *Nicotiana benthamiana*.

2 Materials

2.1 Plant Growth

1. Plastic square pots (10 × 10 × 9 cm).
2. Soil mixture: 3 parts of potting substrate (Pindstrup Mosebrug A/S, Denmark) and 1 part of inorganic substrate (e.g., vermiculite).
3. Fertilizer (20-20-20+TE, Xiamen Vastland Chemical Co., Ltd. China).

4. Controlled plant growth environment (setup at 22–23 °C and 100–150 mE/m^2/s; *see* details below about suitable light cycles depending on the plant species).

2.2 Sample Collection

1. Biopsy punch with plunger (4 mm diameter; Miltex, USA) (*see* **Note 1**).
2. White 96-well plates suitable for chemiluminescence measurement (OptiPlate™-96, PerkinElmer).
3. Distilled water.

2.3 ROS Assay

1. Multichannel pipette (12 channels).
2. Luminol (Sigma) (*see* **Note 2**): 100 mM stock solution in DMSO. Aliquot and store at −20 °C.
3. Horseradish Peroxidase (Sigma): 20 mg/mL stock solution in ultra-pure distilled water. Aliquot and store at −20 °C.
4. Elicitor (e.g., flg22 peptide—QRLSTGSRINSAKDDAAGLQIA, >98% purity; Abclonal, USA): *See* Subheading 3.2 for the preparation of elicitors.
5. Ultra-pure distilled water (Milli-q water, Millipore Corporation, Germany).
6. Microplate reader (Varioskan flash, Thermo Scientific, USA) (*see* **Note 3**).

3 Methods

3.1 Preparing Plant Material

1. Sow *A. thaliana* or *N. benthamiana* seeds in the soil mixture, cover with a transparent plastic lid, and stratify for 2 days at 4 °C in the dark (*see* **Note 4**).
2. Move pots to a controlled environment room or growth chamber. Arabidopsis plants should grow in short-day conditions (8 h light/16 h dark cycle or 10 h light/14 h dark cycle) (*see* **Note 5**). *N. benthamiana* plants should grow in long-day conditions (16 h light/8 h dark cycle). Check every 2 days and water the plants slightly if needed, to ensure that the soil is humid, while carefully avoiding excess water in the tray.
3. When the seedlings begin emerging (5–7 days after sowing), the plastic lid can be removed.
4. 10–14 days after sowing, transplant Arabidopsis or *N. benthamiana* seedlings into individual pots or tray wells (*see* **Note 6**). Cover again with a transparent plastic lid for 1 week to minimize stress after transplanting.
5. 4- to 5-week-old healthy plants are suitable for ROS assays.

3.2 Elicitor Preparation and Storage

1. For simplicity, we will describe the preparation and application of the elicitor peptide flg22, corresponding to a highly conserved domain of bacterial flagellin, which is widely used to trigger immune responses [15].
2. Preparation of peptide stocks from lyophilized peptides should be conducted with extreme care to avoid peptide contaminations (*see* **Note** 7).
3. Carefully resuspend the lyophilized flg22 peptide in ultra-pure distilled water (*see* **Note 8**) to generate a storage stock at a concentration of 10 mM (*see* **Note 9**).
4. Prepare 10 μL aliquots and store at −20 °C. These aliquots can also be freeze-dried and stored at −20 °C to extend their stability, and reconstituted with distilled water before use.

3.3 Sample Collection

1. Add 100 μL of distilled water into each well of a white 96-well plate. If the purpose of the experiment is to test the impact of a given chemical in the ROS burst, it can be dissolved in the water at this step, or later during the incubation period (*see* **Note 10**). Organize your different samples in columns, using at least eight replicates (wells) per sample (*see* **Note 11**).
2. Collect the leaf discs individually by pressing gently the biopsy punch toward the leaf surface supported by your finger (*see* **Note 12**). Eject each disc into a well of the white 96-well plate. Do not submerge the discs. They should be floating on the water (*see* **Note 13**).
3. Cover the plate with a transparent plastic lid or plastic film and place it in the controlled environment room or growth chamber (*see* **Note 14**). Incubate for 14–16 h (*see* **Note 15**). Avoid shaking.

3.3.1 Collection of Leaf Discs from Arabidopsis

1. Use well-expanded rosette leaves from 4- to 5-week-old plants (*see* **Note 16**) (Fig. 1a).
2. We recommend using three leaves per plant (Fig. 1a) and two discs per leaf. The central vein should be avoided when collecting the leaf discs (Fig. 1b). Therefore, four plants are sufficient to prepare a sample of 24 replicates/leaf discs.

3.3.2 Collection of Leaf Discs from N. benthamiana

1. This assay can be performed for *N. benthamiana* leaves after transient expression or silencing of a gene of interest (*see* **Notes 17** and **18**).
2. We recommend using the third and fourth youngest leaves from 4- to 5-week-old plants (see the previous note about using healthy plants; *see* **Note 19**; Fig. 1c).
3. It is possible to collect 24 discs from one half of a well-expanded *N. benthamiana* leaf (*see* **Note 20**). Avoid collecting discs from leaf veins, the edges, or the tip of the leaf (Fig. 1d).

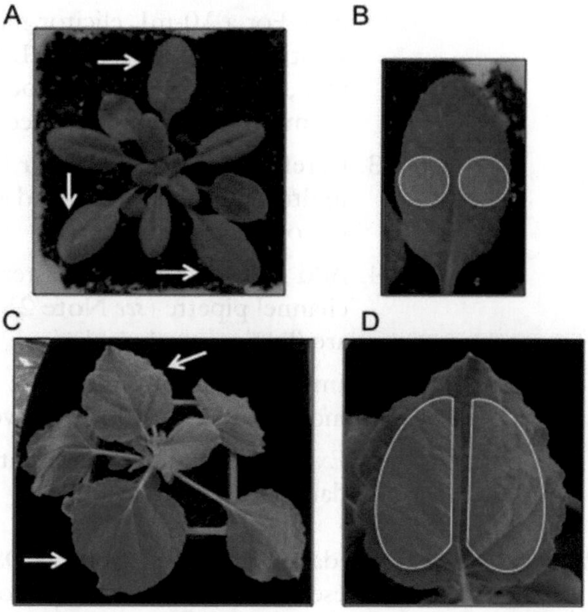

Fig. 1 Representative tissues used to collect leaf discs for ROS assays. (**a**, **c**) *Arrows* indicate the Arabidopsis (**a**) and *N. benthamiana* (**c**) leaves preferred for ROS assays. (**b**, **d**) *Shadowed areas* indicate the preferred tissues for collection of leaf discs in Arabidopsis (**b**) and *N. benthamiana* (**d**) leaves

3.4 Elicitor Treatment and Analysis of Reactive Oxygen Species

3.4.1 Settings of the Microplate Reader (See Note 21)

1. Turn on the microplate reader and set up a kinetics session to measure the luminescence in each well over a total time of 30–60 min (*see* **Note 22**).
2. Measurement duration per well could be 1 s or shorter, depending on the number of samples on one plate and the measuring limits of your machine (*see* **Note 23**).
3. Cycle times can be set according to the duration of one measurement. For example, to read a microplate with 96 samples, we usually record the luminescence in each well during 400 ms, being able to record the luminescence of the full microplate every minute. Therefore, in our experiments, sixty measuring cycles per 96-well microplate are performed over 60 min. The reading direction of the microplate reader should be oriented based on the position of the different replicates of each sample, especially when multiple samples are tested on one microplate (*see* **Note 24**).

3.4.2 Luminescence Measurement

1. Prepare a 100 μM working stock of the elicitor by adding 990 μL of ultra-pure distilled water to a storage stock tube (which contains 10 μL of elicitor at 10 mM). Prepare 100 μL aliquots and keep them at −20 °C (*see* **Note 25**).
2. Prepare the elicitor master mix containing 50 nM flg22 (*see* **Note 26**), 100 μM luminol, and 20 μg/mL HRP (*see* **Note**

27). For a 10-mL elicitor master mix, 10 μL of luminol stock solution (100 mM), 10 μL of HRP stock solution (20 mg/mL), and 5 μL of flg22 stock solution (100 μM) are added to 10 mL of ultra-pure distilled H_2O, and mixed well.

3. Carefully remove the water from each well completely using a multichannel pipette. Avoid any mechanical damage or wounding of the leaf discs.
4. Add 100 μL of elicitor master mix into each well using multiple-channel pipette (*see* **Note 28**). Make sure that all the leaf discs are floating on the solution.
5. Immediately, put the plate into the microplate reader and monitor ROS production over 60 min.
6. Export the luminescence data to an Excel file and analyze the data as described below.

3.5 Data Analysis

The data obtained from the ROS burst assay can be analyzed and represented in different ways to extract as much information as possible. One usual representation shows the relative luminescence units (RLU; *see* **Note 29**) produced per minute upon elicitor treatment, representing the kinetics of ROS production in different samples. Another usual representation shows the total accumulative values of RLU for each sample, representing the total ROS production over the duration of the assay. A calculation template can be made in an Excel file to make calculations easier and faster.

3.5.1 Total ROS Production Induced by the Elicitor During the Experiment

1. Calculate the sum of luminescence values from each leaf disc over the 60 min. Then calculate the average and standard error of the total luminescence values from leaf discs corresponding to the same sample.
2. Represent RLU values for each sample in a column graph, as shown in Fig. 2a.

3.5.2 Kinetics of ROS Production in Response to the Elicitor

1. Collect the luminescence values produced at each cycle from all the leaf discs from the same sample, and calculate mean and standard error of RLU corresponding to each cycle.
2. Represent RLU values for each sample (in the *y* axis) and the time of each cycle (in the *x* axis). The measurement of the oxidative burst over 60 min usually results in a kinetics curve as shown in Fig. 2b. Treatment of samples with 50 nM flg22 usually leads to a ROS-burst curve with a maximum peak at approximately 10–15 min (*see* **Note 30**).
3. Kinetics curves can be compared visually, but can also provide interesting quantitative features to compare ROS production patterns between samples. Some of these features include the maximum value of ROS detected in a sample on a single measurement and the time required to reach the maximum value after elicitation (Fig. 2b).

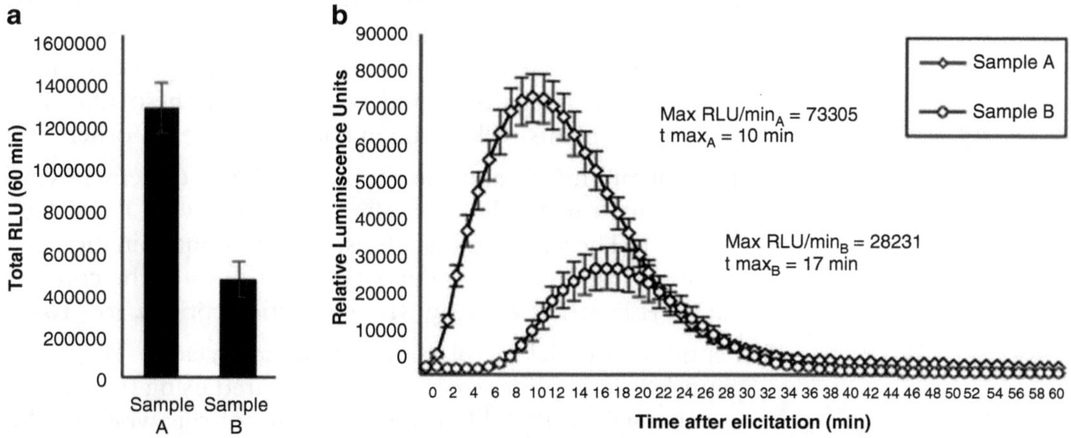

Fig. 2 Common representation of a flg22-triggered ROS assay. (**a**) Total ROS production induced during 60 min after elicitor treatment. (**b**) Kinetics of ROS production in response to the elicitor, including quantitative features obtained from the kinetics data

4 Notes

1. Several companies provide good-quality cork borers suitable for the collection of leaf discs. The cork borer used should be sharp to minimize tissue damage, since this may impact significantly the outcome of the assay.
2. Standard luminol can be used for most applications. More sensitive luminol derivatives (e.g., L-012, Wako Chemicals) can be used if higher sensitivity is required. Luminol is light-sensitive. All solutions containing luminol must be protected from light by wrapping tubes in aluminum foil.
3. Several different instruments can be used to measure the bioluminescence generated in this assay. We use microplate readers such as Promega GloMax™ 96 Microplate Luminometer and Varioskan Flash. High-resolution photon-counting cameras also provide excellent results, but data acquisition and processing may require specialized software.
4. Seeds can also be stratified before sowing.
5. It is extremely important to control carefully the life cycle of the Arabidopsis plants used for ROS assays. Bolting or flowering plants give inconsistent results and are not suitable for these assays. Since 4- to 5-week-old Arabidopsis plants are used, it is important to keep the plants in short-day conditions. Long-day conditions or accidental disruptions of the dark cycles (e.g., opening the growth rooms or chambers during the dark cycles) will cause premature bolting and should be avoided.

6. We use 10 × 10 × 9 cm square pots, placing 1 *N. benthamiana* plant or 4 Arabidopsis plants per pot. Arabidopsis plants can also be transplanted into multiwell trays. The growth surface for Arabidopsis plants should be wide enough so that 4- to 5-week-old plants will not be in contact with each other.
7. The chemiluminescence detection of ROS is extremely sensitive (signal can be detected after elicitation with 1 nM flg22 peptide). Therefore, the slightest elicitor contamination may cause misleading results. For a careful analysis of the contamination risks when working with synthetic peptides, *see* [16].
8. Peptides can stick to plastics and other surfaces. To prevent peptide loss, peptides stocks can be prepared using 0.1 M NaCl and 1 mg/mL BSA [15]. Be also aware of this when reusing reagent reservoirs and other materials.
9. The stability of the flg22 peptide decreases at low concentrations. Therefore, it is recommended to keep peptides in concentrated stocks for long-term storage. Our 10 mM flg22 stocks prepared in ultra-pure distilled water retain similar elicitor activity for at least 1 year stored at −20 °C.
10. We have tested this with a variety of water-soluble chemicals and inhibitors, providing a very powerful assay for pharmacological or chemical genetics approaches [11, 17, 18].
11. We recommend using 24 replicates (three columns) per sample, allowing the simultaneous analysis of four samples per plate. For screening purposes, the number of replicates can be reduced to 16 (two columns) or 8 (one column) per sample, although this will increase the variation in the experiment. To reduce variation, the number of replicates per sample can be increased to 32 (four columns).
12. Always use gloves when handling leaf tissues. Take special care when using fresh sharp biopsy punches to avoid injuries. Biopsy punches can be reused hundreds of times with satisfactory results. Replace if a significant loss of sharpness is detected.
13. We have compared results obtained with discs floating with the abaxial or adaxial side facing the water, and did not find significant differences.
14. Plates can be incubated overnight at room temperature on the top of the laboratory bench, which does not impact the result in most cases. However, doing side-by-side comparisons, we have found that results are slightly more accurate and reproducible after incubation in controlled conditions.
15. This incubation is important so that the leaf discs recover from the mechanical damage. We normally collect the leaf discs in the evening and perform the assays the next morning.

16. To obtain reliable and reproducible results, only healthy plants should be used. Stressed, bolting, or flowering plants are not suitable for ROS assays.
17. When using *Agrobacterium tumefaciens*-based transient expression, make sure to optimize the conditions so that there is no tissue necrosis at the time of the assay. The leaf discs can be carefully collected with forceps after the assay to analyze the transient gene expression or protein accumulation.
18. The infiltration with *A. tumefaciens* will change the ROS burst in comparison to naïve or mock-treated leaves. Therefore, we recommend using leaf tissue infiltrated with *A. tumefaciens* as control.
19. Different leaves in *N. benthamiana* are likely to produce quantitatively different ROS results. Therefore, we recommend including the internal control on one side of the leaf, in comparison to the other half expressing the gene of interest.
20. Intra-experimental variation when using *N. benthamiana* for ROS burst assays is significantly higher than when using Arabidopsis. If this causes a problem, we suggest increasing the number of discs per sample (from the same leaf) or the number of leaves tested.
21. These settings can be used on a Varioskan Flash microplate reader or similar instruments. The settings when using a photon-counting camera or different instruments may vary significantly.
22. Luminescence can be measured for longer time to detect additional features of the elicitation pattern.
23. Measurement duration per well should be set as short as possible to minimize the variation introduced by the time difference between the measurement of the first and the last sample during each cycle. Therefore, if you want to measure the luminescence of the full plate every minute, one round of measurement should be less than 60 s.
24. When multiple samples are tested in one plate, it is important to avoid delays in the measurement of the different samples, which may cause misleading results. Therefore, the reading direction should be set crossing through the different samples (e.g., if the replicates are placed in columns, the reading direction of the microplate reader should be horizontal).
25. Avoid thawing/freezing cycles of the working stock solutions. Solutions at 100 μM should not be stored for longer than 2–3 months.
26. A range of elicitor concentrations can be used to detect slight differences between samples. Dose-response curves are very useful to optimize the sensitivity of this assay.

27. A master mix containing 50 nM flg22, 100 μM luminol, and 20 μg/mL HRP is sufficient to trigger a consistent ROS burst in our experimental conditions. If you are testing a range of elicitor concentrations or stronger elicitors, you should make sure that the concentrations of luminol and HRP are not limiting in your assays.
28. When multiple samples are tested in one plate, if the replicates are organized in columns, the master mix should be added horizontally (i.e., using a 12-channel pipette) to allow the comparison of different samples elicited at the same time.
29. The luminescence monitored in this assay corresponds to the light emitted due to the HRP-mediated oxidation of luminol. Therefore, RLU is an arbitrary unit that can vary between different assays and different machines. When comparing different samples, it is very important to relate to internal controls within the same plate, instead of comparing absolute values from different experiments.
30. The shape of the curve may vary from assay to assay, and it may be significantly different depending on the plant used, the nature of the elicitor, and the elicitor concentration.

Acknowledgments

We thank Rosa Lozano-Durán for critical reading and suggestions on this manuscript. We also thank members of the Macho laboratory, past and present members of Cyril Zipfel's group (The Sainsbury Laboratory, UK), and several other groups for their contribution to the optimization of this protocol over the years. Research in the Macho laboratory is supported by the Shanghai Center for Plant Stress Biology (Chinese Academy of Sciences), and the Chinese 1000 Talents Program.

References

1. Macho AP, Zipfel C (2014) Plant PRRs and the activation of innate immune signaling. Mol Cell 54:263–272
2. Gilroy S, Suzuki N, Miller G, Choi WG, Toyota M, Devireddy AR, Mittler R (2014) A tidal wave of signals: calcium and ROS at the forefront of rapid systemic signaling. Trends Plant Sci 19(10):623–630
3. Nathan C, Cunningham-Bussel A (2013) Beyond oxidative stress: an immunologist's guide to reactive oxygen species. Nat Rev Immunol 13(5):349–361
4. Suzuki N, Koussevitzky S, Mittler R, Miller G (2012) ROS and redox signalling in the response of plants to abiotic stress. Plant Cell Environ 35(2):259–270
5. Marino D, Dunand C, Puppo A, Pauly N (2012) A burst of plant NADPH oxidases. Trends Plant Sci 17(1):9–15
6. Kadota Y, Shirasu K, Zipfel C (2015) Regulation of the NADPH oxidase RBOHD during plant immunity. Plant Cell Physiol 56(8):1472–1480
7. Nühse TS, Bottrill AR, Jones AME, Peck SC (2007) Quantitative phosphoproteomic analysis of plasma membrane proteins reveals regulatory mechanisms of plant innate immune responses. Plant J 51(5):931–940

8. Zhang J, Shao F, Li Y, Cui H, Chen L, Li H, Zou Y, Long C, Lan L, Chai J, Chen S, Tang X, Zhou J-M (2007) A Pseudomonas syringae effector inactivates MAPKs to suppress PAMP-induced immunity in plants. Cell Host Microbe 1(3):175–185
9. Boutrot F, Segonzac C, Chang KN, Qiao H, Ecker JR, Zipfel C, Rathjen JP (2010) Direct transcriptional control of the Arabidopsis immune receptor FLS2 by the ethylene-dependent transcription factors EIN3 and EIL1. Proc Natl Acad Sci U S A 107(32): 14502–14507
10. Hann DR, Dominguez-Ferreras A, Motyka V, Dobrev PI, Schornack S, Jehle A, Felix G, Chinchilla D, Rathjen JP, Boller T (2014) The Pseudomonas type III effector HopQ1 activates cytokinin signaling and interferes with plant innate immunity. New Phytol 201(2): 585–598
11. Macho AP, Boutrot F, Rathjen JP, Zipfel C (2012) ASPARTATE OXIDASE plays an important role in Arabidopsis stomatal immunity. Plant Physiol 159(4):1845–1856
12. Gimenez-Ibanez S, Hann DR, Ntoukakis V, Petutschnig E, Lipka V, Rathjen JP (2009) AvrPtoB targets the LysM receptor kinase CERK1 to promote bacterial virulence on plants. Curr Biol 19(5):423–429
13. Monaghan J, Matschi S, Shorinola O, Rovenich H, Matei A, Segonzac C, Malinovsky FG, Rathjen JP, MacLean D, Romeis T, Zipfel C (2014) The calcium-dependent protein kinase CPK28 buffers plant immunity and regulates BIK1 turnover. Cell Host Microbe 16(5): 605–615
14. Smith JM, Heese A (2014) Rapid bioassay to measure early reactive oxygen species production in Arabidopsis leave tissue in response to living Pseudomonas syringae. Plant Methods 10(1):6
15. Felix G, Duran JD, Volko S, Boller T (1999) Plants have a sensitive perception system for the most conserved domain of bacterial flagellin. Plant J 18(3):265–276
16. Mueller K, Chinchilla D, Albert M, Jehle AK, Kalbacher H, Boller T, Felix G (2012) Contamination risks in work with synthetic peptides: flg22 as an example of a pirate in commercial peptide preparations. Plant Cell 24(8):3193–3197
17. Lozano-Duran R, Bourdais G, He SY, Robatzek S (2014) The bacterial effector HopM1 suppresses PAMP-triggered oxidative burst and stomatal immunity. New Phytol 202(1):259–269
18. Macho AP, Schwessinger B, Ntoukakis V, Brutus A, Segonzac C, Roy S, Kadota Y, Oh MH, Sklenar J, Derbyshire P, Lozano-Duran R, Malinovsky FG, Monaghan J, Menke FL, Huber SC, He SY, Zipfel C (2014) A bacterial tyrosine phosphatase inhibits plant pattern recognition receptor activation. Science 343(6178):1509–1512

Chapter 12

MAPK Assays in *Arabidopsis* MAMP-PRR Signal Transduction

Hoo Sun Chung and Jen Sheen

Abstract

Activation of MAPK (Mitogen-Activated Protein Kinase) cascades after MAMP (Microbe-Associated Molecular Pattern) perception through PRR (Pattern Recognition Receptor) is one of the first conserved responses when plants encounter microbial organisms. Phosphorylation of various cellular factors in the MAMP-PRR pathway by MAPK cascades is critical for broad-spectrum plant innate immunity. Measurement of MAPK activation and identification of MAPK phosphorylation targets in the MAMP-PRR signal transduction pathway are essential to understand how plants reprogram their cellular processes to cope with unfavorable microbial attack. Here, we describe detailed protocols of three assays measuring MAPK activity after MAMP perception: (1) immune-blotting analysis with anti-phospho ERK1/2 antibody; (2) in-gel kinase assay using a general substrate myelin basic protein (MBP); (3) an in vitro kinase assay to evaluate phosphorylation of MAPK substrate candidates during MAMP-PRR signaling based on a protoplast expression system.

Key words *Arabidopsis*, Signal transduction, MAMP, PRR, MAPK cascades, Protoplast transient expression system, Anti-phospho-ERK1/2 antibody, Myelin basic protein (MBP), Kinase assay

1 Introduction

In the immediate environment, plants are able to recognize highly conserved features of microbial organisms known as MAMPs such as flagellin, elongation factor Tu (EF-Tu), and oligosaccharides from microbial cell walls [1–4]. Remarkably sensitive recognition of those non-self-molecules by plants is achieved by specific PRRs such as FLS2 and EFR for flagellin and EF-Tu, respectively, located on the plasma membrane [1, 2]. As this self-surveillance system gets the alert, one of the immediate cellular processes triggered by MAMP-PRR signaling is the activation of MAPK cascades composed of three-tiered protein kinases, MAPKKK (MTK), MAPKK (MKK), and MAPK (MPK) ([4, 5], Fig. 1a). In both plants and animals, MAPK cascades play central roles in innate immune responses by modulating functions of downstream targets critical for broad-spectrum immunity [3–6]. Current knowledge in the

Libo Shan and Ping He (eds.), *Plant Pattern Recognition Receptors: Methods and Protocols*, Methods in Molecular Biology, vol. 1578, DOI 10.1007/978-1-4939-6859-6_12, © Springer Science+Business Media LLC 2017

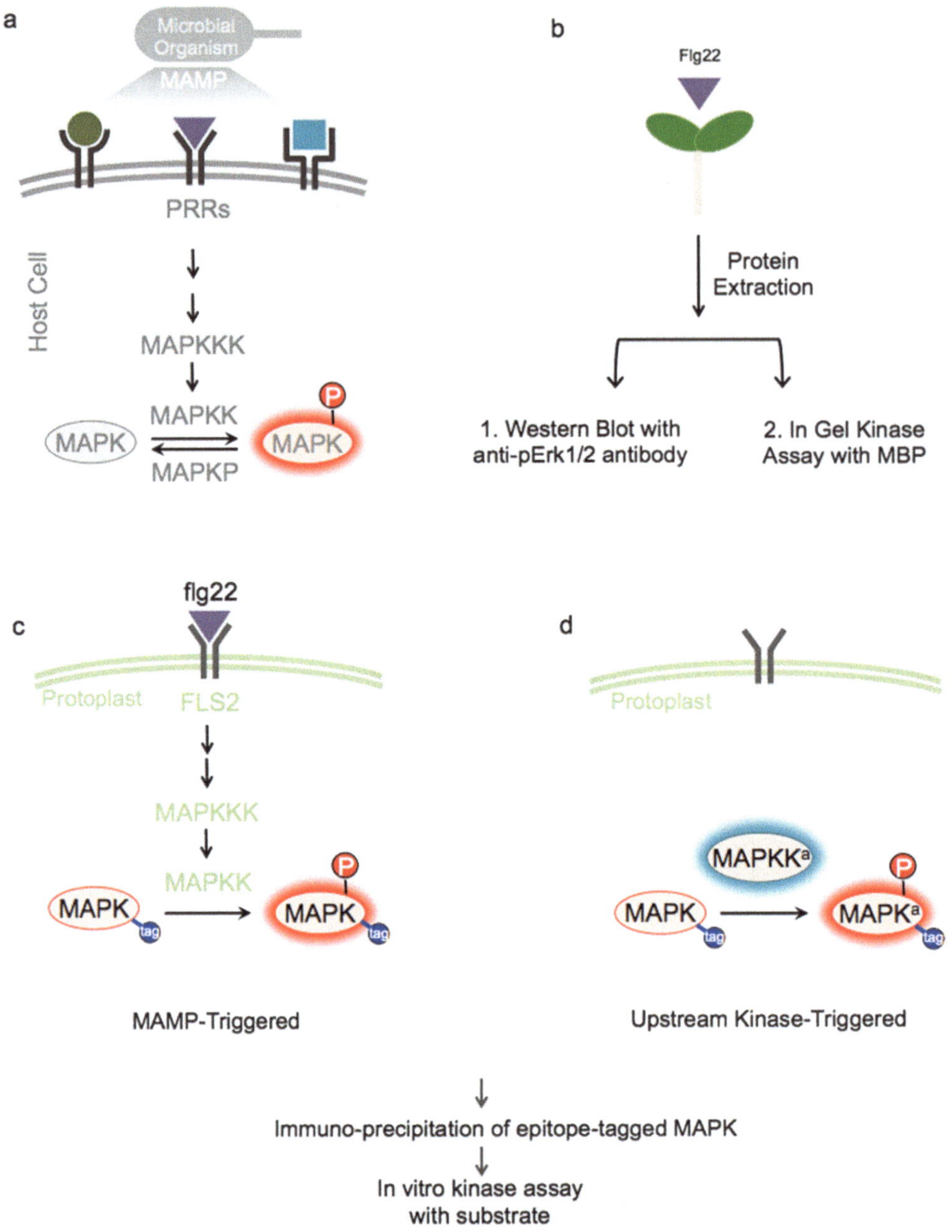

Fig. 1 Scheme of MAPK activation assays in MAMP-PRR signaling. (**a**) Activation of MAPK cascade in MAMP-PRR signal transduction pathway. Various MAMPs from microbial organism are recognized by specific PRRs at the cell membrane and triggers immune signal transduction pathway including activation of MAPK cascade. MAPK phosphatase (MAPKP) reverses MAPK activity by de-phosphorylation. (**b**) Experimental scheme of measuring endogenous MAPK activity after MAMP perception in Arabidopsis seedlings. (**c, d**) Underlying molecular basis of testing target phosphorylation by active MAPK using protoplast system. Ectopically expressed epitope-tagged MAPK in protoplasts is activated after MAMP perception through the PRR at the cell membrane (**c**). Co-expression of constitutively active upstream kinase (MKK^{a}) triggers MAPK activation without MAMP treatment (**d**)

field indicates that different MAMPs and bacterial proteases may trigger different members of MAPKs through distinct mechanisms, exemplified by flg22, LPS, and PrpL [7, 8]. It is also essential to study the signaling consequences of differential dynamics, amplitude, duration, and specificity of MAPK activation triggered by individual MAMP of various microbial organisms to understand sophisticated mechanisms that plants develop for their immune system.

All plant MAPKs identified so far share the highest sequence similarity and general phosphorylation and activation features to the ERK1/2 type of MAPKs in mammals. The development of an antibody specific to phosphorylated ERK1/2 (p42/p44) and identification MBP as a general substrate of ERK1/2, established valuable and convenient assays to study MAPK activities in plants for various signaling pathways including the MAMP-PRR pathway [9, 10]. In this chapter, we first describe experimental details of two protocols for the analysis of endogenous MAPK activation in the MAMP-PRR pathway, including immuno-blot analysis using an anti-phospho ERK1/2 antibody and in-gel kinase assay with MBP as a general substrate (Fig. 1b).

A critical question in studying the MAMP-PRR signaling pathway is to identify downstream targets that are phosphorylated by MAPKs, which may directly control reprograming of various cellular processes to support plant immunity [3, 11]. Validating phosphorylation of target candidates of a specific MAPK is crucial to further understand detailed molecular mechanisms of plant immune responses. A relatively straightforward and high-throughput way to answer the question would be the use of a biochemical assay to directly determine protein phosphorylation with a signal-activated kinase in the tube. *Arabidopsis* mesophyll cell protoplasts have been proved to serve as a great tool to study signal transduction pathways [12–14]. Combined with a facile and efficient DNA transfection technique, it could be an excellent system to express epitope-tagged signaling proteins such as kinases, activate them by desired signals such as MAMPs, and immunoprecipitate the proteins for further biochemical and functional assays. We introduce the third detailed protocol for using *Arabidopsis* protoplast system in a biochemical assay to evaluate substrate phosphorylation by the MAMP-PRR activated MAPKs (Fig. 1c, d).

2 Materials

2.1 Plant Growth and MAMP Treatment

1. Seeds of *Arabidopsis thaliana* Columbia-0.
2. Jiffy-7 peat soil pellets (Jiffy group) or Fafard soil mix (Fafard).
3. Liquid growth media: 0.5× Murashige and Skoog basal salt mixture, 0.5% Sucrose, pH 5.8 with KOH.

4. Falcon™ 6–12 well tissue culture plates.
5. Growth chamber: 12 h light/12 h dark under low light (75 μmol/m^2.s) at a 23/20 °C light/dark temperature, 65% relative humidity.
6. Flg22 peptide: the conserved 22 amino acids of flagellin, chemically synthesized according to the published peptide sequence [1].

2.2 Immuno-Detection of MAPK Activity

1. Protein Isolation (PI) buffer: 50 mM Tris–HCl, pH 7.5, 150 mM NaCl, 5 mM EDTA, 1 mM DTT, 2 mM NaF, 2 mM Na_3VO_4, protease inhibitor (Roche, one tablet of Complete™ Mini for 10 mL of PI buffer, prepare fresh), 1% v/v Triton X-100.
2. 10% SDS-polyacrylamide gel.
3. 10× Tris-Glycine SDS-PAGE running buffer.
4. 10× Tris buffered saline with 0.05%? Tween®20 (TBST).
5. Blocking buffer: 1× TBST, 5% w/v nonfat dry milk.
6. Primary antibody incubation buffer: 1× TBST, 5% w/v Bovine Serum Albumin (BSA), 1:1000 dilution of anti-phospho ERK1/2 antibody (Cell Signaling, #9101).
7. Secondary antibody incubation buffer: 1× TBST, 5% w/v nonfat dry milk, 1:10,000 dilution of HRP-conjugated anti-rabbit IgG (Pierce, #31460).

2.3 In-Gel MAPK Assay

1. Protein Isolation buffer: same as in Subheading 2.4.
2. Myelin basic protein (MBP from Invitrogen).
3. Wash buffer: 25 mM Tris–HCl, pH 7.5, 0.5 mM DTT, 5 mM NaF, 0.1 mM Na_3VO_4, 0.5% w/v BSA (Sigma, A7906), 0.1% v/v Triton X-100.
4. Renaturation buffer: 25 mM Tris–HCl, pH 7.5, 0.5 mM DTT, 5 mM NaF, 0.1 mM Na_3VO_4.
5. Reaction buffer: 25 mM Tris–HCl, pH 7.5, 2 mM EDTA, 12 mM $MgCl_2$, 1 mM DTT, 0.1 mM Na_3VO_4.
6. Stop/Wash buffer: 5% w/v trichloroacetic acid, 1% w/v sodium pyrophosphate.
7. 50 μCi γ-^{32}P-ATP per gel (Perkin Elmer).
8. Gel dryer (Bio-Rad).

2.4 Isolation of Protoplasts and DNA Transfection

1. Enzyme solution: 0.4 M mannitol. 20 mM KCl. 20 mM MES, pH 5.7, 1% w/v cellulase R10 (Yakult Pharmaceutical Industry Co., Ltd, Japan), 0.2% w/v macerozyme R10 (Yakult Pharmaceutical Industry Co., Ltd, Japan), 10 mM $CaCl_2$, 1 mM β-mercaptoethanol, 0.1% w/v BSA. Heat the enzyme solution including everything except $CaCl_2$ and BSA at 55 °C

for 10 min to dissolve enzymes and inactivate proteases and DNases. Cool the solution to room temperature and then add $CaCl_2$ and BSA. Filter the solution through a 0.45 μm filter. Prepare the enzyme solution fresh.

2. W5 solution: 154 mM NaCl, 125 mM $CaCl_2$, 5 mM KCl, 2 mM MES, pH 5.7.
3. MMg solution: 0.4 M mannitol, 15 mM $MgCl_2$, 4 mM MES, pH 5.7.
4. Polyethylene glycol (PEG) solution: 40% w/v PEG 4000 (Sigma-Aldrich, #81240), 0.2 M mannitol, 100 mM $CaCl_2$. Prepare the PEG solution fresh.
5. WI solution: 0.5 M mannitol, 20 mM KCl, 4 mM MES, pH 5.7.
6. Bovine calf serum (Sigma, A7906).
7. Nylon mesh (35–70 μm, Lab-Line Instruments, #190-158-00).
8. Hemocytometer (Hausser Scientific, #1492).
9. Falcon™ tissue culture plates, 6–12 wells.
10. DNA plasmid of epitope tagged MAPKs and the constitutively activated version of MKK[a].

2.5 In Vitro MAPK Assay

1. Protein Isolation buffer: same as in Subheading 2.4.
2. Monoclonal anti-FLAG M2 antibody (Sigma, F3165).
3. Protein G Sepharose beads (GE Healthcare, #17-0618-02).
4. MAPK phosphorylation buffer: 20 mM Tris–HCl, pH 7.5, 20 mM $MgCl_2$, 10 mM $MnCl_2$, 5 mM EDTA, 1 mM DTT.
5. MAPK phosphorylation mix (15 μL per sample): MAPK phosphorylation buffer, 1–5 μg of desired substrate (e.g., MBP), 2 μCi of γ-^{32}P-ATP.
6. Fixation buffer: 10% v/v ethanol, 10% v/v acetic acid.
7. Gel dryer (Bio-Rad).

3 Methods

3.1 Arabidopsis Seedling Growth and MAMP Treatment

1. Sterilize *Arabidopsis* seeds with 30% v/v bleach solution and stratify them at 4 °C for 3 days.
2. Add 0.5 mL of 0.5× MS media into each well of the 12-well Falcon™ Tissue Culture Plate.
3. Place six seeds in each well of the plate.
4. Seal the plate with 3 M micropore™ tape.
5. Grow plants for 7 days (*see* Subheading 2.1 for growth condition).

6. Prepare 10 nM flg22 in 0.5× MS liquid media.
7. Carefully take out liquid medium remaining in the plate and add 0.5 mL of fresh media containing 10 nM flg22 (*see* **Note 1**).
8. Incubate seedlings at room temperature for desirable periods (*see* **Note 2**).
9. Collect seedlings into the 1.5-mL microcentrifuge tube after gently blot them on Kimwipes® to remove residual liquid, and immediately freeze them with liquid nitrogen (*see* **Note 3**).
10. Store samples at −80 °C or proceed to the next step immediately.

3.2 Protein Isolation from Whole Seedlings (*see* Note 4)

1. Grind seedlings to a fine powder using a pellet pestle.
2. Add 100 μL of PI buffer and then homogenize well by vortexing vigorously for 30 s.
3. Centrifuge the samples at 18,000 × *g* at 4 °C for 5 min.
4. Transfer 80 μL of the clear supernatant to the new tubes.
5. Quantify total protein extract.
6. Store samples at −80 °C or proceed to the next step immediately.

3.3 Western Blot Analysis with Anti-Phospho ERK1/2 Antibody

1. Prepare 10% SDS-polyacrylamide gel (*see* **Note 5**).
2. Take 20 μg of total protein prepared from Subheading 3.2, and add appropriate amount of 4× sample buffer.
3. Boil the samples for 5 min and ice bath them.
4. After brief spin down, load the samples and run a gel until 20- to 100-kDa protein markers are well resolved.
5. Electrotransfer to PVDF membrane.
6. Wash membrane with 1× TBST for 5 min.
7. Incubate membrane in 10 mL of blocking buffer for 1 h at room temperature with gentle shaking.
8. Wash membrane three times for 5 min each with 15 mL of 1× TBST at room temperature.
9. Incubate membrane with of 10 mL of primary antibody incubation buffer at 4 °C for overnight with gentle shaking.
10. Wash membrane three times for 5 min each with 15 mL of 1× TBST three times at room temperature.
11. Incubate membrane with 10 mL of secondary antibody incubation buffer for 1 h at room temperature.
12. Wash membrane three times for 5 min each with 15 mL of 1× TBST at room temperature.
13. Proceed detection with ECL reagent (Fig. 2).

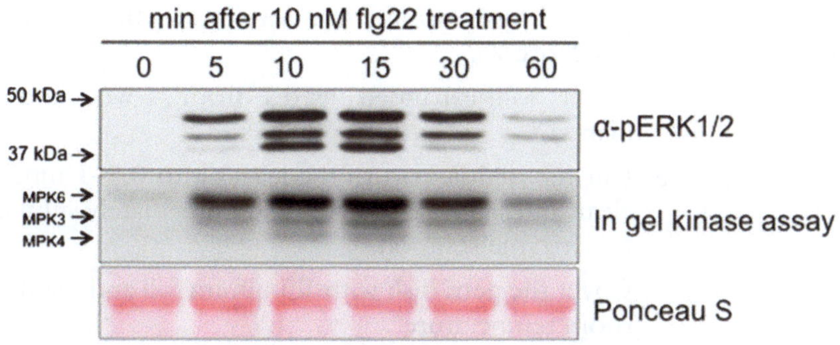

Fig. 2 Activation of endogenous MAPK after MAMP perception. Seven-day-old seedlings grown in liquid medium were treated with 10 nM flg22 peptide for indicated time length (0–60 min) and activation of endogenous MAPK proteins was measured by western blot analysis using anti-phospho Erk1/2 antibody (*top panel*) and by in-gel kinase assay using a general substrate MBP (*middle panel*). Twenty microgram of total protein was loaded in each well. Ponceau S staining of Rubisco large subunit (rbcL) shows equal amount of protein loading

3.4 In-Gel MAPK Assay

1. Prepare 10% SDS-polyacrylamide gel containing 0.25 mg/mL of MBP.
2. Take 20 μg of total protein prepared from Subheading 3.2, and add appropriate amount of 4× sample buffer.
3. Boil the samples for 5 min and ice bath them.
4. After brief spin down, load samples and run a gel until 20- to 100-kDa protein markers are well resolved.
5. Wash the gel three times for 30 min each with 100 mL of the wash buffer at room temperature with gentle agitation.
6. Incubate the gel in renaturation buffer at 4 °C overnight with three changes of the buffer during the period.
7. Equilibrate the gel with 25 mL of reaction buffer (without ATP) at room temperature for 1 h with gentle agitation.
8. Bring the gel to the designated radioactive work area.
9. Add 50 μCi of γ-^{32}P-ATP to the remaining 25 mL of reaction buffer.
10. Replace buffer with ATP-containing reaction buffer and incubate the gel for 1 h with gentle agitation at room temperature.
11. Stop reaction by replacing buffer with 100 mL of stop/wash buffer and wash the gel with gentle agitation at room temperature for 6 h. Change buffer at least six times (*see* **Note 6**).
12. Vacuum dry the gel at 80 °C for 1 h.
13. Proceed detection with autoradiography film or any phosphor imaging system.

3.5 Isolation of Protoplasts

1. Prepare fresh enzyme solution into a 100 × 20 mm petri dish (*see* **Note 7**).
2. Select well-expanded leaves from 4-week-old *Arabidopsis* plants.
3. Cut the middle part of the leaves into 0.5–1 mm strips with a clean and sharp razor blade and immediately submerge them in the enzyme solution.
4. Cover the petri dish with aluminum foil and incubate for 3 h at room temperature.
5. Gently agitate the petri dish to release protoplasts.
6. Add 10 mL of W5 solution and filter the slurry through a 35–70 μm nylon mesh into a 30-mL round-bottom tube.
7. Centrifuge the flow-through at 100 × *g* and remove the supernatant.
8. Gently resuspend protoplasts with 10 mL of W5 solution and rest the slurry on ice for at least 30 min.
9. The protoplasts should settle at the bottom of the tube. Pipet out W5 solution and resuspend the pellet in MMg solution to 2×10^5 cells/ml.

3.6 Protoplast Transfection and MAPK Activation

1. Prepare fresh PEG transfection solution.
2. Add 10 μL of plasmids DNA of MAPK-FLAG (2 μg/μL) into a 2-mL round-bottom tube (*see* **Notes 8** and **9**).
3. If desired, add 5 μL of plasmid DNA of MKK^a for co-transfection (*see* Fig. 3).
4. Add 200 μL of protoplasts in MMg solution into the tube and mix well by gentle tapping.
5. Add 220 μL of PEG transfection solution and mix by gentle tapping.
6. Incubate at room temperature for 5 min.
7. Add 440 μL of W5 solution and mix gently by inverting tubes several times to stop transfection.
8. Spin at 100 × *g* for 1 min and remove the supernatant.
9. Resuspend the protoplasts with 1 mL of WI solution and place the cell in a 6-well Falcon plate.
10. Incubate protoplasts for 5 h at room temperature.
11. Treat the protoplasts with MAMPs (e.g., flg22 peptide) for desirable period (*see* **Note 10**).
12. Harvest protoplasts by centrifugation at 100 × *g* for 2 min.
13. Remove the supernatant and freeze immediately with liquid nitrogen.
14. Store the samples at −80 °C or proceed to the next step.

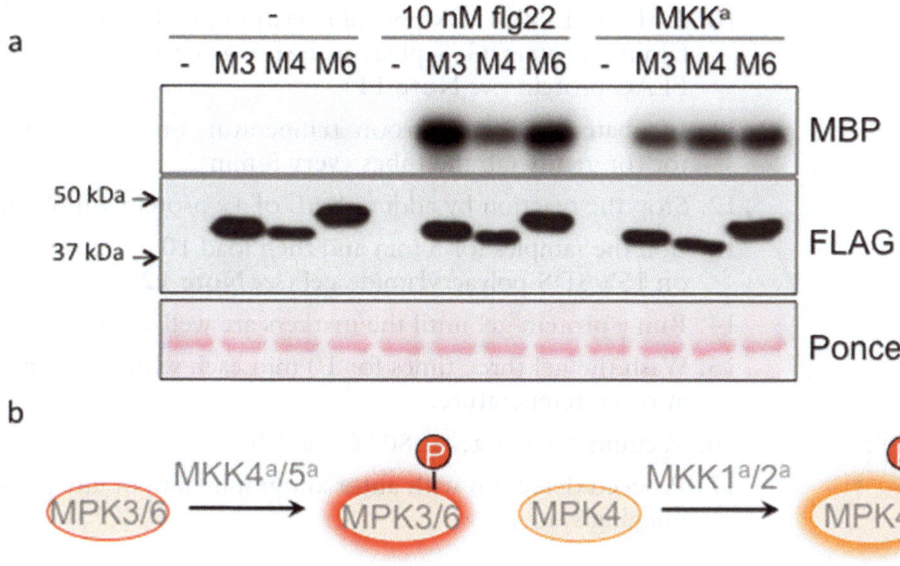

Fig. 3 In vitro phosphorylation of MBP by MAPK triggered by MAMP or upstream kinase in protoplast system. (**a**) Isolated protoplasts from 4-week-old *Arabidopsis* were transfected with indicated DNA plasmid to express epitope-tagged MAPK (e.g., MPK3-FLAG, MPK4-FLAG, and MPK6-FLAG). Protoplast with no DNA transfection (–) was served as a negative control. To activate MAPK, cells were treated with 10 nM flg22 for 10 min after 5 h of transfection, or co-transfected with constitutively active upstream kinase (MKK5[a] for MPK3 and MPK6, and MKK1[a] for MPK4) and incubated for 5 h before harvest. After immuno-purification of individual MAPK-FLAG, kinase activity of each protein was estimated by in vitro kinase assay using a general substrate MBP (*top panel*). Western blot analysis using anti-FLAG antibody indicates expression of individual MAPK-FLAG in protoplast (*middle panel*). Ponceau S staining of Rubisco large subunit (rbcL) shows equal amount of protein loading. (**b**) Different upstream kinases of MPK3/6, and MPK4. Without MAMP, MPK3/6, and MPK4 can be triggered by active MKK4/5 and MKK1/2, respectively

3.7 In Vitro MAPK Assay

1. Add 200 μL PI buffer to the frozen protoplasts and gently tab the tube to lysate the cells in ice.
2. Centrifuge at 18,000 × *g* at 4 °C for 5 min and transfer all supernatant to the new tube.
3. Add 1 μL of FLAG m2 antibody to the tube and incubate in a rotating wheel for 2 h at 4 °C.
4. Wash Protein G Sepharose beads twice with 1-mL PI buffer.
5. Add 10 μL of Protein G Sepharose beads into the tube and incubate for 1 h at 4 °C.
6. Centrifuge at 3,000 × *g* for 30 s and carefully remove the supernatant.
7. Wash the beads with 1-mL PI buffer and then with 1 mL of MAPK phosphorylation buffer.
8. Bring samples to the designated radioactive work area.
9. Prepare MAPK phosphorylation mix.

10. Add 15 μL of MAPK phosphorylation mix into the tube containing Protein G Sepharose beads associated with MAPK-FLAG protein (*see* **Note 11**).
11. Incubate for 30 min at room temperature on a shaking incubator (or gently tab the tubes every 5 min).
12. Stop the reaction by adding 8 μL of 4× protein sample buffer.
13. Boil the samples for 5 min and then load 10 μL of supernatant on 15% SDS-polyacrylamide gel (*see* **Note 12**).
14. Run a protein gel until the markers are well separated.
15. Wash the gel three times for 10 min each with Fixation buffer at room temperature.
16. Vacuum dry the gel at 80 °C for 1 h.
17. Proceed detection with autoradiography film or any phosphor imaging system.

4 Notes

1. After seedling growth for 7 days, remaining liquid medium amount in each well of the plate is varying. To have identical concentration of MAMP treatment in each well, it is recommended to remove all remaining media and add freshly made MAMP containing media to the plants in each well.
2. Individual MAMP triggers MAPK cascade with specificity. Design experiment accordingly to cover temporal dynamics of MAPK activation.
3. MAPK cascade could be activated by numerous stress signals including mechanical stress. Be cautious with handling of samples to keep the MAMP-specific effect on MAPK activation.
4. It is critical to keep the sample in ice during the whole procedure to keep the proteins intact.
5. 10% SDS-polyacrylamide gel gives you best separation of MPK6, MPK3, and MPK4, as they are resolved between 50 and 37 kDa.
6. First round of used wash buffer is highly radioactive. It must be treated accordingly.
7. Prepare 10 mL of enzyme solution to digest up to 40 leaves with a yield of around 4.0×10^6 protoplasts.
8. The quality of the plasmid DNA is critical to achieve high transfection efficiency. We use CsCl gradient for maxi preparation of plasmid DNA. The protocol is available at the Sheen Lab web site (http://molbio.mgh.harvard.edu/sheenweb/protocols_reg.html).

9. The protoplast system is ideal to prepare proteins for kinase assay when the antibody against endogenous protein is not available. Add an epitope tag sequence (e.g., FLAG, HA, or MYC) to the gene of interest, express in the protoplast, and then immune-precipitate with an appropriate antibody.
10. Dynamics of MAPK activation varies by MAMPs. Incubation time with each MAMP to activate MAPK must be decided based on empirical data. If you coexpress MAPK with constitutively active MKK (MKK^a), MAMP treatment is not necessary.
11. Immuno-purified MAPK proteins prepared from 200 μL protoplast (4×10^4 cells) could be used up to five reactions. After washing the beads, aliquot them into upto five tubes in case to test multiple substrates.
12. Prepare 10–15% SDS-polyacrylamide gel according to the size of the substrate. Size of MBP is 18 kDa. Most of the unbound γ-^{32}P-ATP migrates with blue dye (bromophenol blue) of the loading buffer. Cut out the dye part of the gel before you fix the gel.

Acknowledgment

The authors thank the former and current members of the Sheen Laboratory for their efforts to develop and improve kinase assays and the *Arabidopsis* mesophyll cell transient expression system. This work was supported by the Gordon and Betty Moore Foundation fellowship to H.S.C through Life Science Research Foundation, and the National Science Foundation grant IOS-0618292 and the National Institute of Health grant R01 GM070567 to J.S.

References

1. Felix G et al (1999) Plants have a sensitive perception system for the most conserved domain of bacterial flagellin. Plant J 18:265–276
2. Kunze G et al (2004) The N terminus of bacterial elongation factor Tu elicits innate immunity in Arabidopsis plants. Plant Cell 16:3496–3507
3. Tena G et al (2011) Protein kinase signaling networks in plant innate immunity. Curr Opin Plant Biol 14:519–529
4. Tena G et al (2001) Plant mitogen-activated protein kinase signaling cascades. Curr Opin Plant Biol 4:392–400
5. Asai T et al (2002) MAP kinase signaling cascade in *Arabidopsis* innate immunity. Nature 415:977–983
6. Arthur J et al (2013) Mitogen-activated protein kinases in innate immunity. Nat Rev Immunol 13:679–692
7. Cheng Z et al (2014) Pathogen-secreted protease activate a novel plant immune pathway. Nature 521:213–216
8. Ranf S et al (2015) A lectin S-domain receptor kinase mediates lipopolysaccharide sensing in *Arabidopsis thaliana*. Nat Immunol 16:426–433
9. Chow S et al (2001) Measurement of MAP kinase activation by flow cytometry using phospho-specific antibodies to MEK and ERK: potential for pharmacodynamic monitoring of signal transduction inhibitors. Cytometry 46:72–78

10. Atkins C et al (1999) Regulation of myelin basic protein phosphorylation by mitogen-activated protein kinase during increased action potential firing in the hippocampus. J Neurochem 73:1090–1097
11. Popescu S et al (2009) MAPK target networks in Arabidopsis thaliana revealed using functional protein microarrays. Genes Dev 23: 80–92
12. Cocking EC (1960) A method for the isolation of plant protoplasts and vacuoles. Nature 187:927–929
13. Sheen J (2001) Signal transduction in maize and Arabidopsis mesophyll protoplast. Plant Physiol 127:1466–1475
14. Yoo S et al (2007) *Arabidopsis* mesophyll protoplasts: a versatile cell system for transient gene expression analysis. Nat Protoc 2:1565–1572

Chapter 13

LeEIX2 Interactors' Analysis and EIX-Mediated Responses Measurement

Meirav Leibman-Markus, Silvia Schuster, and Adi Avni

Abstract

Plant-pathogen interactions involve a large number of wide regulatory systems, necessary for plant defense responses against pathogen attack. The fungal protein ethylene-inducing xylanase (EIX) elicits defense responses in specific cultivars of tobacco and tomato. The response to EIX is controlled by a single locus encoding for LeEIX2, a leucine-rich-repeat receptor-like-protein (LRR-RLP). As an RLP, LeEIX2 does not possess an obvious cytoplasmic signaling moiety such as a kinase domain. To study LeEIX2 mode of action, it is essential to identify the potential interactors involved after EIX perception. Here, we describe the in vivo co-IP methodology used for protein interaction verification and ethylene and ROS (reactive oxygen species) measurements used for physiological effects assessment.

Key words LRR-RLP, Ethylene-inducing xylanase, MAMP, Defense responses, Co-IP, ROS

1 Introduction

Plants are constantly being exposed to a wide variety of microorganisms such as bacteria and fungi. As plants lack a network of circulating cells, their defense relies on the innate immunity of each cell and systemic signals emanating from infection sites [1]. Plant cells exploit various types of cell surface receptor molecules to perceive extracellular signals. These signals can be molecular-associated microbial patterns (MAMPs) or self-derived [2]. Often, these receptors contain extracellular leucine-rich repeats (LRRs) that mediate protein-protein interaction [3].

Two types of LRR receptors have been reported in plants: RLKs composed of an LRR domain, a single-pass transmembrane domain, and a cytoplasmic kinase domain. RLPs, which differ from RLKs by carrying a short cytoplasmic tail without a kinase domain. RLKs and RLPs were shown to play roles in innate immunity and development in several plant species [4, 5]. A total of 647 RLKs were identified in the tomato genome [6]. The number of RLP-type receptors predicted from genomic sequences varies according

Libo Shan and Ping He (eds.), *Plant Pattern Recognition Receptors: Methods and Protocols*, Methods in Molecular Biology, vol. 1578, DOI 10.1007/978-1-4939-6859-6_13, © Springer Science+Business Media LLC 2017

to the plant species studied. Arabidopsis has 57 [7] and the number of RLPs for the tomato genome can be expected to be much higher. Among the tomato RLPs involved in defense responses the best characterized members include the Cf family against *Cladosporium fulvum* [8, 9], VeI against *Verticillium* [10], and LeEIX2 which recognizes EIX [11].

Since RLPs lack a kinase domain it is crucial to identify the interactors responsible for downstream signaling upon elicitation. Different screening procedures such as yeast two hybrid screens and proteomics analysis will yield numerous putative interactors. Such interactions probably involve protein-protein interaction or even complex formation [2]. Co-IP methodology can be used to verify and characterize such interactions. Here, we describe an efficient procedure to detect co-IP of membrane integrated proteins and their interactors both cytoplasmatic and membranal.

Once a putative interactor is verified, its relevance to defense responses induction must be investigated. Several defense responses are known and can be monitored. Some of them occur rapidly upon elicitation such as the production of reactive oxygen species (ROS) burst. ROS production promotes cell wall strengthening and H_2O_2 triggers local PCD [12, 13]. Other responses develop at a slower rate such as ethylene production. Ethylene is a gaseous plant hormone whose level increases after elicitation [14]. We will describe the methodology to measure and quantify both these parameters.

All the methods described refer to the LeEIX2/EIX system. Other systems may require different concentrations or incubation times. The methodology can apply both to transiently expressing and stably transgenic plants.

2 Materials

2.1 Co-IP of Membrane Proteins

Prepare all solutions using ultrapure water and analytical grade reagents. Prepare and store all reagents at 4 °C (unless indicated otherwise).

1. Extraction buffer (EB): 50 mM Tris–HCl, pH 7.5, 2 mM $MgCl_2$, 150 mM NaCl.

   Immediately prior to use add the following: 140 mM β-mercaptoethanol, 2 mM PMSF, 1 mM protease inhibitor cocktail (EDTA free).
2. Wash buffer 1: 50 mM Tris–HCl, pH 7.5, 1 mM EDTA, 150 mM NaCl, 0.1% Nonidet P40.
3. Wash buffer 2: 50 mM Tris–HCl, pH 7.5, 1 mM EDTA, 500 mM NaCl, 0.1% Nonidet P40.
4. Wash buffer 3: 10 mM Tris–HCl, pH 7.5, 0.1% Nonidet P40.

5. 4× Sample buffer: 8% SDS, 40% glycerol, 200 mM Tris-Cl, pH 6.8, 388 mM dithiothreitol (DTT), and 0.1 mg/mL bromophenol blue dye.

2.2 ROS Burst from Leaf Discs

Prepare all solutions using ultrapure water and analytical grade reagents.

1. Luminol 17 mg/mL in DMSO (100 mM) prepare freshly prior to use. Keep in the dark at room temperature.
2. HRP 10 mg/mL in 0.1 M Phosphate buffer pH 6.0. Store at −20 °C in the dark (*see* **Note 1**).
3. EIX 1 mg/mL.
4. White 96-well multiplate suitable for optical reading.
5. Luminometer suitable for 96-well multiplate.

2.3 Ethylene Induction from Leaf Discs

Prepare all solutions using ultrapure water and analytical grade reagents at room temperature.

1. Ethylene assay buffer: 10 mM MES pH 6.0, 250 mM sorbitol.
2. EIX 1 mg/mL.
3. 10-mL erlenmeyer flasks tightly septa sealed with septum rubber caps sleeve type that can be needle perforated.
4. 5-mL disposable syringes with 21G needles (to pump out the gaseous sample) and rubber stoppers (to hold the sample till injecting into the gas chromatographer; GC).

3 Methods

3.1 Co-IP of Membrane Proteins

Carry out all procedures at room temperature on ice unless otherwise specified.

1. Using precooled mortar and pestle; grind approximately 200 mg (*see* **Note 2**) plant tissue expressing the tagged membranal protein with liquid nitrogen to a fine powder.
2. When powder starts to melt, add three volumes of EB and continue grinding until a homogeneous liquid solution has been achieved.
3. Transfer solution to a precooled microfuge tube.
4. Centrifuge for 16 min at 20,000 × *g*, 4 °C.
5. Discard the supernatant (*see* **Note 3**).
6. Suspend pellet in two volumes EB containing 0.5% Triton X-100 (*see* **Note 4**).
7. Homogenize thoroughly inside the microfuge tube using a small plastic or glass pestle. Leave on ice for 20 min mixing every 5 min (*see* **Note 5**).

8. Centrifuge for 16 min at 20,000 × *g*, 4 °C.
9. Vortex for 5 s and collect the supernatant (Triton soluble Membrane (TSM) fraction).
10. Remove 5–50 μL (Input) according to the expected protein expression level, add to the input sample buffer ×4 (*see* **Note 6**).
11. Add to the TSM fraction three volumes of EB so as to dilute the Triton X-100 to 0.2% to avoid unspecific binding, and 5 μL antibody bound agarose beads previously washed twice with 1000 μL EB.
12. Incubate the samples 1 h (*see* **Note 7**) on a rotating wheel at 4 °C.
13. Centrifuge at 2500 × *g* for 3 min and discard supernatant.
14. Resuspend beads with 500 μL wash buffer 1 and transfer to a dolphin microfuge tube wash previous tube with another 500 μL of wash buffer 1 and add to dolphin tube, Final wash volume will be 1 mL. Return to rotating wheel at 4 °C for 5–7 min.
15. Centrifuge at 2500 × *g* for 3 min, discard supernatant.
16. Wash twice with 1-mL wash buffer 1 and centrifuge at 2500 × *g* for 2 min. Incubate between each wash the samples on the rotating wheel for 5 min at 4 °C.
17. Wash twice with 1-mL wash buffer 2 and once with 1-mL wash buffer 3. Incubate between each wash the samples on the rotating wheel for 5 min at 4 °C.
18. Resuspend beads with 500 μL wash buffer 3 and transfer to a new dolphin microfuge tube wash previous dolphin tube with another 500 μL of wash buffer 3 and add to the new dolphin tube.
19. Centrifuge at 2500 × *g* for 2 min and discard supernatant leaving beads as dry as possible.
20. Add to the beads 20 μL 2× sample buffer.

3.2 ROS Burst from Leaf Discs

Carry out all procedures at room temperature unless otherwise specified.

1. Sample leaf discs (LD) with a cork borer (5 mm inside diameter; *see* **Note 8**).
2. Place one LD in each well of a white 96-well multiplate containing 250 μL of ddH_2O water, abaxial side facing water. Handle LD carefully to prevent wounding and avoid taking any curvy or curly LD.
3. Incubate the LDs with a transparent lid on, for 4–24 h at room temperature with gentle agitation.
4. After incubation, remove water. Take care not to damage the LDs and make sure that they lay at the bottom of the well

(*see* **Note 9**). Immediately add 50 μL ddH_2O water to each well to prevent LDs drying up (*see* **Note 10**).

5. Prepare ROS assay solution concentrated 1.5 fold: Luminol 150 μM, HRP 15 μg/mL, EIX 1 μg/mL or control.
6. Add 100 μL to each well and immediately start measuring the light emission in repeating cycles till you observe emission decay (*see* **Note 11**).

3.3 Ethylene Induction from Leaf Discs

Carry out all procedures at room temperature unless otherwise specified.

1. Dispense 1-mL ethylene assay buffer with or without EIX 1 μg/mL in each erlenmeyer flask.
2. Sample leaf discs (LD) with a cork borer (7–8 mm inside diameter).
3. Place six LDs in each erlenmeyer flask, abaxial side facing solution. Avoid taking any curvy or curly LDs. Take care to distribute equally LDs from all sampled leaves.
4. Tightly cap flasks and incubate with gentle agitation for 4 h.
5. For each sample inject the needle connected to a 5-mL syringe, pump up and down gaseous sample two to three times and finally extract a 3-mL sample and immediately stick the syringe into a rubber stopper (*see* **Note 12**).
6. Inject and read samples in GC.

4 Notes

1. HRP is sensitive to repeated defrosting cycles, therefore prepare solution and divide into small aliquots before freezing.
2. For genes with low expression level, protocol can be scaled up to 1 g tissue.
3. This is the cytosolic soluble fraction. If one of the proteins is present also in the cytosolic fraction, keep this fraction on ice and use it to dilute the Triton X-100 concentration in **step 11**.
4. It is useful to prepare 10–25% Triton X-100 stock solution and add the required amount to EB to avoid pipetting concentrated viscous Triton X-100.
5. Alternatively, the samples can be incubated on the rotating wheel at 4 °C at a slow speed.
6. Input samples can be stored at −20 °C after boiling.
7. For low protein expression and/or weak interaction, incubation may be prolonged up to overnight.
8. When using *Agrobacterium* for transient expression, sample LD 24–44 h after infiltration.

9. To thoroughly remove all water, use a blunt needle connected to a vacuum source.
10. To avoid LD drying up, remove water from one row at a time.
11. Longer integration time will yield fewer readings with higher level of emission. It is crucial to match the time elapsed between adding the solution to each well to the integration time between well readings.
12. Samples can be stored for 1–2 h before injecting them into the GC if required.

Acknowledgments

This work was partly supported by the Israel Science Foundation administered by the Israel Academy of Science and Humanities no. 388/12 and Research Grant Award no. IS-4842-15R from BARD, the United States–Israel Binational Agriculture Research and Development Fund and by the United States-Israel Binational Science Foundation no. 2013227.

References

1. Chisholm ST, Coaker G, Day B, Staskawicz BJ (2006) Host-microbe interactions: shaping the evolution of the plant immune response. Cell 124(4):803–814
2. Bohm H, Albert I, Fan L, Reinhard A, Nurnberger T (2014) Immune receptor complexes at the plant cell surface. Curr Opin Plant Biol 20(1):47–54
3. Kobe B, Kajava AV (2001) The leucine-rich repeat as a protein recognition motif. Curr Opin Struct Biol 11(6):725–732
4. De Smet I, Voss U, Jurgens G, Beeckman T (2009) Receptor-like kinases shape the plant. Nat Cell Biol 11(10):1166–1173
5. Wang G, Fiers M (2010) CLE peptide signaling during plant development. Protoplasma 240(1–4):33–43
6. Sakamoto T, Deguchi M, Brustolini OJ, Santos AA, Silva FF, Fontes EP (2012) The tomato RLK superfamily: phylogeny and functional predictions about the role of the LRRII-RLK subfamily in antiviral defense. BMC Plant Biol 12:229
7. Fritz-Laylin LK, Krishnamurthy N, Tor M, Sjolander KV, Jones JD (2005) Phylogenomic analysis of the receptor-like proteins of rice and Arabidopsis. Plant Physiol 138(2): 611–623
8. Jones DA, Thomas CM, Hammond-Kosack KE, Balint-Kurti PJ, Jones JD (1994) Isolation of the tomato Cf-9 gene for resistance to Cladosporium fulvum by transposon tagging. Science 266(5186):789–793
9. Thomma BP, VANEsse HP, Crous PW, DE Wit PJ (2005) Cladosporium fulvum (syn. Passalora fulva), a highly specialized plant pathogen as a model for functional studies on plant pathogenic Mycosphaerellaceae. Mol Plant Pathol 6(4):379–393
10. Fradin EF, Zhang Z, Juarez Ayala JC, Castroverde CD, Nazar RN, Robb J, Liu CM, Thomma BP (2009) Genetic dissection of Verticillium wilt resistance mediated by tomato Ve1. Plant Physiol 150(1):320–332
11. Ron M, Avni A (2004) The receptor for the fungal elicitor ethylene-inducing xylanase is a member of a resistance-like gene family in tomato. Plant Cell 16(6):1604–1615
12. Hammond-Kosack KE, Jones JD (1996) Resistance gene-dependent plant defense responses. Plant Cell 8(10):1773–1791
13. Tenhaken R, Levine A, Brisson LF, Dixon RA, Lamb C (1995) Function of the oxidative burst in hypersensitive disease resistance. Proc Natl Acad Sci U S A 92(10):4158–4163
14. Felix G, Grosskopf DG, Regenass M, Basse CW, Boller T (1991) Elicitor-induced ethylene biosynthesis in tomato cells: characterization and use as a bioassay for elicitor action. Plant Physiol 97(1):19–25

Chapter 14

CDPK Activation in PRR Signaling

Heike Seybold, Marie Boudsocq, and Tina Romeis

Abstract

Calcium-dependent protein kinases undergo a rapid biochemical activation in response to an intracellular Ca increase induced by the PRR-dependent perception of a pathogen-related stimulus. Based on SDS gel resolution, the in-gel kinase assay allows the analysis of multiple in vivo protein samples in parallel, combining the advantage of protein separation according to molecular mass with the activity read-out of a protein kinase assay. It thus enables to follow the transient CDPK activation and inactivation in response to in vivo elicitation with a time-wise resolution. In addition, changes of CDPK phosphorylation activity often correlate with slight shifts in the enzyme's apparent molecular mass, indicating posttranslational modifications and a conformational change of the active enzyme compared to its inactive resting form. These band shifts can be detected by a simple immunoblotting to monitor CDPK activation.

Key words CDPK, Signal transduction, In-gel kinase assay, Electrophoretic band shift, Transient expression, Protoplast, Phosphorylation target

1 Introduction

The technique of in-gel kinase assays had first been introduced into the field of plant signaling research from yeast with the objective to study stimulus-dependent activation of MAP kinases [1]. Based on the SDS gel resolution, the advantage of this method is the analysis of multiple in vivo protein samples (even within crude protein extracts) in parallel, for example following an in vivo response to pathogen stimulation of plant cell cultures or plants. Compared to the alternative method of antibody-based immune-complex kinase assays, where the protein kinase of interest is immunoprecipitated prior to kinase activity measurement, the in-gel kinase assay allows the analysis of several protein kinases simultaneously without the requirement of specific antibodies. Thus, by analyzing protein kinases in crude extracts directly after harvesting, each sample represents a "frozen snap-shot" of kinase in vivo modification/activation, which is particularly important for the study of dynamic signaling processes.

Libo Shan and Ping He (eds.), *Plant Pattern Recognition Receptors: Methods and Protocols*, Methods in Molecular Biology, vol. 1578, DOI 10.1007/978-1-4939-6859-6_14,

An important prerequisite is, however, that the type of kinase "survives" the denaturing condition of the SDS-gel procedure, and the kinase's intrinsic folding triggers the renaturation into the kinase native active state. So far, in the context of PRR perception, in-gel kinase assays have been successfully applied to investigate MAPKs and CDPKs.

In the context of immune responses, endogenous CDPKs have been detected in tobacco cell cultures carrying the tomato R-gene Cf9 upon elicitation with the fungal-derived elicitor peptide Avr9 [2]. Using histone as a synthetic protein substrate cast into the gel, kinases *Nt*CDPK2 and *Nt*CDPK3 showed predominant differential in-gel kinase activity. In addition, kinase activity correlated with a stimulus-dependent mobility shift in the SDS gel toward a slower migrating form of the kinase protein, indicating a stimulus-induced post-translation modification (for example, phosphorylation) of the CDPK protein.

The in-gel kinase assay has subsequently been further developed and applied to investigate single CDPK protein isoforms within the CDPK gene family using specific protein tags. The activation of a tagged CDPK isoform can be followed either in transgenic plant lines expressing an appropriate fusion protein, as performed with CPK5-YFP lines to investigate flg22-mediated responses [3]), or upon transient expression in *N. benthamiana* [4, 5]. A further and significant improvement of the in-gel kinase assay has been provided with the use of Arabidopsis protoplast system that allows to monitor the kinase activity of both endogenous and transiently expressed protein kinases either in wild-type or mutant plant backgrounds [6, 7]. This approach has then been extended to the research of CDPKs [3, 8–10] providing the following advantages for the study of PRR-induced kinase-mediated signaling: (1) The transient expression of single isoforms enables the dissection of PRR-components required for CDPK activation by comparing protoplasts originating from different plant mutant backgrounds such as *fls2* or *bak1* mutant lines [3, 11]. (2) Based on a transient expression in protoplasts, differences in kinase activation even among close homologs of CDPK family members can be addressed, thus providing a valuable and important tool to decipher "apparent functional redundancy" on a biochemical level. (3) A selection of different peptides or protein substrates can be cast into the polyacrylamide gel. In the past, synthetic proteins with a basic phosphorylation motif (such as histone) have been frequently chosen, but also other generic substrates such as casein or even the preferred MAPK substrate MBP (myelin basic protein). However, now that increasingly more genuine in vivo phosphorylation target proteins of CDPKs are identified, the in-gel kinase method will proof even more powerful, because peptides, domains or entire proteins representing (potential or already proven) in vivo phosphorylation targets can be included in the gel. (4) In the absence of a substrate protein cast into the gel, the assay visualizes enzyme auto-phosphorylation, which may also yield important information about PRR signaling. Furthermore, by incubating the gel in reaction buffer either in presence or absence of Ca^{2+}, calcium-dependent properties of the enzymes can be addressed. (5) On the molecular mechanistic level of CDPK activation, the transient in-gel kinase assay allows a parallel investigation of CDPK protein kinase variants, which either lack certain consensus domains or carry site-specific point mutations, for example in sites required for calcium or ATP binding. A compromised in-gel kinase activity in response to PRR-mediated stimulation thus identifies single amino acids which may be required for CDPK signal transmission in plant immunity.

The protocols provided in this chapter describe how to monitor CDPK activation by in-gel kinase assay and immunoblotting, using Arabidopsis mesophyll cell protoplasts (Fig. 1).

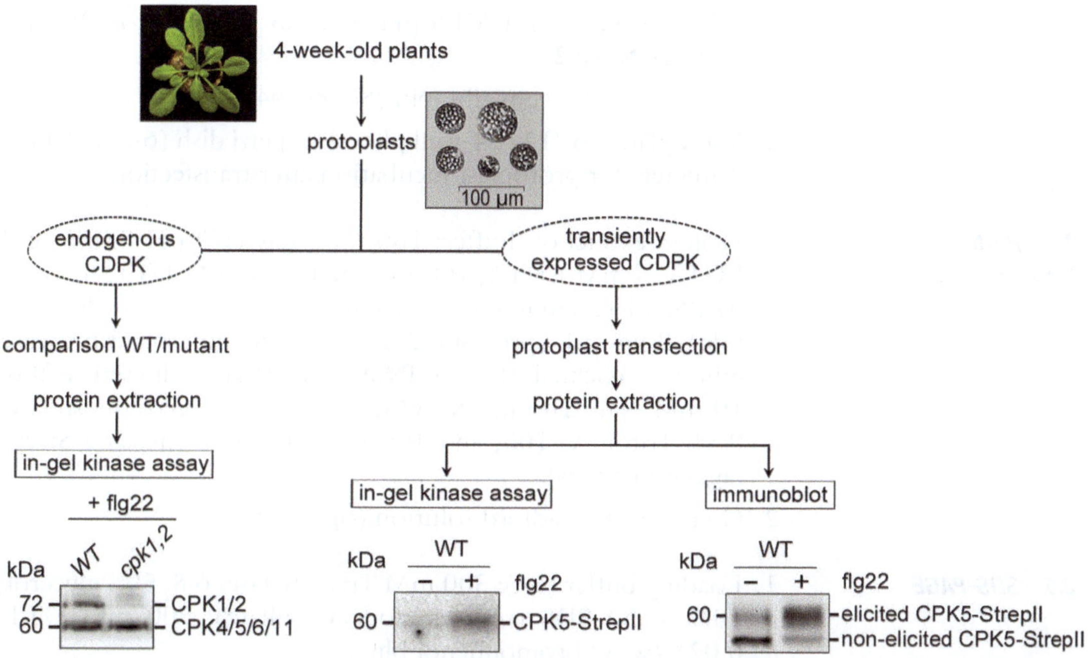

Fig. 1 Workflow of the method. Protoplasts isolated from 4-week-old plants can be directly used to monitor CDPK activity by in-gel kinase assay, allowing the comparison of several endogenous CDPKs in various mutant backgrounds (*left panel*). Protoplasts can also be transiently transformed with CDPK-tagged constructs to monitor the activity of individual CDPK isoforms in WT or mutant backgrounds, either by in-gel kinase assay or immunoblot (*right panel*)

2 Materials

2.1 Arabidopsis Plants

1. Arabidopsis soil: jiffy-7® pellets.
2. 0.1% agar type E or M (plant cell tested, Sigma).

2.2 Mesophyll Protoplast Isolation

1. Enzyme solution: 20 mM MES pH 5.7, 0.4 M mannitol, 20 mM KCl, 1.5% (w/v) cellulase R10 (Yakult, Japan), 0.4% macerozyme (Yakult, Japan), 0.1% bovine serum albumin (BSA), 10 mM $CaCl_2$ (*see* **Note 1**).
2. W5 solution: 2 mM MES pH 5.7, 154 mM NaCl, 125 mM $CaCl_2$, 5 mM KCl (*see* **Note 2**).
3. Desiccator.
4. 35–75 μm nylon mesh.
5. Swing centrifuge.

2.3 Mesophyll Protoplast Transfection

1. MMg solution: 4 mM MES pH 5.7, 0.4 M mannitol, 15 mM $MgCl_2$ (*see* **Note 2**).
2. PEG solution: 200 mM mannitol, 50 mM $CaCl_2$, 40% PEG (w/v) (Fluka) (*see* **Note 3**).
3. WI solution: 4 mM MES pH 5.7, 0.5 M mannitol, 20 mM KCl (*see* **Note 2**).
4. 5% (w/v) calf serum albumin (Sigma N4637).
5. Microplates (6/12/24-well plates) or petri dish (6-cm/10-cm diameter) for protoplast incubation after transfection.

2.4 Protein Extraction

1. Protein extraction buffer: 100 mM Tris-HCl pH 8.0, 5 mM EGTA, 5 mM EDTA, 100 mM NaCl, 20 mM DTT, 0.5 mM AEBSF (4-(2-aminoethyl)benzenesulfonyl fluoride hydrochloride), 2 μg/mL aprotinin, 2 μg/mL leupeptin, plant protease inhibitor cocktail (Sigma P9599, Germany) diluted 1:200, 10 mM NaF, 10 mM Na_3VO_4, 10 mM β-Glycerolphosphate, 0.5% Triton X–100, and 100 μg/mL avidin (if using Strep-tagged proteins).
2. Commercial Bradford solution (optional).

2.5 SDS-PAGE

1. Loading buffer (5×): 300 mM Tris–HCl pH 6.8, 50% glycerol, 10% (w/v) SDS (sodium dodecyl sulfate), 500 mM DTT, 0.02% (w/v) bromophenol blue.
2. Resolving gel: 10% (v/v) acrylamide/bis, 375 mM Tris–HCl pH 8.8, 0.04% (v/v) TEMED, 0.1% (w/v) APS.
3. Histone III-S (*see* **Note 4**): Addition of 0.25 mg/mL Histone III-S (final concentration) to mixture before pouring the resolving gel.

4. Stacking gel: 5% (v/v) acrylamide/bis, 125 mM Tris–HCl pH 6.8, 0.1% (v/v) TEMED, 0.1% (w/v) APS.
5. SDS-PAGE running buffer: 25 mM Tris-HCl pH 6.8, 0.1% (w/v) SDS, 19.2 mM glycine.
6. Protein marker, prestained.
7. Gel running apparatus.

2.6 In-Gel Kinase Assay

1. Washing buffer: 25 mM Tris–HCl pH 7.5, 0.5 mM DTT, 5 mM NaF, 0.1 mM Na_3VO_4, 0.5 mg/mL BSA, 0.1% Triton X–100.
2. Renaturation buffer: 25 mM Tris–HCl pH 7.5, 0.5 mM DTT, 5 mM NaF, 0.1 mM Na_3VO_4.
3. Reaction buffer: 25 mM Tris–HCl pH 7.5, 1 mM DTT, 0.1 mM Na_3VO_4, 12 mM $MgCl_2$, optional (*see* **Note 5**): 1 mM $CaCl_2$ or 2 mM EGTA.
4. Cold ATP and [γ-^{32}P]-ATP.
5. Stop solution: 5% TCA, 1% phosphoric acid (*see* **Note 6**).
6. Horizontal shaker.
7. Reusable detection screens for radioactivity (phosphor screen) and a phosphor imager.

2.7 Immunoblot

1. Transfer buffer: 25 mM Tris, 192 mM glycine, (optional: 0.0124% (w/v) SDS), 20% (v/v) Methanol.
2. Nitrocellulose membranes.
3. Transfer blotting apparatus.
4. Tris-buffered saline (TBS): 20 mM Tris–HCl pH 7.6, 150 mM NaCl.
5. TBST: TBS containing 0.1% (v/v) Tween 20.
6. Blocking solution: 5% skimmed milk powder in TBST (*see* **Note 7**).
7. Anti-tag antibody and corresponding secondary antibody.
8. ECL detection kit.

3 Methods

3.1 Protoplast Isolation

1. Arabidopsis seeds are stratified in 0.1% agar at 4 °C for 2 days in the dark before being sown on watered Jiffy-7® pellets. The plants are grown for 4 weeks in an environment-controlled growth chamber with a medium photoperiod (12 h light at 22 °C/12 h dark at 18 °C) under low-light conditions (75 μE/m^2/s) and 50–70% relative humidity (*see* **Note 8**).

2. Freshly prepare 10 mL of enzyme solution (*see* **Note 1**) and pour it in a 10-cm petri dish. Using a clean razor blade, cut 0.5–1 mm leaf strips in the middle part of the leaf. Transfer the leaf strips gently in the enzyme solution by capillarity using a bacterial loop and immerge them completely (*see* **Note 9**).
3. Cap the petri dish, place it in a desiccator and apply vacuum for maximum 30 min, in the dark. It will help the enzyme solution to penetrate the plant tissues to increase the digestion yield. Recover the plate and incubate in the dark in a controlled environment for 2.5 h (*see* **Note 10**).
4. To release protoplasts, gently shake the petri dish with a circle motion. Add 10 mL of W5 solution and continue gentle shaking to homogenization. Pipet off the protoplast solution and filter through a wetted 35/75-μm nylon mesh, in a 30-mL round-bottomed tube. Collect the cells on ice. Centrifuge at 100 × *g* in a swing centrifuge at RT, without any break, for 2 min (*see* **Note 11**). Pipet off the supernatant. Wash the cells with 10 mL W5, gently mix and centrifuge again. Discard flow-through. Resuspend the protoplasts in 4–5 mL W5 and count them under a microscope with a hemacytometer. Adjust the concentration to 2 × 10^5 cells/mL in cold MMg solution. Keep on ice for 30 min before transfection.

3.2 Protoplast Transfection

1. Prepare the plates for protoplast incubation: add 5% calf serum to each well, shake the plate until calf serum has covered the wells and remove it. Let the plate dry before adding WI in each well (*see* **Note 12**). The coating will prevent the protoplasts to stick to the well.
2. Add 20 μL DNA (20–40 μg) to each 2-mL microcentrifuge tube. One by one, add 200 μL protoplasts and mix gently. One by one, add 200 μL PEG solution and mix gently by tapping until the suspension is homogenous. Incubate at RT for 5–15 min (*see* **Note 13**).
3. One by one, add 440 μL W5 solution and mix gently by inverting the tube until the solution is homogenous.
4. Centrifuge at 100 × *g* in a swing centrifuge at RT, without any break, for 1 min. Discard supernatant. Resuspend protoplasts in 200 μL WI solution and transfer them to the culture plate. When the plate is done, gently shake it with a circle motion to homogenize the protoplast suspension. Incubate the cells on the bench for 6–14 h. The protoplasts will settle down on the periphery of the well.
5. To apply PAMP, gently resuspend the cells with a circle motion. Add PAMP (for example, 200 nM flg22), gently homogenize and incubate for 10 min. To stop treatment, transfer the cells to a 1.5 mL microcentrifuge tube by pipetting.

Centrifuge at 100 × *g* in a swing centrifuge at RT, without any break, for 2 min. Discard supernatant and freeze pellets in liquid nitrogen.

3.3 Protein Extraction

1. Add 25 μL protein extraction buffer to the frozen protoplast pellet and vortex to thaw and lyse the cells.
2. Incubate the samples on ice for 10 min and, if desired, centrifuge the samples for 10 min at 20,000 × *g* to remove cell debris.
3. Protein quantification is optional for protoplast crude extracts. Protoplast transformation samples were adjusted to equal cell amount, so protein levels in crude extracts are estimated as equal (*see* **Note 14**).
4. Add 5× Loading buffer to each sample, incubate for 3 min at 70 °C (NOT 95 °C!!).
5. Load samples on gels (with or without matrix proteins) to continue with SDS-PAGE (*see* **Note 15**).

3.4 In-Gel Kinase Assay

Gels in liquid are incubated on a horizontal shaker if not stated otherwise. Carry out all steps at room temperature unless specified otherwise.

3.4.1 Non-Radioactive Part

1. Immediately after SDS-PAGE, the gel plates are separated with the help of a spatula or a similar tool. The stacking gel is removed and the resolving gel is rinsed 2–3 times with washing buffer (~25 mL each time) followed by incubation in washing buffer (~75 mL) for 1 h. This washing procedure is repeated twice.
2. After washing, the gel is rinsed with renaturation buffer, followed by incubation in renaturation buffer for 1 h.
3. Rinse the gel with renaturation buffer and incubate in renaturation buffer at 4 °C for at least 14 h or overnight.
4. Back at RT, rinse the gel with renaturation buffer and incubate the gel in renaturation buffer for 1 h.
5. Equilibrate the gel for 30–60 min in reaction buffer without ATP (but either with Ca^{2+} or EGTA).

3.4.2 Radioactive Part

1. Incubate the gel for 1.5 h in 10 mL reaction buffer supplemented with 50 μCi [γ-^{32}P] ATP and 0.1 nM nonradioactive ATP (*see* **Note 16**).
2. Replace the reaction buffer with stop solution to wash away radioactivity. Replace the stop solution in washing steps until the liquid has less than 0.5–1 count per second (*see* **Note 17**).
3. Place the gel on a whatman paper and cover it with saran foil.

4. Dry the gel on the whatman paper on a gel dryer until it is completely dry (*see* **Note 18**).
5. Cover gel with a second layer of saran foil before exposure to a phosphor screen.

3.5 Immunblotting

The same protein samples (used for in-gel kinase assay) can be loaded on SDS-gels without any embedded protein substrate to control and quantify protein expression via common immunoblotting.

1. After separating the gel plates, remove the stacking gel, rinse the resolving gel with deionized water to remove traces of SDS-PAGE running buffer, and place it on a nitrocellulose membrane to electrotransfer the proteins onto the membrane via semi-dry or wet/tank blotting.
2. Place the gel sandwich in the transfer apparatus and transfer the proteins at 40 V or 100 mA overnight at 4 °C. Alternatively, the transfer can be performed at 100 V or 350 mA in about 60 min at 4 °C or at room temperature with additional cooling of the system (*see* **Note 19**).
3. Remove the membrane from the transfer unit and rinse with deionized water. Optional: The efficiency of the protein transfer as well as an equal protein loading can be controlled by reversible staining of the membrane for example with Ponceau S.
4. Incubate the membrane in blocking solution for 1 h at room temperature or overnight at 4 °C to prevent unspecific binding of the antibodies.
5. Incubate the membrane with the primary antibody (according to the tag/fusion protein used) diluted in the blocking solution, at 4 °C overnight with gentle shaking.
6. Wash the membrane 3 times in TBST, each 10 min at RT.
7. Incubate the membrane in the appropriate secondary antibody diluted in the blocking solution, at RT for 1 h with gentle shaking.
8. Wash the membrane twice in TBST, each 10 min, and twice in TBS, each 5 min. Continue with the detection of the tagged proteins using an ECL detection kit following the manufacturer's instructions.

4 Notes

1. Mix together H_2O, mannitol, KCl, and MES solutions and pre-heat at 70 °C for 3–5 min. Then, add cellulase and macerozyme powder, and incubate at 55 °C for 10 min to help the enzyme dissolution. Cool down the enzyme solution to room

temperature (5–10 min on ice). Add $CaCl_2$ and BSA before filtering the solution through a 0.45 μm syringe filter device. The enzyme solution should be clear light brown.

2. W5 and WI solutions are filtered sterilized and kept at room temperature but the protoplast isolation and transformation do not need to be carried out under the hood. The MMg solution is stored at 4 °C to be ready for protoplast transformation.
3. The PEG solution must be prepared freshly, usually at the end of the leaf digestion. Combine all the components in a falcon tube and let them dissolve gently on an orbital shaker. Do not vortex the PEG solution nor adjust the pH. The PEG quality is critical and we recommend the Fluka product.
4. Our experience is that CDPK can also phosphorylate MBP, presumably in part depending on the enzyme tag. The advantage is that MAPK activity can be investigated in parallel. Like histone, MBP is used at 0.25 mg/mL final concentration. Moreover, the auto-phosphorylation of the kinase can be investigated by omitting the protein substrate from the resolving gel.
5. Prepare 0.5 L washing and renaturation buffer and 100 mL reaction buffer per gel. Buffers containing DTT must always be prepared fresh. Use $CaCl_2$ to increase kinase activity or EGTA to inhibit CDPK activity.
6. The stop solution can be prepared in advance and stored at room temperature or used for the next experiment.
7. Depending on the protein tag, BSA should be used instead of milk powder if the tag interferes with biotinylated proteins.
8. Healthy plants are critical for the transfection efficiency and physiological responses. Plants should not show any stressed phenotype (like small dark green leaves, thick leaves, or anthocyanes) and they should still be in the vegetative phase (no visible inflorescence buds).
9. We select the well-expanded leaves (usually leaves 5–8) one by one. We cut the leaves on a clean bench on the top of a clean white paper to identify crushed material that we discard. The cut must be clean to avoid wounding cells. Each cut leaf must be quickly transferred to the enzyme solution. The use of a bacterial loop (0.5 mm-diameter) allows to catch the leaf strips by capillarity to avoid wounding. Cutting 20 leaves in 10 mL enzyme solution should give around 10^6 protoplasts.
10. The digestion must be performed in the dark to avoid light-induced oxidative stress. We use foil to cover the desiccator and then, the petri dish, during the incubation in the growth chamber. Although the digestion can be performed on the

bench, the use of the growth chamber guarantees constant temperature.

11. The protoplasts are fragile and must be handled gently. The use of round-bottomed tubes with a swing centrifuge allows to pellet the cells carefully. The break must be removed otherwise it will put back the protoplasts in suspension.
12. The final volume in each well is critical to avoid hypoxia of the cells. We select the plate according to the amount of cells: 6-well plate for 200 μL cells in 1 mL final volume, 12-well plate for 100 μL cells in 0.5 mL final, 24-well plate for 50 μL cells in 0.25 mL final, 6-cm diameter petri dish for 400 μL cells in 2 mL final, 10-cm diameter petri dish for 2 mL cells in 5 mL final.
13. The transfection can be performed in 2 mL or 15 mL round-bottomed centrifuge tube, depending on the required volume. The incubation time in PEG solution should not exceed 15 min, so we usually handle maximum ten tubes at the same time. If more transfections are planned, they must be performed successively.
14. Using similar protein amounts is important to be able to compare the kinase activity in different samples. Protein quantification with commercial Bradford solutions is always an option here. Indeed, when working with crude extracts, adjusting the cell density (in the case of protoplast transfection) or measuring fresh weight (in the case of leaf material) can replace the Bradford assay.
15. Use a fresh gel but make sure it is fully polymerized. To save time, the gels can also be poured the evening before you run the assay. Run the SDS-PAGE at low voltage (100–120 V) to obtain best results concerning the resolution between elicited and non-elicited CDPK isoforms.
16. Concentration of cold ATP can be adjusted to increase the signal or lower the background intensities. We tested a concentration range from 0 to 25 μM cold ATP and we obtained the best results with 0.1 nM. The gels must be covered with liquid. So, the volume of the reaction buffer should be adjusted depending on the container.
17. This procedure usually takes 3–4 h if you replace the stop solution every 20–30 min.
18. Start with 2 h at 80 °C with a vacuum pump.
19. Small molecular weight proteins may pass through the membrane while high molecular weight proteins are not transferred.

Acknowledgments

This work was supported by the Agence Nationale de la Recherche to M.B. (grant number ANR-15-CE20-0003-01) and by German Science Foundation DFG within SFB 973 and FOR 964 to T.R. The IPS2 benefits from the support of the LabEx Saclay Plant Sciences-SPS (grant number ANR-10-LABX-0040-SPS).

References

1. Suzuki K, Shinshi H (1995) Transient activation and tyrosine phosphorylation of a protein kinase in tobacco cells treated with a fungal elicitor. Plant Cell 7:639–647
2. Romeis T, Piedras P, Jones JD (2000) Resistance gene-dependent activation of a calcium-dependent protein kinase in the plant defense response. Plant Cell 12:803–816
3. Dubiella U, Seybold H, Durian G, Komander E, Lassig R, Witte C-P, Schulze WX, Romeis T (2013) Calcium-dependent protein kinase/NADPH oxidase activation circuit is required for rapid defense signal propagation. Proc Natl Acad Sci U S A 110:8744–8749
4. Romeis T, Ludwig AA, Martin R, Jones JD (2001) Calcium-dependent protein kinases play an essential role in a plant defence response. EMBO J 20:5556–5567
5. Ludwig AA, Saitoh H, Felix G, Freymark G, Miersch O, Wasternack C, Boller T, Jones JDG, Romeis T (2005) Ethylene-mediated cross-talk between calcium-dependent protein kinase and MAPK signaling controls stress responses in plants. Proc Natl Acad Sci U S A 102:10736–10741
6. Asai T, Tena G, Plotnikova J, Willmann MR, Chiu W-L, Gomez-Gomez L, Boller T, Ausubel FM, Sheen J (2002) MAP kinase signalling cascade in Arabidopsis innate immunity. Nature 415:977–983
7. Kovtun Y, Chiu WL, Tena G, Sheen J (2000) Functional analysis of oxidative stress-activated mitogen-activated protein kinase cascade in plants. Proc Natl Acad Sci U S A 97: 2940–2945
8. Boudsocq M, Willmann MR, McCormack M, Lee H, Shan L, He P, Bush J, Cheng S-H, Sheen J (2010) Differential innate immune signalling via Ca^{2+} sensor protein kinases. Nature 464:418–422
9. Gao X, Chen X, Lin W, Chen S, Lu D, Niu Y, Li L, Cheng C, McCormack M, Sheen J, Shan L, He P (2013) Bifurcation of Arabidopsis NLR immune signaling via Ca^{2+}-dependent protein kinases. PLoS Pathog 9:e1003127
10. Monaghan J, Matschi S, Romeis T, Zipfel C (2015) The calcium-dependent protein kinase CPK28 negatively regulates the BIK1-mediated PAMP-induced calcium burst. Plant Signal Behav 10:e1018497
11. Li L, Li M, Yu L, Zhou Z, Liang X, Liu Z, Cai G, Gao L, Zhang X, Wang Y, Chen S, Zhou J-M (2014) The FLS2-associated kinase BIK1 directly phosphorylates the NADPH oxidase RbohD to control plant immunity. Cell Host Microbe 15:329–338

Chapter 15

Chitin and Stress Induced Protein Kinase Activation

Chandra Kenchappa*, Raquel Azevedo da Silva*, Simon Bressendorff, Sabrina Stanimirovic, Jakob Olsen, Morten Petersen, and John Mundy

Abstract

The assays described here are pertinent to protein kinase studies in any plant. They include an immunoblot phosphorylation/activation assay and an in-gel activity assay for MAP kinases (MPKs) using the general protein kinase substrate myelin basic protein. They also include a novel in-gel peptide substrate assay for Snf1-related kinase family 2 members (SnRK2s). This kinase family-specific assay overcomes some limitations of in-gel assays and permits the identification of different types of kinase activities in total protein extracts.

Key words Acryloylated peptide substrates, In-gel assays, Myelin basic protein

1 Introduction

MAP kinases (MPKs) regulate responses to developmental and stress signals including pathogen infection [1–3]. They are activated by phosphorylation on conserved Thr and Tyr residues in a regulatory site (TEY) by MAP kinases kinases which are activated by MAP kinase kinase kinases. MPK phosphorylation is assayed by immunoblot with α-pTEpY antibodies. MPK cascades amplify stimuli to responses by phosphorylating their substrates, including transcription factors, on Ser/Thr followed by Pro (S/TP sites). MPK activities in total or purified protein extracts are routinely assayed in gels following separation by SDS-PAGE using Myelin Basic Protein (MBP) as substrate [4]. MBP contains four S/TP sites phosphorylated in vivo [5].

SnRK2s kinases are required for responses to abiotic stress and the osmoregulatory hormone abscisic acid (ABA) [6, 7]. In the ABA pathway, hormone binding to PYL receptors represses the activity of protein phosphatases 2C which otherwise dephosphorylate and

*These authors contributed equally to this work.

Libo Shan and Ping He (eds.), *Plant Pattern Recognition Receptors: Methods and Protocols*, Methods in Molecular Biology, vol. 1578, DOI 10.1007/978-1-4939-6859-6_15, © Springer Science+Business Media LLC 2017

inactivate SnRK2s. SnRK2 autophosphorylation on Ser residues [8, 9] then permits their phosphorylation of substrates including ABA-responsive transcription factors. Like other SnF1 related enzymes, SnRK2s preferentially phosphorylate substrates on Ser in the consensus ΦXRXXSXXXΦ (Φ = hydrophobic) [10, 11]. Peptides containing such sequences have been used to assay purified SnF1-related and SnRK2 activities in vitro [12, 13]. MBP is phosphorylated by SnRK2s [9] and may contain three SnRK2 phosphorylation sites.

Although in-gel kinase assays are useful, they have limitations. First, during gel casting substrates are copolymerized via their Lys residues to acryl gel groups to prevent substrates from being eluted during electrophoresis and subsequent processing steps. Second, proximal Lys-acryl linkages may mask substrate phosphorylation sites. Thus, protein substrates like MBP, with multiple phosphorylation sites and/or relatively distal Lys residues, are commonly used. Third, MBP is phosphorylated by different kinases which may confound their identification.

For example, immunoblotting with α-pTEpY antibodies identifies at least two MPKs of 40–42 kD that are phosphorylated in the moss *Physcomitrella patens* in response to treatment with the fungal elicitor chitin (Fig. 1a, b, lane 3). This immunoblot assay also shows that moss MPKs are not phosphorylated in response to osmotic stress (NaCl, lanes 4–8). Similarly, an in-gel kinase activity assay with MBP as substrate shows that one or more kinases are activated in response to chitin (Fig. 1c, lane 3). However, this activity assay also detects one or more kinases of ~39 kD that are activated in response to osmotic stress (500 mM NaCl, lanes 4–8). While these two assays indicate that the 40–42 kD bands are MPKs, the nature of the ~39 kD activities are unclear.

These assays illustrate the limitations of in-gel activity assays using MBP as substrate. These limitations have also precluded the use of shorter peptide substrates lacking Lys residues for gel linking, or peptides necessarily containing Lys proximal to phosphorylation sites. We have recently shown that this limitation may be circumvented by synthesizing substrate peptides lacking Lys residues but containing an acryloylated *N*-terminus for acryl gel linking (Fig. 2) [14, 15].

This is shown in Fig. 1d with an *N*-acryloylated peptide with the consensus SnRK2 phosphorylation site ALARAASAAALARRR [13]. It identifies SnRK2s activated in response to osmotic stress (NaCl or mannitol, lanes 3–8) which are not responsive to chitin (lane 3). This indicates that peptides containing phosphorylation sites for specific kinase types can be used to study their activation and to identify them. For example, multiple substrate proteins are phosphorylated on lysine-less peptides by different kinase types in response to various cues in *Arabidopsis* [16, 17]. The specificity of in-gel peptide substrate assays could confirm and extend their

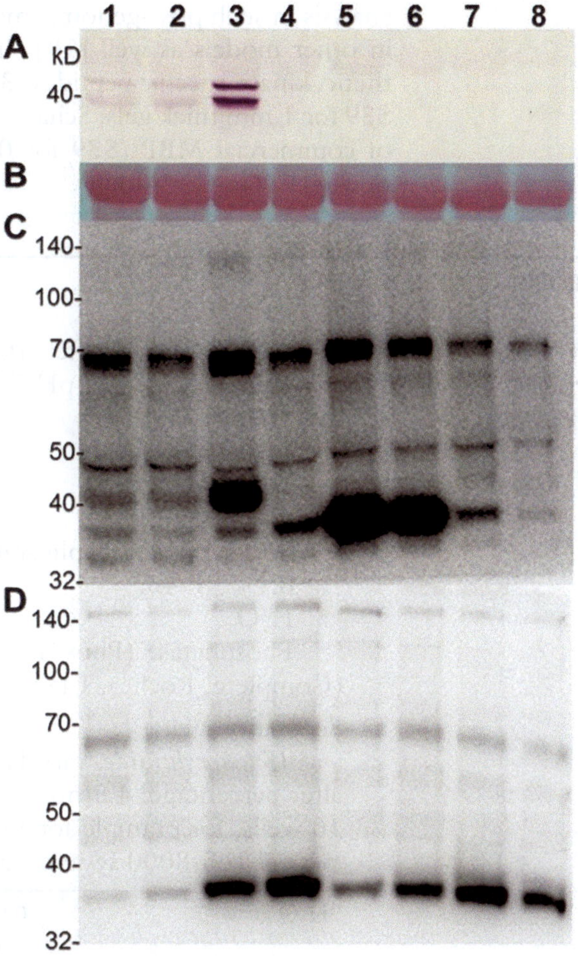

Fig. 1 Protein kinase assays. (**a**) Immunoblot with α-pTEpY antibodies; (**b**) Ponceau-stained gel as loading control; (**c**) In-gel kinase activity assay with MBP as substrate. **a–c** are same samples of wild type moss treated as follows in lanes: 1, 5 min H_2O control; 2, 10 min H_2O control; 3, 5 min 100 μg/mL chitin; lanes 4–8, 2, 5, 10, 20, 60 min 500 mM NaCl. (**d**) In-gel kinase activity assay with acryloylated peptide ALARAASAAALARRR containing the SnRK2 consensus phosphorylation site ΦXRXXSXXXΦ. Lanes are wild type moss treated with: 1, 5 min H_2O; 2, 5 min 100μg/mL chitin; 3–5, 5, 10, and 20 min 500 mM NaCl; 6–8, 5, 10, and 20 min 800 mM mannitol. Protein molecular weight markers in kDa at *left*

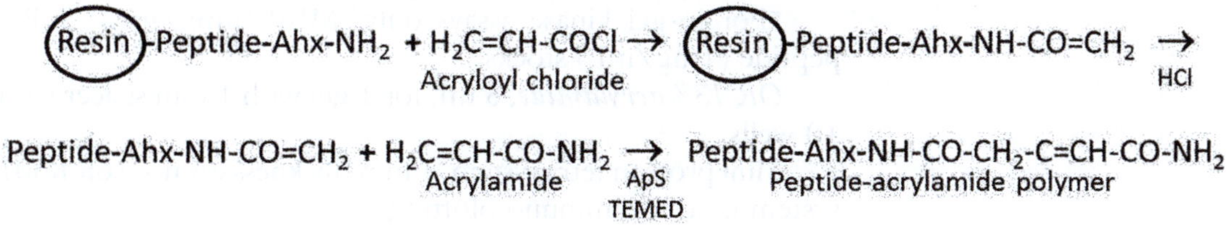

Fig. 2 Peptide synthesized on standard resin with *N*-terminal 6-aminohexanoic acid (Ahx) is acryloylated with acryloyl chloride. Acryloylated peptide released from resin by HCl hydrolysis can be copolymerized as usual by ammonium persulfate (ApS) and tetramethylethylenediamine (TEMED) with acrylamide (and/or bisacrylamide, not shown)

analysis in such post-genomic models and would aid kinase studies in other models as well [18]. In addition, the cost of such synthetic, kinase-specific peptides (30 mer = $31 for 0.75 mm thick or $39 for 1 mm thick gels; Schaefer N, Copenhagen) is less than that of commercial MBP ($39 for 0.75 mm thick or $49 for 1 mm thick gels; Sigma-Aldrich).

2 Materials

2.1 MPK Phosphorylation by Immunoblotting

1. Protein extraction buffer, 10 mL.
   50 mM Tris–HCl pH 7.5.
   10 mM $MgCl_2$.
   15 mM EGTA.
   100 mM NaCl.
   2 mM DTT.
   30 mM β-glycerophosphate.
   0.1% NP-40.
   H_2O.
   Phosphatase (PhosSTOP, Roche) and protease inhibitors (Complete, Roche), 1 tablet each for 10 mL buffer.
2. Separation gel.
   15% acrylamide, vary H_2O and 30% acryl mix volumes to alter percentage. 4.5 mL for 1 gel with 0.75 mm spacer with 10 wells, for example for Bio-Rad Mini Protean Tetra system (Cat. # 165-8000 system), gel size 10 × 8 cm.

H_2O	0.90 mL (1.15 mL without substrate for immunoblot).
30% acrylamide–bis mix	2.25 mL (30/0.8 w/w acrylamide–bis, Sigma Cat. # A3699).
1.5 M Tris pH 8.8	1.00 mL
Substrate*	250 μL
10% SDS	45 μL
10% APS	45 μL
TEMED	5 μL

   *For in-gel kinase assays only: MBP (5mg/mL), SnRK peptide (1mg/mL) stocks.
   OR 13% acrylamide, 5 mL for 1 gel with 1 mm spacer with 10 wells.
   Mini protean tetra system, 1 mm thickness cat no. 165-8001 system used for immuno blotting.

H_2O	1.41 mL (1.72 mL without substrate for immunoblot)
30% acrylamide–bis mix	2.17 mL

1.5 M Tris pH 8.8	1.0 mL
Substrate *	312 μL
10% SDS	50 μL
10% APS	50 μL
TEMED	8 μL.

*MBP (5 mg/mL), SnRK peptide (1 mg/mL) stocks.

- After pouring the acrylamide separation solution between the plates, CAREFULLY overlay with water-saturated isopropanol.
- Rinse with water after the gel has polymerized (20 min).
- Pour the stacking gel (below) and insert the comb without air bubbles! Allow the gel to polymerize (minimum10 min).

3. Stacking gel (5% acrylamide) 6 mL for 3 gels 2 mL/gel.

H_2O	3.5 mL
30% acrylamide–bis mix	1.0 mL
1.0 M Tris pH 6.8	1.5 mL
10% SDS	60 μL
10% APS	60 μL
TEMED	10 μL

4. 6× SDS loading buffer.
   60% glycerol.
   12% SDS.
   375 mM Tris pH 6.8.
   60 mM DTT.
   0.5% bromophenol blue.
5. Broad range marker—Spectra multicolour broad range ladder, Cat. # 26634, Thermofisher, SE).
6. Electrophoresis buffer.
   25 mM Tris.
   250 mM glycine.
   0.1% SDS.
7. Transfer buffer.
   48 mM Tris.
   39 mM glycine.
   0.037% SDS.
   20% methanol.
8. Immunoblot 2ashing buffer (TBST)
   1× Tris-buffered saline (TBS, 50 mM Tris, 150 mM NaCl).
   0.1% Tween.
9. Immunoblot blocking buffer.

1× TBST.
5% dry milk.
0.1% Tween.

10. Immunoblot alkaline phosphatase buffer (AP)
100 mM NaCl.
5 mM MgCl2.
100 mM Tris–HCl (pH 9.5, important).

2.2 In-Gel Kinase Activity Assay with MBP or Acryloylated Peptide Substrates

1. PAGE and separation gel—Essentially the same as for immunoblotting (Subheading 2.1 above) EXCEPT for the addition of kinase substrate to the separation gel.
2. Kinase activity assay exchange buffer.
25 mM Tris–HCl pH 7.5.
0.5 mM DTT.
0.1 mM Na3VO4.
5 mM NaF.
0.5 mg/mL BSA.
0.1% Triton X−100.
pH to 7.5
3. Kinase activity assay denaturing buffer.
50 mM Tris–HCl pH-7.5.
6 M guanidine hydrochloride (Gua-HCl).
20 mM DTT.
4. Kinase activity assay renaturing buffer.
25 mM Tris–HCl pH 7.5.
1 mM DTT.
0.1 mM Na3VO4.
5 mM NaF.
5. Kinase activity assay reaction buffer.
25 mM Tris–HCl pH 7.5.
2 mM EGTA.
12 mM MgCl2.
1 mM DTT.
0.1 mM Na3VO4.
After washing with this buffer, add to 20 mL volume for kinase reaction add cold ATP to 100 nM and 3.7 MBq (100 μCi) γ-ATP32.
6. Kinase activity assay termination solution.
5% (w/v) trichloroacetic acid (TCA).
1% (w/v) sodium pyrophosphate (Na4O7P2).
7. PhosphorImaging—we use the following but other systems also work: MultiSensitive Phosphor Screens, Medium, Size, 12.5 × 25.2 cm Product No. 7001723, PerkinElmer Cyclone Phosphor Imager: http://macro.lsu.edu/psrgroup/PersonalPages/Derek%20Dorman/SAXS%20Webpage/TurboCad%20Drawings/PSI/Phosphorimager%20manual.pdf.

3 Methods

3.1 MPK Phosphorylation by Immunoblotting

Phosphorylated MPKs are detected by immunoblot using α-pTEpY antibody following SDS-PAGE.

3.1.1 Sample Treatment and Extraction

Samples—For moss, three 14-day-old colonies on BCDAT medium of a given genotype/sample.

For *Arabidopsis*, 3–6 seedlings grown for 10–14 days in 6-well plates with 4 mL liquid MS.

1. Label Eppendorf tubes according to samples.
2. For moss: open the petri dish with three colonies and spray them evenly with 100 μg/mL chitin in H_2O.

   For *Arabidopsis*: Add chitin to a final concentration of 100 μg/mL.
3. For moss: After 5 min., carefully and quickly pick the three colonies, gently blot them on tissue paper to remove excess liquid, place them in a labelled Eppendorff tube and freeze in liquid nitrogen.

   For *Arabidopsis*: After 10 min, carefully and quickly pick the 4–5 seedlings, gently blot them on tissue paper to remove excess liquid, place them in a labelled Eppendorff tube and freeze in liquid nitrogen.
4. Add 250 μL of Extraction Buffer to each sample and grind with a pestle while the sample thaws.
5. Centrifuge at 13,500 × *g* for 10 min at 4–8 °C.
6. Transfer 200 μL of the supernatant to a new tube without disturbing the pellet (*see* **Note 1**).
7. Add 40 μL of 6× SDS loading buffer. Mix well and incubate for 5 min in heating block at 95 °C.

3.1.2 SDS-PAGE and Blotting

The samples prepared above generally contain 2–3 μg/μL protein. ~30 μg of protein per lane may give good results in gels of 1 mm (for western blotting) and 0.75 mm (for in-gel kinase assay) thickness.

1. Carefully and rapidly load equal amounts of protein in max. 20 μL of each sample on the polyacrylamide gel. Include a well with 10 μL of protein marker. Freeze store the remaining sample at −80 °C.
2. Run samples at 90 V for 30 min to allow proteins and dye to enter the stacking gel.
3. Run the gel at 130 V. As MPKs are 40–50 kDa, stop electrophoresis when ~32 kDa marker runs out. This takes approx. 4 h at RT.

4. Carefully dismount the gel, then incubate it in Transfer Buffer for 2 min.
5. Immobilize the proteins on a nitrocellulose membrane (Amersham Hybond Nitrocellulose blotting membrane, GE Healthcare, Cat # RPN303D) by wet transfer overnight or 2 h at 70 V (see http://www.bio-rad.com/webroot/web/pdf/lsr/literature/4006190b.pdf).

3.1.3 Blocking and Incubation in 1° Antibody

1. Stain membrane for total protein loading by adding Ponceau stain (Ponceau S: 0.1% (x/v) Ponceau S in 1% (v/v) acetic acid) for 2 min.
2. Remove the Ponceau stain by washing 3× in TBST (approximately 5 min each wash with gentle shaking)
3. Incubate the membrane in 5% milk TBST for 30 min.
4. Wash 3× in TBST (approximately 5 min each wash with gentle shaking).
5. Add 10 mL 1° antibody (α-pTEpY—anti-p42/p44-erk) 1/2000 in 5% BSA-TBST). Incubate for 2 h at RT or at 4 °C overnight.

3.1.4 2° Antibody and Developing

1. Collect 1° antibody and wash blot 3× in TBST (approximately 5 min per wash with gentle shaking).
2. Add 10 mL 2° antibody (anti-rabbit AP 1/5000 in 5% BSA-TBST). Incubate for 1 h at RT.
3. Collect 2° antibody and wash blot 3× (approximately 5 min per wash with gentle shaking).
4. Wash once with AP buffer for 5 min, then develop blot by adding 10 mL BCIP/NBT substrate (Cat. # 11697471001, Roche: dissolve 1 tablet in 10 mL H_20). Dark bands should appear after 2–10 min. Stop reaction by washing membrane 3× in water.

3.2 In-Gel Kinase Activity Assay with MBP or Acryloylated Peptide Substrates

The protocol is slightly modified from an earlier one [19]. Sample treatments and protein extractions are the same as for ASSAY 1—MPK phosphorylation by immunoblotting. The polyacrylamide gels are cast with the substrate in 0.75 mm gels and run as in Subheading 3.1. When SDS-PAGE is finished:

1. Exchange the SDS buffer by rinsing gel 4–6× in 50 mL Washing Buffer, 30 min each rinse at RT with gentle shaking.
2. Fully denature proteins by rinsing the gel 2× in 50 mL chilled Denaturing Buffer, 30 min each with shaking at RT.
3. Renature the proteins by rinsing the gel 4× in 50 mL at 4 °C to remove Gua-HCl in chilled Renaturing Buffer, with shaking at RT, followed by 1 longer rinse in Renaturing Buffer for 3–12 h at 4 °C without shaking.

4. Wash gel 2× at RT in 50 mL Reaction Buffer, 30 min each with shaking.
5. Incubate gel for 1–2 h at RT in 20 mL Reaction Buffer containing 3.7 MBq (100 μCi) γ-ATP^{32} with shaking.
6. Carefully discard the radioactive waste.
7. Quickly stop the reaction by adding 50 mL Termination Solution with shaking.
8. Rinse 6× in Termination Solution for a total of 3–6 h to eliminate unincorporated γ-ATP^{32}.
9. To avoid shrinking/brittleness, treat gels with 10% glycerol/5% ethanol/7% methanol for 30 min prior to drying at 80 °C under vacuum for 1 h.
10. Expose gels to Phosphor image screen overnight.

4 Notes

1. *Optional determination of sample protein concentration for SDS-PAGE:
   (a) Add 795 μL water to an eppendorff tube.
   (b) Add 5 μL protein extract. Mix. Remember a control with 5 μL Extraction Buffer.
   (c) Add 200 μL commercial Bradford Reagent. Mix.
   (d) Incubate for 10 min and measure optical density at 595 nm (OD595).
   (e) Calculate [protein] using: A595 = 0.0556 c + 0.0245 where c = μg protein in 1 mL Bradford reaction.

Acknowledgments

This work was supported by grants from the Danish Basic Research Foundation (Center for Comparative Genomics) and the Danish Research Council for Nature and the Universe (Comparative & functional genomics of plant innate immunity, 1323-00267A) to JM.

References

1. Fiil BK, Petersen K, Petersen M, Mundy J (2009) Gene regulation by MAP kinase cascades. Curr Opin Plant Biol 12:615–621
2. Rodriguez M, Petersen M, Mundy J (2010) Plant MAP kinase cascades. Annu Rev Plant Biol 61:621–649
3. Rasmussen M, Roux M, Petersen M, Mundy J (2012) Map kinase cascades in plant innate immunity. Front Plant Sci 3(169):1–6
4. Zhang S, Klessig DF (1997) Salicylic acid activates a 48-kD MAP kinase in tobacco. Plant Cell 9:809–824

5. Martenson RE, Law MJ, Deibler GE (1983) Identification of multiple in vivo phosphorylation site in rabbit myelin basic protein. J Biol Chem 258:930–937
6. Weiner JJ, Peterson FC, Volkman BF, Cutler SR (2010) Structural and functional insights into core ABA signaling. Curr Opin Plant Biol 13:495–502
7. Fujii H, Zhu JK (2012) Osmotic stress signaling via protein kinases. Cell Mol Life Sci 69:3165–3173
8. Belin C, de Franco PO, Bourbousse C, Chaignepain S, Schmitter JM, Vavasseur A, Giraudat J, Barbier-Brygoo H, Thomine S (2006) Identification of features regulating OST1 kinase activity and OST1 function in guard cells. Plant Physiol 141:1316–1327
9. Boudsocq M, Droillard M-J, Barbier-Brygoo H, Laurière C (2007) Different phosphorylation mechanisms are involved in the activation of sucrose non-fermenting 1 related protein kinases 2 by osmotic stresses and abscisic acid. Plant Mol Biol 63:491–503
10. Huang JZ, Huber SC (2001) Phosphorylation of synthetic peptides by a CDPK and plant SNF1-related protein kinase. Influence of proline and basic amino acid residues at selected positions. Plant Cell Physiol 42:1079–1087
11. Sirichandra C, Davanture M, Turk BE, Zivy M, Valot B, Leung J, Merlot S (2010) The *Arabidopsis* ABA-activated kinase OST1 phosphorylates the bZIP transcription factor ABF3 and creates a 14-3-3 binding site involved in its turnover. PLoS One 5:e13935
12. Dale S, Wilson WA, Edelman AM, Hardie DG (1995) Similar substrate recognition motifs for mammalian AMP-activated protein kinase, higher plant HMG-CoA reductase kinase-A, yeast SNF1, and mammalian calmodulin-dependent protein kinase I. FEBS Lett 361: 191–195
13. Halfter U, Ishitani M, Zhu JK (2000) The *Arabidopsis* SOS2 protein kinase physically interacts with and is activated by the calcium-binding protein SOS3. Proc Natl Acad Sci USA 97:3735–3740
14. O'Brien-Simpson NM, Ede NJ, Brown LE, Swan J, Jackson DC (1997) Polymerization of unprotected synthetic peptides: a view toward synthetic peptide vaccines. J Am Chem Soc 119:1183–1188
15. Bressendorff S, Azevedo R, Kenchappa CS, Ponce de León I, Olsen JV, Rasmussen MW, Erbs G, Newman MA, Petersen M, Mundy J (2016) An innate immunity pathway in the moss Physcomitrella patens. Plant Cell 28:1328–1342
16. Yang Z, Guo G, Zhang M, Liu CY, Hu Q, Lam H, Cheng H, Xue Y, Li J, Li N (2013) Stable isotope metabolic labeling-based quantitative phosphoproteomic analysis of Arabidopsis mutants reveals ethylene-regulated time-dependent phosphoproteins and putative substrates of constitutive triple response 1 kinase. Mol Cell Proteomics 12:3559–3582
17. Roitinger E, Hofer M, Köcher T, Pichler P, Novatchkova M, Yang J, Schlögelhofer P, Mechtler K (2015) Quantitative phosphoproteomics of the ataxia telangiectasia-mutated (ATM) and ataxia telangiectasia-mutated and rad3-related (ATR) dependent DNA damage response in Arabidopsis thaliana. Mol Cell Proteomics 14:556–571
18. Jiménez C, Berl T, Rivard CJ, Edelstein CL, Capasso JM (2004) Phosphorylation of MAP kinase-like proteins mediate the response of the halotolerant alga *Dunaliella viridis* to hypertonic shock. Biochim Biophys Acta 1644:61–69
19. Romeis T, Piedras P, Zhang S, Klessig DF, Hirt H, Jones DG (1999) Rapid Avr 9- and Cf-9–dependent activation of MAP kinases in tobacco cell cultures and leaves: convergence of resistance gene, elicitor, wound, and salicylate responses. Plant Cell 11:273–288

Chapter 16

Measuring Callose Deposition, an Indicator of Cell Wall Reinforcement, During Bacterial Infection in Arabidopsis

Lin Jin and David M. Mackey

Abstract

The plant cell wall responds dynamically during interaction with various pathogens. Upon recognition of "nonself" components, plant cells deploy a variety of immune responses including cell wall fortification. Callose, a β-(1, 3)-D-glucan polymer, is a component of the material deposited at the site of infection between the plasma membrane and the preexisting cell wall that is hypothesized to serve as a physical barrier and platform for directed antimicrobial compound deposition. The defense-associated function of callose deposition is supported by its induction during pathogen-associated molecular patterns (PAMP)-triggered immunity (PTI) and its inhibition by defense suppressing virulence effectors. Thus, callose deposition is a commonly monitored read-out in plant defense. This protocol describes the use of aniline blue staining and fluorescent microscopy to measure callose deposition in bacteria-infected or elicitor-challenged Arabidopsis leaf tissues.

Key words Callose, PTI, Aniline blue, Arabidopsis, *Pseudomonas syringae*, Flg22

1 Introduction

Callose is a high molecular weight, β-(1, 3)-D-glucan polymer that is normally found in a number of plant tissues/structures, including sieve plates, pollen tubes, microsporocytes, cell plates, and plasmodesmata [1]. Callose deposition can also be triggered upon pathogen attack. It was first characterized as a major constituent of the papillae which are cell wall fortifications induced at the site of fungal infection. Additional components of papillae include phenolics, reactive oxygen species, peroxidases, cell wall structural proteins, and other cell wall polymers [2]. It is hypothesized that the formation of papillae at sites of infection restricts hyphae invasion by fortifying the cell wall and facilitating the directed deposition of antimicrobial compounds. A recent nanoscale study revealed the 3D structure of callose, including interaction between callose and the preexisting cellulosic cell wall. Callose fibrils migrate to and permeate the cell wall nanopore structure at the site of infection,

Libo Shan and Ping He (eds.), *Plant Pattern Recognition Receptors:Methods and Protocols*, Methods in Molecular Biology, vol. 1578,
DOI 10.1007/978-1-4939-6859-6_16,

supporting a hypothesis that callose may restrict the ingression of pathogen-secreted cell wall-degrading enzymes [3]. However, the role of callose as an effective physical barrier against pathogen invasion remains a topic of open discussion. For example, study of the Arabidopsis *PMR4* (*Powdery Mildew Resistance 4*) gene, which encodes the stress-induced callose synthase, yielded contrasting results in which a *PMR4*-overexpressing Arabidopsis line completely restricted fungal penetration [4], whereas mutation of *PMR4* resulted in higher resistance to adapted fungal pathogens that was linked to hyper-activation of salicylic acid-signaling [5, 6]. Therefore, functional study of callose in plant–microbe interaction remains an active area of research.

Cell wall reinforcements, such as callose deposition, are induced during plant-microbe interactions and can be readily induced by administration of PAMPs such as chitosan (fungal cell wall component) and flg22 (bacterial flagellin epitope) [7]. As such, callose deposition is widely used as a measure of cell wall defense during plant PTI (PAMP-triggered immunity). Compared to other commonly used PTI read-outs such as reactive oxygen species (ROS) production or mitogen-associated protein kinase (MAPK) phosphorylation, an advantage of callose measurement is that it is a relatively late defense output (occurs hours after infection). Therefore, it is especially useful to assess defense suppressing function of virulence effectors secreted by pathogens via their natural delivery mechanism (e.g., the type-III secretion system, T3SS, of gram negative bacteria) [8–12].

Aniline blue is commonly used to stain callose in plant tissues [13]. Commercial aniline blue is a heterogeneous blend of several fluorophores. A few characterized fluorophores, including the chemically defined sirofluor, when bound to callose, are excited by UV light and emit wavelengths within the blue light range. The fluorescent intensity peaks under alkaline conditions [14–16]. The number of stained callose deposits is commonly used as a quantitative measurement of plant basal immunity. Other characteristics of the stained callose such as size of the deposits [17], and pixel intensity [7] have also been followed to reflect resistance level. This protocol outlines a procedure for aniline blue staining of callose in Arabidopsis leaf tissues following infection with the bacterial pathogen *Pseudomonas syringae* or treatment with a PAMP elicitor (flg22), and details procedures of callose visualization using fluorescent microscopy as well as image processing steps to quantify the number of callose deposits. This protocol can be used to measure callose under other types of defense elicitor challenges, and is likely adaptable to other plant species as well.

2 Materials

2.1 Bacterial Infection and Elicitor Treatment of Arabidopsis Leaves

1. Five- to six-week-old Arabidopsis plants grown on peat moss-based potting mix (e.g., Metro Mix 360, Sun Gro Horticulture, Vancouver, Canada), at 22 °C/16 °C under an 8 h light (115 $\mu molm^{-2}s^{-1}$)/16 h dark cycle.
2. King's B (KB) agar plate (per 1 L): 20 g peptone, 15 mL glycerol (100%), 1.5 g K_2HPO_4, 1.5 g $MgSO_4 \cdot 7H_2O$, 18 g agar, containing appropriate antibiotics.
3. 10 mM $MgCl_2$, filter-sterilized.
4. Purified flg22 (custom synthesized by EZBiolab, Indiana, USA), 22 amino acid peptide (QRLSTGSRINSAKDDAAGLQIA) derived from the conserved amino-terminus of flagellin from *Pseudomonas aeruginosa*.
5. Sterile inoculation loop or toothpick.
6. One-milliliter needleless syringe.
7. Spectrophotometer set to 600 nm and disposable plastic cuvettes.

2.2 Callose Staining with Aniline Blue

1. Lactophenol solution (per 600 mL): 50 mL distilled water, 50 mL glycerol (100%), 50 mL lactic acid (85%), 50 mL phenol (pH 6.4–6.6, water saturated), 400 mL ethanol (95%) (*see* **Note 1**).
2. 50% ethanol in distilled water.
3. Staining solution: 0.01% (w/v) aniline blue (Methyl blue, Sigma-Aldrich, Missouri, USA. Cat. # M-6900) in 150 mM K_2HPO_4 (pH 9.5).
4. Six-well cell culture plates (approximately 9.6 cm^2 per well).
5. Heating incubator set to 65 °C.
6. Vacuum pump and liquid trap for liquid aspiration.
7. Platform shaker.

2.3 Callose Visualization by Fluorescent Microscopy

1. 50% glycerol (v/v) in distilled water.
2. Microscope slide glasses (75 × 25 × 1.0 mm) and cover glasses (22 × 50 mm).
3. Nail polish.
4. Epifluorescent microscope equipped with UV lamp, UV filters (370 nm excitation, 420–530 nm emission), and mounted camera.

2.4 Callose Quantification

Image J software: http://rsbweb.nih.gov/ij/.

3 Methods

3.1 Bacteria and Elicitor Preparation

1. Revive frozen glycerol stocks of *P. syringae* strains by streaking the bacteria onto a fresh King's B agar plate containing appropriate antibiotics and grow for 24 h at 28 °C.
2. Scrape up the overnight bacterial culture with a sterile inoculation loop or toothpick and resuspend in 5 mL of 10 mM $MgCl_2$.
3. Measure the optical density at 600 nm (OD_{600}) of the bacterial cell suspension using a spectrometer. Then dilute to a final OD_{600} of 0.2 (approximately 1×10^8 colony-forming units (CFU)/mL). Approximately 100 μL of bacterial inoculum is sufficient for infiltrating a single leaf.
4. For elicitors, such as flg22, dissolve the peptide in sterilized distilled water at a final concentration of 100 μM (*see* **Note 2**). Approximately 100 μL of elicitor solution is sufficient for infiltrating a single leaf.

3.2 Leaf Infiltration

1. Use 5–6-week-old Arabidopsis plants grown vegetatively under short-day conditions (8 h day length). Mark leaves of approximately the same size (no more than 2 cm in length) with an indelible ink pen. Six leaves from two plants are used for each treatment (three leaves from each plant, *see* **Note 3**).
2. Mist the plants with water prior to infiltration while preparing bacterial inoculum. High moisture will open stomata and facilitate infiltration.
3. Infiltrate the abaxial (dorsal) side of the marked leaves with a 1-mL needleless syringe. The syringe is placed away from the central vein (*see* **Note 4**) opposite to gentle pressure from a finger on the adaxial (ventral) leaf surface, and the bacteria inoculum or elicitor solution is pressure-infiltrated into the intercellular space of the leaf. Make sure the whole leaf is infiltrated, this is easily identifiable by observing water-soaking of the infiltrated area.
4. Tap off residual inoculum or solution from the leaf surface with absorbent paper, allow the infiltrated leaves to dry for approximately 30 min, and return the plants to normal growth environment.

3.3 Clearing Leaves of Chlorophyll

For this section, all procedures should be carried out in a chemical fume hood and with appropriate personal protective equipment when handling phenol-containing solutions.

1. Infiltrated leaves are collected at 14–16 h after infiltration and placed in six-well plates. Leaves from different treatments are placed into different wells and identified by marking the bottom of the well. A single well can accommodate up to six leaves.

2. Add 5 mL lactophenol solution into each well. Gently shake the plates on a platform shaker for 15 min. The marker ink will be washed off the leaves by lactophenol, and the syringe infiltrated spots may turn yellow to clear after this step while the rest of the leaf stays green.
3. Remove lactophenol solution by aspiration. Care should be taken to avoid touching leaf tissues (*see* **Note 5**).
4. Arrange the leaves to be separated from each other using forceps, use the wall of the wells if necessary (*see* **Note 6**).
5. Incubate the leaf tissue in a heating incubator at 65 °C for 60 min with lids removed from the six-well plates. Leaves will appear brownish yellow after the incubation, and should remain moist.
6. Add 5 mL lactophenol solution into each well, and gently shake on a platform shaker at room temperature. Change lactophenol solution after 2–3 h, then shake overnight.

3.4 Callose Staining and Slide Preparation

1. Remove lactophenol solution by aspiration.
2. Add 5 mL 50% ethanol solution to each well, and gently shake the plates for 5 min. Aspirate out the ethanol solution. At this step, leaves should turn transparent and appear yellowish white.
3. Add 5 mL distilled water into each well, and gently shake for 10 min (*see* **Note 7**).
4. Remove water by aspiration, and add 5 mL of freshly made staining solution to each well. Samples are gently shaken for 1 h (*see* **Note 8**).
5. Remove staining solution by aspiration (*see* **Note 9**).
6. Add mounting solution (50% glycerol) into each well. The samples are then ready to be mounted.
7. Leaves are carefully transferred onto the slide glass (*see* **Note 10**). Seal the cover glass with nail polish. Wipe off residual solution from the slides. The samples can be viewed right after mounting or can be stored in dark for up to 2 weeks.

3.5 Callose Visualization by Fluorescent Microscopy

The slides are viewed with an epifluorescent microscope equipped with UV lamp, UV filters (370 nm excitation, 420–530 nm emission, *see* **Note 11**), and mounted camera. Microscope and image acquisition settings will differ by equipment make and model and the software used. Outlined here are general settings using a Nikon® eclipse 80i microscope, a Mercury arc lamp, a Nikon® DS-Qi1Mc camera, and an NIS-Element Basic Research software (Nikon Instruments Inc., Tokyo, Japan).

1. Select a 4× or 10× objective lens and a 10× eyepiece to visualize samples.

2. Secure the specimen on the microscope stage. Focus on the sample under bright-field, then switch to UV filter (340–380 nm excitation filter, 400 nm long pass dichromatic mirror, 435–485 nm bandpass barrier filter) and turn on fluorescence to visualize the callose. No additional filters (such as chlorophyll filter) are needed for callose observation.
3. The general image acquisition set-up is as follows: format: 1280 × 1024 no binning; exposure: 300–400 ms; analog gain: 2–3×; Filter: DAPI; Objective lens: 4×. Settings should be adjusted depending on the microscope and camera and should allow sufficient visualization while avoiding oversaturated signals. The same setting is used across all samples.
4. Images can be saved as TIFF (8-bit) or JPEG file formats (*see* **Note 12**).
5. Scale bar is stamped in the corner of the picture for area measurement purpose (*see* **Note 13**).
6. Images are acquired from 3–5 random areas on a single leaf (*see* **Note 14**).
7. Examples of the microscopic images are shown in Fig. 1.

3.6 Image Processing and Callose Quantification

A number of software can be used for particle quantification from the acquired images, including NIS-Element software (Nikon Instruments Inc., Tokyo, Japan) and Photoshop (Adobe Systems Inc, California, USA). Batch processing software such as CalloseMeasurer [18], and AutoSPOTs [19] are also available specifically for callose quantification purposes. Outlined here is a procedure for use of the freely available Image J software (http://rsbweb.nih.gov/ij/) (Fig. 2).

1. Determine image area. Open the image in Image J software. Draw a straight line over the scale bar. Under "Analysis", select "Set Scale". Enter the numerical length of the scale bar in the "Known distance" box and the type of unit in "Unit of length" box. Click "OK". Remove the line drawn over the scale bar by clicking at the blank area of the image. Under "Analysis", select "Measure". The number under "Area" column is the area of the entire image in units entered into the scale bar, e.g., μm^2 (*see* **Note 15**). The image area needs to be determined only once as long as all the images are acquired under the same magnification.
2. Image processing to remove background by adjusting brightness and contrast. Aniline blue will also enhance fluorescence coming from the non-elicited cell walls and the vascular system. To remove these less intense sources of background, go to "Image"—"Adjust"—"Brightness/Contrast" and dial the "Minimum" bar gradually to the right until the veins just

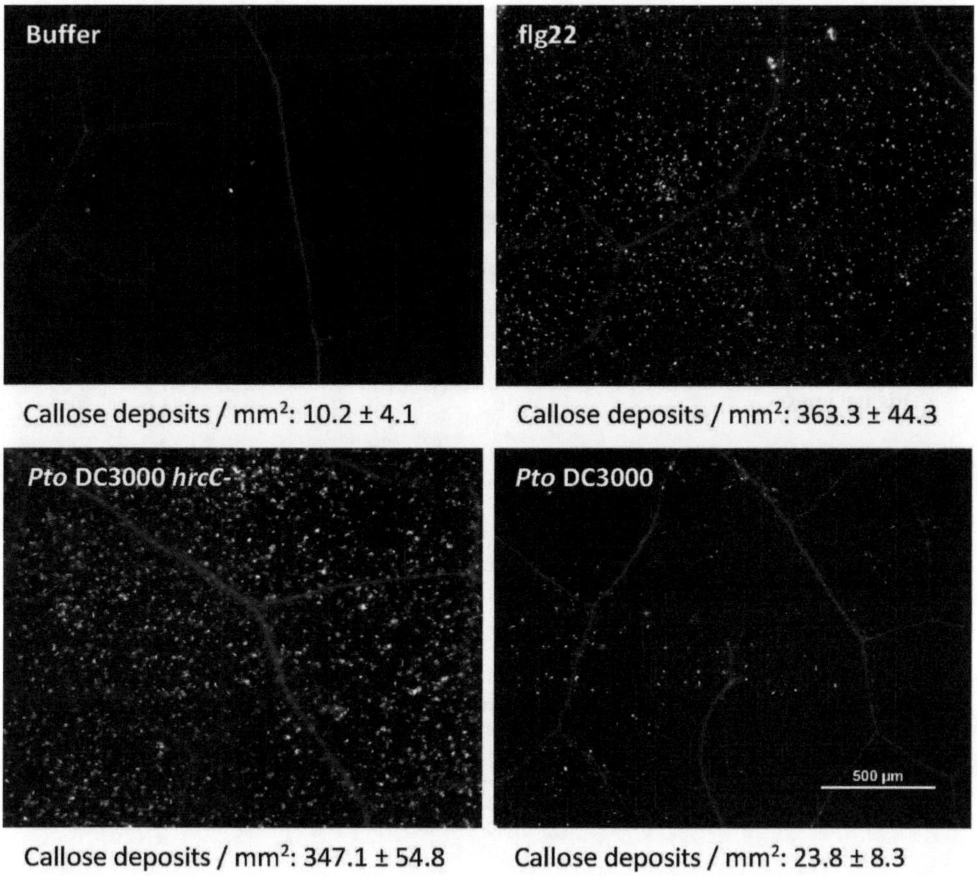

Fig. 1 Callose deposition is triggered by PAMPs and suppressed by virulence effector proteins from Pto DC3000 in leaves of Arabidopsis. Five-week-old Arabidopsis leaves were infiltrated with buffer (10 mM $MgCl_2$), 100 μM flg22, or suspensions of indicated bacterial strains at $OD_{600} = 0.2$ (1×10^8 CFU/mL). Callose deposition was assessed at 16 h after infiltration. Callose deposition is elicited by purified flg22. Effector proteins from the virulent strain Pseudomonas syringae pv. tomato strain DC3000 (Pto DC3000) suppresses callose deposition that is induced by the T3SS-deficient mutant Pto DC3000 hrcC- [23]. Numbers shown are mean ± SEM of callose per mm^2 from five images

become invisible. Click "Apply". This step will remove background signal coming from the veins, but the staining of defense-associated callose deposits should be much brighter and still be scored in subsequent steps.

3. Image processing to convert to a binary image. Go to "Image"—"Adjust"—"Threshold". Choose "Default" and "B&W" from the drop-down menus. Check "Dark background" box. If background removal was insufficient from the last step, you should be able to see veins very easily once the threshold box is opened. If this occurs, close the threshold box, and repeat **step 2** until veins are invisible.

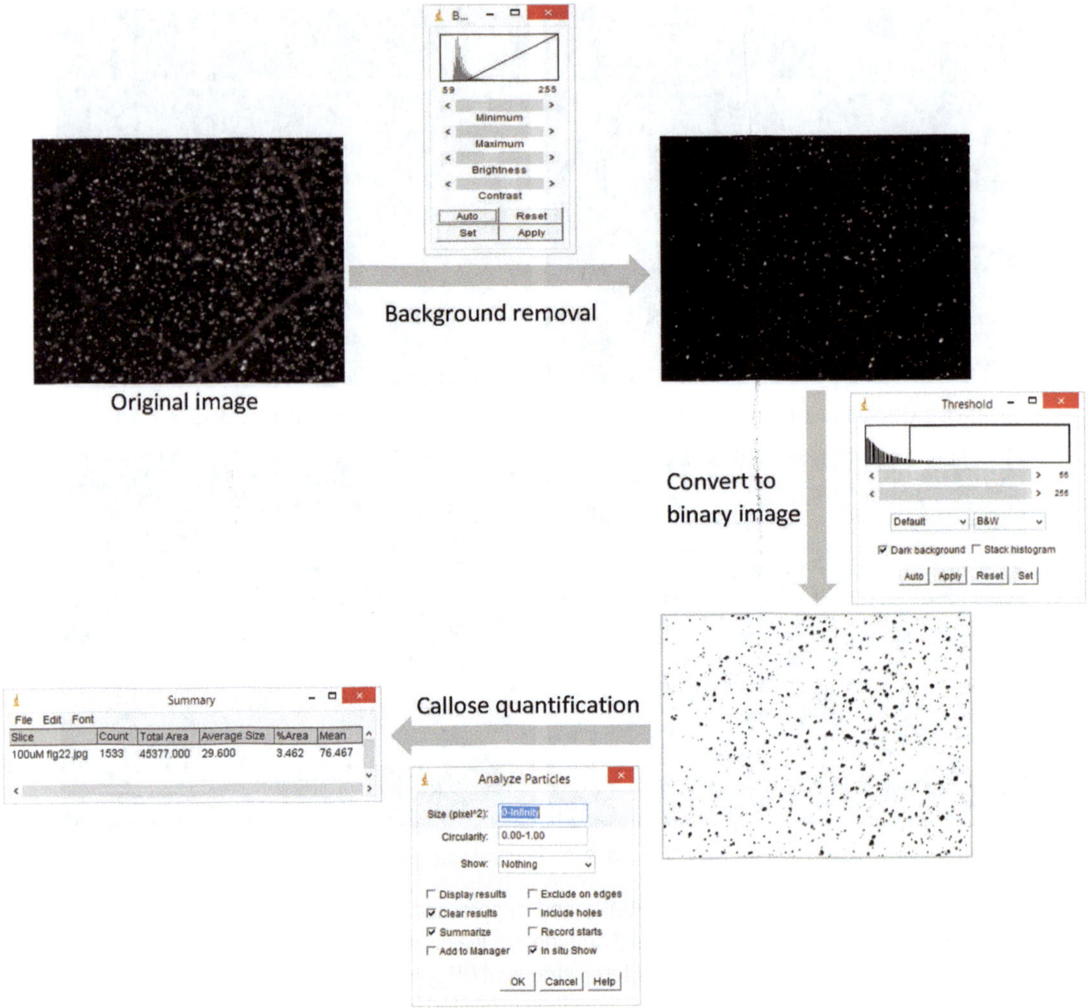

Fig. 2 General workflow of image processing and callose quantification. Shown is a representative image of Arabidopsis leaf infiltrated with 100 μM flg22. The image goes through background removal, binary conversion, and partical analysis processes. Shown are images following each processing procedure and the final partical count summary. Images shown next to arrows are the Image J software interface for each step

4. Callose quantification. Go to "Analyze"—"Analyze particles" and choose "Size: 0-infinity", "Circularity: 0.00–1.00", "Show: Outline" (optional, this will outline the particles used for the analysis and can be used to verify that background has been set low enough to avoid scoring the veins), check "summarize", and click "OK". The number shown under "Count" column in the "Summary" box is the number of particles (callose deposits) present in this particular image (*see* **Note 16**).
5. Repeat **steps 2–4** for every image, record particle numbers, and calculate the average and standard deviation/standard error for each treatment. Data is usually presented as number of callose deposits/mm^2.

4 Notes

1. Mix the lactophenol solution thoroughly before use to allow proper mixing of glycerol with the rest of the components. Store in dark and in a chemical fume hood. It can be used for up to 6 months, if stored properly. Experimental procedures involving handling lactophenol should be performed within a fume hood and with appropriate personal protective equipment. Dispose of phenol-containing chemical waste properly according to local institutional guidelines.
2. Use of a higher concentration of flg22 will ensure robust elicitation of callose deposition. However, syringe infiltration into Arabidopsis leaves of flg22 at concentrations as low as 0.5–10 nM, can reveal more subtle differences in callose deposition [20, 21].
3. Using leaves of approximately the same size (age) will help avoid age-associated difference in the plant defense response. This is especially important if subtle differences in callose deposition need to be assessed.
4. The central vein (midrib) of the leaf should be avoided during infiltration as this will greatly affect the viability of the leaf tissue.
5. It is essential to remove lactophenol solution prior to incubation. Failure to do so results in no callose staining, presumably due to interference with chlorophyll removal.
6. Folded or overlapping leaf area will cause incomplete chlorophyll clearing by lactophenol solution, which will interfere with subsequent callose observation.
7. Leaves may float on top of water, but this does not significantly affect washing or staining.
8. Callose staining using aniline blue achieves optimal fluorescence under alkaline conditions. Staining solution needs to be made fresh right before staining. The solution appears purplish blue initially, and the color will fade away during the staining procedure. This will, however, not affect staining. It is not recommended to stain the leaves for extended period of time as this will further soften the tissue and increase the difficulty of subsequent mounting.
9. Special care needs to be taken at this step as the leaves become very tender and easily lost during aspiration. Use tools (such as forceps) to stop the leaves from flowing with the stream.
10. Wetting the slide glass with plenty of mounting solution will minimize friction and aid the transfer. Slant the slide glass into mounting solution, and carefully drag leaf onto the slide glass using forceps. It is easier to handle the leaf from the base (petiole).
11. Optimal excitation wavelength for aniline blue is 370 nm, and maximum emission around 509 nm (Nikon Microscopy U).

However, commercial aniline blue is a heterogeneous blend of several fluorophores that have a wide range of excitation and emission wavelengths. For callose visualization, researchers have used UV excitation filters with peak excitation wavelength in the 350–390 nm range (excitation range includes 370 nm), and bandpass barrier filters with emission wavelength falls in the 420–530 nm range, or long pass UV barrier filters of 420 nm up to 460 nm.

12. Images saved in JPEG format will appear in color, while TIFF format (8-bit) images will appear in black and white. We find it easier to process images in greyscale for downstream image processing and quantification purposes. For colored JPEG images, convert to greyscale by "Image"—"Type"—"8-bit" in the ImageJ software.
13. Scale bar needs to be stamped onto only one image if the same magnification is used throughout. For callose quantification, the images need to be free of scale bars.
14. Try to avoid the infiltration spot when acquiring images. Mechanical wounding can also induce callose deposition, more callose may be observed at this area.
15. Image area can typically be determined in the image acquisition software. The ROI (region of interest) function in NIS-Element software allows one to select the entire image, and the area measurement can be calculated automatically.
16. Pixel range can also be arbitrarily set to distinguish "big" and "small" callose deposits that can be elicited by different treatments [as reported in [17].

Acknowledgments

This protocol is adapted from Kim and Mackey, 2008, MiMB [22]. Funding for this work was provided by the US Department of Agriculture (National Institute of Food and Agriculture, grant #2015-11870612), the Korean Rural Development Administration Next-Generation BioGreen Program (System and Synthetic Agro-Biotech Center and grant nos. PJ009088 and PJ011091), and the Ohio Agricultural Research and Development Center of the Ohio State University.

References

1. Ellinger D, Voigt CA (2014) Callose biosynthesis in Arabidopsis with a focus on pathogen response: what we have learned within the last decade. Ann Bot 114(6):1349–1358
2. Underwood W (2012) The plant cell wall: a dynamic barrier against pathogen invasion. Front Plant Sci 3:85
3. Eggert D, Naumann M, Reimer R, Voigt CA (2014) Nanoscale glucan polymer network causes pathogen resistance. Sci Rep 4:4159
4. Ellinger D, Naumann M, Falter C, Zwikowics C, Jamrow T, Manisseri C, Somerville SC, Voigt CA (2013) Elevated early callose deposition results in complete penetration resistance

to powdery mildew in Arabidopsis. Plant Physiol 161(3):1433–1444
5. Jacobs AK, Lipka V, Burton RA, Panstruga R, Strizhov N, Schulze-Lefert P, Fincher GB (2003) An arabidopsis callose synthase, GSL5, Is required for wound and papillary callose formation. Plant Cell 15(11):2503–2513
6. Nishimura MT, Stein M, Hou BH, Vogel JP, Edwards H, Somerville SC (2003) Loss of a callose synthase results in salicylic acid-dependent disease resistance. Science 301(5635):969–972
7. Luna E, Pastor V, Robert J, Flors V, Mauch-Mani B, Ton J (2011) Callose deposition: a multifaceted plant defense response. Mol Plant Microbe Interact 24(2):183–193
8. Geng X, Cheng J, Gangadharan A, Mackey D (2012) The coronatine toxin of *Pseudomonas syringae* is a multifunctional suppressor of Arabidopsis defense. Plant Cell 24(11): 4763–4774
9. DebRoy S, Thilmony R, Kwack YB, Nomura K, He SY (2004) A family of conserved bacterial effectors inhibits salicylic acid-mediated basal immunity and promotes disease necrosis in plants. Proc Natl Acad Sci USA 101(26): 9927–9932
10. Ham JH, Majerczak D, Ewert S, Sreerekha MV, Mackey D, Coplin D (2008) WtsE, an AvrE-family type III effector protein of *Pantoea stewartii* subsp. *stewartii*, causes cell death in non-host plants. Mol Plant Pathol 9(5):633–643
11. Fabro G, Steinbrenner J, Coates M, Ishaque N, Baxter L, Studholme DJ, Korner E, Allen RL, Piquerez SJ, Rougon-Cardoso A, Greenshields D, Lei R, Badel JL, Caillaud MC, Sohn KH, Van den Ackerveken G, Parker JE, Beynon J, Jones JD (2011) Multiple candidate effectors from the oomycete pathogen *Hyaloperonospora arabidopsidis* suppress host plant immunity. PLoS Pathog 7(11):e1002348
12. Guo M, Tian F, Wamboldt Y, Alfano JR (2009) The majority of the type III effector inventory of *Pseudomonas syrinage* pv. tomato DC3000 can suppress plant immunity. Mol Plant Microbe Interact 22(9):1069–1080
13. Currier HB (1957) Callose substance in plant cells. Am J Bot 44(6):478–488
14. Smith MM, McCully ME (1978) A critical evaluation of the specificity of aniline blue induced fluorescence. Protoplasma 95(3): 229–254
15. Wood PJ, Fulcher RG (1983) Dye interactions. A basis for specific detection and histochemistry of polysaccharides. J Histochem Cytochem 31(6):823–826
16. Stone BA, Evans NA, Bonig I, Clarke AE (1984) The application of Sirofluor, a chemically defined fluorochrome from aniline blue for the histochemical detection of callose. Protoplasma 122(3):191–195
17. Ham JH, Kim MG, Lee SY, Mackey D (2007) Layered basal defenses underlie non-host resistance of Arabidopsis to Pseudomonas syringae pv. phaseolicola. Plant J 51(4):604–616
18. Zhou J, Spallek T, Faulkner C, Robatzek S (2012) CalloseMeasurer: a novel software solution to measure callose deposition and recognise spreading callose patterns. Plant Methods 8(1):49
19. Cumbie JS, Pankow RC, Thomas WJ, JH, C (2010) AutoSPOTs: automated image analysis for enumerating callose deposition. In: Akimitsu K et al. (eds) 10th Japan-US Seminar: genome-enabled integration of research in plant pathogen systems, 2010. APS press: St Paul, MN
20. Kim MG, da Cunha L, McFall AJ, Belkhadir Y, DebRoy S, Dangl JL, Mackey D (2005) Two Pseudomonas syringae type III effectors inhibit RIN4-regulated basal defense in Arabidopsis. Cell 121(5):749–759
21. Daudi A, Cheng Z, O'Brien JA, Mammarella N, Khan S, Ausubel FM, Bolwell GP (2012) The apoplastic oxidative burst peroxidase in Arabidopsis is a major component of pattern-triggered immunity. Plant Cell 24(1):275–287
22. Kim MG, Mackey D (2008) Measuring cell-wall-based defenses and their effect on bacterial growth in Arabidopsis. Methods Mol Biol 415:443–452
23. Yuan J, He YH (1996) The *Pseudomonas syringae* Hrp regulation and secretion system controls the production and secretion of multiple extracellular proteins. J Bacteriol 178:6399–6402

Chapter 17

Quantitative Evaluation of Plant Actin Cytoskeletal Organization During Immune Signaling

Yi-Ju Lu and Brad Day

Abstract

High spatial and temporal resolution microscopy-based methods are valuable tools for the precise real-time imaging of changes in cellular organization in response to stimulus perception. Here, we describe a quantitative method for the evaluation of the plant actin cytoskeleton during immune stimulus perception and the activation of defense signaling. As a measure of the biotic stress-induced changes in actin filament organization, we present methods for analyzing changes in actin filament organization following elicitation of pattern-triggered immunity and effector-triggered immunity. Using these methods, it is possible to not only quantitatively evaluate changes in actin cytoskeletal organization following biotic stress perception, but to also use these protocols to assess changes in actin filament organization following perception of a wide range of stimuli, including abiotic and developmental cues. As described herein, we present an example application of this method, designed to evaluate changes in actin cytoskeletal organization following pathogen perception and immune signaling.

Key words Actin cytoskeleton, Confocal microscopy, Quantitative evaluation, ETI, PTI, *Pseudomonas syringae*, Immune signaling, PAMP

1 Introduction

The actin cytoskeleton is required for the function of numerous cellular processes, including those required for growth, development, response to the environment, and immunity [1]. For example, in plants, abiotic stress, as well as a multitude of pathogen-derived elicitors, has been demonstrated to induce rapid and transient changes in actin organization, including polymerization and bundling [2–6]. However, our understanding of the role of actin, including the purpose of the incessant remodeling to the host cytoskeleton during stress signaling, is largely corollary and lacking a mechanistic definition of the role of actin as a required component of the signaling cascade. In plants, a multitude of immune-eliciting bacterial strains and purified pathogen elicitors cause rapid and transient increases in actin polymerization [2]. These changes have been demonstrated to be specific, and only occur when the

Libo Shan and Ping He (eds.), *Plant Pattern Recognition Receptors: Methods and Protocols*, Methods in Molecular Biology, vol. 1578, DOI 10.1007/978-1-4939-6859-6_17,

plant has its full complement of corresponding pattern recognition receptors and known components of the pattern receptor complex. To fully define the role of the plant actin cytoskeleton, including the ability to rapidly and accurately associate changes in cytoskeletal organization with the precise activation of specific signaling cascades, requires a comprehensive, and parallel, analysis of the genetic, biochemical, and cellular processes during stimulus perception. In recent years, this approach has been successfully employed to define the plant cell response, including the activation of immunity, in response to pathogen perception.

Recent advances in live cell imaging technologies have made it possible to define the temporal and spatial changes in cytoskeletal organization, and to accurately associate these changes with concomitant changes in host physiology and cell signaling [2, 7–9]. Of these advanced technologies, fluorescent imaging of the cytoskeleton has emerged as one of the most reliable tools for monitoring changes in the microtubule and actin filament array [10]. In this Chapter, we describe confocal microscopy-based methods for imaging of the plant actin cytoskeleton, including quantification of changes in filament organization, during host infection with phytopathogenic microbes. As a highly robust method to define and associate changes in actin filament organization with concomitant changes in the host cell physiology during pathogen infection, the procedure presented herein is a reliable, highly quantitative method to visualize and describe changes in host cell organization during stress response and signaling.

2 Materials

All liquid reagents described herein should be prepared using ultrapure, sterile deionized water (dH_2O). Freshly prepared solutions should be stored at room temperature, unless otherwise indicated.

2.1 General Reagents and Disposables

1. Polystyrene sterile six-well flat bottom plates.
2. Sterile 100 × 100 mm Petri dishes.
3. Fine-tip forceps.
4. Incubator, 28 °C.

2.2 Bacterial Culture and Media Preparation

1. NYGA (nutrient yeast glucose agar; [11]) media: 0.5% Bacto peptone, 0.3% yeast extract, 2% (v/v) glycerol, 15% Bacto agar. Autoclave media at 121 °C for 45 min. Cool NYGA media to 55 °C, and then add antibiotics, if necessary, mix, and pour 25 mL of media into 100 × 100 mm petri dishes. Cool to room temperature.

2. Antibiotic stock concentrations: rifampicin (100 mg/mL) and kanamycin (50 mg/mL). Antibiotic working concentrations: rifampicin (100 μg/mL) and kanamycin (25 μg/mL). Antibiotic stock solutions should be stored at −20 °C.

2.3 Arabidopsis Thaliana Growth and Treatment

The methods described herein have been optimized for the visualization of Arabidopsis epidermal cells following treatment with purified pathogen elicitors and/or dip-inoculation with *Pseudomonas syringae* pv. *tomato* DC3000 (*Pst* DC3000). Based on the sum of our experiences, the methods described below are suitable for the evaluation of any number of plant pathogenic bacteria, including those harboring heterologously expressed type three effectors (T3Es).

1. Arabidopsis plants expressing GFP-fABD2 [12].
2. Fiberglass window screen (Fig. 1a).
3. Potting soil mix (*see* **Note 1**).
4. Rubber bands.
5. Growth chamber conditions: 20 °C, 16 h light–8 h dark cycle with 60% relative humidity and a light intensity of 120 μmol photons/m^2s.
6. Resuspension solution: 10 mM $MgCl_2$.
7. Dip inoculation solution: 10 mM $MgCl_2$ + 0.02% Silwet.
8. Spectrophotometer.
9. Conical Falcon tubes, 15 mL.

2.4 Pathogen Elicitors

In the case presented herein, we broadly refer to pathogen elicitors as pathogen-associated molecular patterns (PAMPs; e.g., flg22, chitin, elf18; [13]). Peptide elicitors can be synthesized and purchased from a number of companies, and should be synthesized at >99% purity (*see* **Note 2**). For the generation of the PAMP chitin, refer to the methods presented in Kaku et al. [14]. Stock solutions of purified pathogen elicitors should be prepared in sterile dH_2O and stored at −20 °C.

The amino acid sequences of two commonly used *Pst* DC3000 PAMPs are as follows:

1. Flg22 peptide: Q-R-L-S-T-G-S-R-I-N-S-A-K-D-D-A-A-G-L-Q-I-A [15].
2. Elf18 peptide: ac-S-K-E-K-F-E-R-T-K-P-H-V-N-V-G-T-I-G [16].

2.5 Preparing Samples for Laser Scanning Confocal Microscopy

1. Confocal microscope with 60× objective.
2. Glass slides (Pre-cleaned, 1″ × 3″, 1.2 mm).
3. Cover slides: 18 × 18 mm, 1.5 oz.

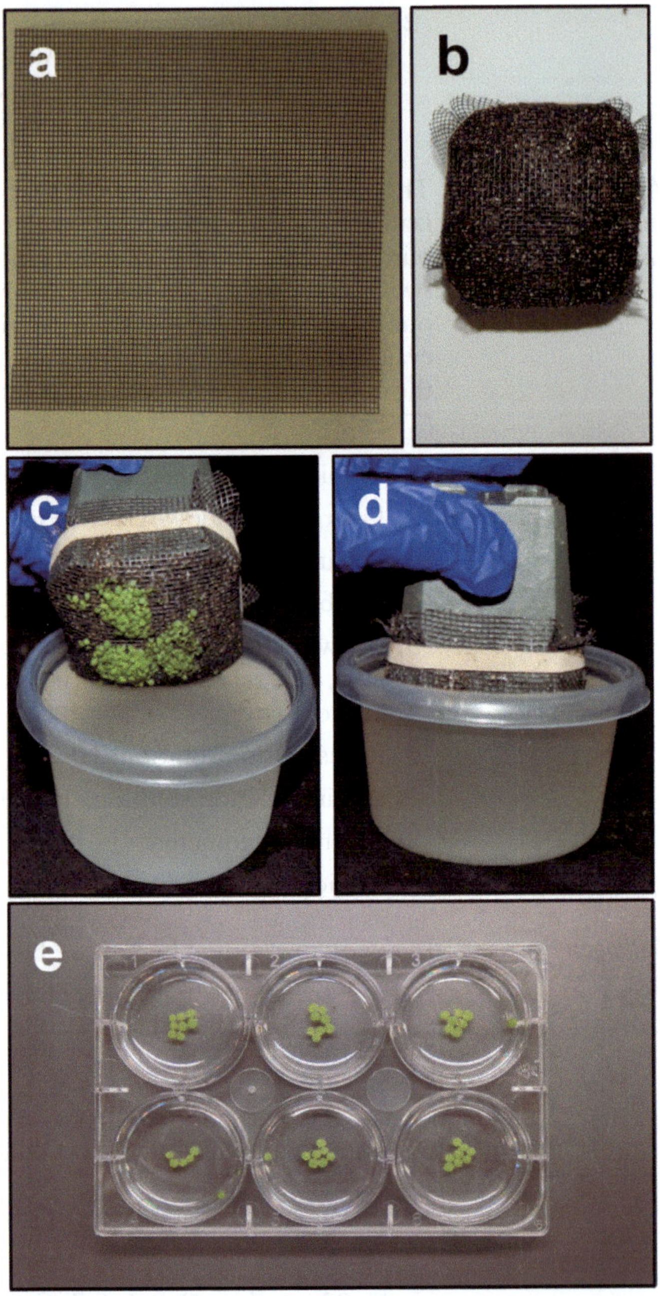

Fig. 1 Treatment of Arabidopsis seedlings with pathogen and pathogen elicitors for induction of immunity and elicitation of changes in host actin cytoskeletal organization. (**a**) Fiberglass insect screen. (**b**) 10–12-day-old seedling in pots covered with fiberglass insect screen. (**c**, **d**) 10–12-day-old Arabidopsis are dip-inoculated in a suspension of *Pseudomonas syringae.* (**e**) Pathogen elicitor treatment of detached Arabidopsis cotyledons

2.6 Image Analysis Software

1. Standard confocal microscopy image collection and analysis software.
2. Fiji image analysis software package (formerly called ImageJ). Free download at: http://fiji.sc/Fiji.
3. The scripts (.txt files) for actin skewness and density analysis, first published by [8], can be downloaded at:

   Skewness:
   http://www.daylab.plp.msu.edu/wp-content/uploads/2015/10/Skewness.txt.
   Density:
   http://www.daylab.plp.msu.edu/wp-content/uploads/2015/10/Density.txt.

3 Methods

All plant inoculation procedures and microscopy analyses should be performed at room temperature (ca. 25 °C).

3.1 Plant Growth

1. Sow GFP-fABD2-expressing Arabidopsis seeds in pots filled with soil and fitted with fiberglass screen (Fig. 1b; *see* **Note 3**). Seeds should be sewn evenly across the pot; after seed germination, plants can be thinned, if necessary, using fine-tip forceps.
2. Stratify seeds at 4 °C in the dark for 3 days.
3. After stratification, transfer pots to an environmentally controlled growth chamber, and water plants as necessary.
4. For actin cytoskeleton imaging during immune elicitation, 10–12-day-old Arabidopsis seedlings are used (Fig. 1c).

3.2 Bacteria Growth and Plant Inoculation

1. Remove bacteria from the −80 °C and streak onto NYGA plates containing appropriate antibiotics, if required, and grow *Pst* DC3000 strains for 2 days at 28 °C.
2. After 2 days growth, re-streak *Pst* DC3000 strains onto fresh NYGA plates containing antibiotics and incubate an additional 16–20 h at 28 °C.
3. After 16–20 h of additional growth, pipette 5 mL of $MgCl_2$ buffer on NYGA plates containing bacterial (i.e., *Pst* DC3000).
4. Gently scrape bacteria from the plate and collect the $MgCl_2$-*Pst* DC3000 solution, transferring into a fresh (sterile) 15 mL Falcon tube.
5. To determine the concentration of the bacterial solution, transfer 100 μL of the bacterial suspension into an Eppendorf tube containing 900 μL $MgCl_2$ buffer (i.e., a 1:10 dilution of your stock solution).

6. Determine concentration of bacteria by measuring the optical density (OD) of the suspension at 600_{nm} using a standard laboratory spectrophotometer.
7. Adjust the concentration of the bacterial suspension to 3×10^7 colony-forming units/mL (CFU/mL) in the required volume of dip inoculation buffer. $OD_{600nm} = 1.0 = 1 \times 10^8$ CFU/mL.
8. For plant inoculation, for example, prepare 200 mL of the bacterial suspension. This is an amount sufficient to dip-inoculate three pots of 10- to 12-day-old seedlings.
9. Invert pots containing Arabidopsis seedlings into the bacterial solution, submerging for 30 s. Gently swirl pots to ensure uniform treatment of the seedlings (Fig. 1c, d)
10. After inoculation, return pots to plant growth trays and place a clear plastic dome lid on flats for 1 h to maintain high relative humidity. After 1 h, remove the dome and return flats to the growth chamber.
11. Periodically observe inoculated seedlings over the course of the experiment. This is most important beginning at around 24–28 h post inoculation (hpi) to ensure that tissue collapse is not occurring.
12. For evaluation of actin cytoskeletal changes during ETI, image seedlings at 24–28 hpi (*see* **Note 4**).

3.3 Inoculation with PAMP Elicitors (PTI)

1. Peptide elicitors, from −20 °C stock solutions, are dissolved in 10 mL of 10 mM $MgCl_2$ to yield working stock solutions. The final working concentration of each elicitor (flg22, elf26, and chitin) ranges from 1 to 25 μM; for our experiments, we typically use a final concentration of 10 μM. For other elicitors, the concentration may vary, depending on the previously defined bioactive concentration.
2. Pipette 5 mL of the elicitor solution into a sterile six-well flat bottom plate.
3. The cotyledons from 10- to 12-day-old Arabidopsis seedling are then floated in the elicitor solution contained in the six-well plate. Approximately 10–15 cotyledons can be placed in each well containing the elicitor solution (Fig. 1e).
4. Return the cover (to maintain humidity), and store plates under ambient light at room temperature. *The incubation time may vary, depending on the elicitor.* For the above noted PAMPs (i.e., flg22, elf26, chitin), incubate over a short time course of 0, 0.5, 1, 2, 4, and 6 h, imaging at each time point. In most cases, optimal time for imaging is approximately 1–2 hpi [2].

3.4 Laser Scanning Confocal Microscopy: Image Collection

1. Images are collected using a laser confocal scanning microscope equipped with a 60× objective. Sample excitation at 488 nm (GFP), emission detection at 495–525 nm. For image

acquisition, 2× magnification is used at a resolution of 800 × 800 pixels.

2. Collect 4–5 images of the lower epidermal pavement cells from one cotyledon (*see* **Note 5**).
3. For each image, collect 25 serial stacks, with an interval of 0.5 μm from the plasma membrane (Fig. 2; *see* **Notes 6** and 7).
4. Save the original image files for skewness and density analyses, described below. For skewness analysis, collect maximum projections of 25 serial stacks each image in the stack as a TIFF file format for skewness analysis.

3.5 Image Analysis

Recent studies have described the imaging and quantification of changes in actin filament organization using two key parameters: Skewness and Density [2, 8]. In short, Skewness is an indirect measure of the degree of actin filament bundling, while Density, also referred to as 'Percent Occupancy', indicates the relative abundance of actin filament signal within the cell.

3.5.1 Skewness Analysis

1. Perform Skewness analysis using the Fiji software package.

   The source files for Skewness analysis are the merged image files, representing 25 stacks, collected as described above. As noted above, these files should be in TIFF format.
2. Open the Fiji software package.
3. From the Fiji 'Menu', select **File ► Import ► Image Sequence**, and select the folder that contains the TIFF image files to be analyzed.
4. In the **Sequence Options** check the box to select **'Convert to 8-bit Grayscale'** and **'Sort names numerically'**. All of the remaining options should be left at the default settings. Click **'OK'**.
5. A new image stack, containing all images (i.e., 25 images, as noted in **'step 4'**, above) will appear (Fig. 2).
6. From the Fiji 'Menu' bar, select **Image ► Adjust ► Threshold**. In the Threshold window, check **'Dark Background'**. Click **'Set'**. In the **'Set Threshold Levels'** window, set the **'Lower Threshold Level'** to **20**, and the **'Upper Threshold Level'** to **255**. Click **'OK'**. Leave the **'Threshold'** window open, and click on image window.
7. From the Fiji 'Menu' bar, select **Plugin ► Macros ► Install ► Skewness script**. At this step, locate the Skewness script (.txt file), downloaded as described above, and select **'Open'**.
8. From the Fiji **'Menu'**, select **Plugin ► Macros ► Skewness**.

At this point, the processing script will being running, and the output of this will yield skewness values for each image, presented in a separate, 'Results', window.

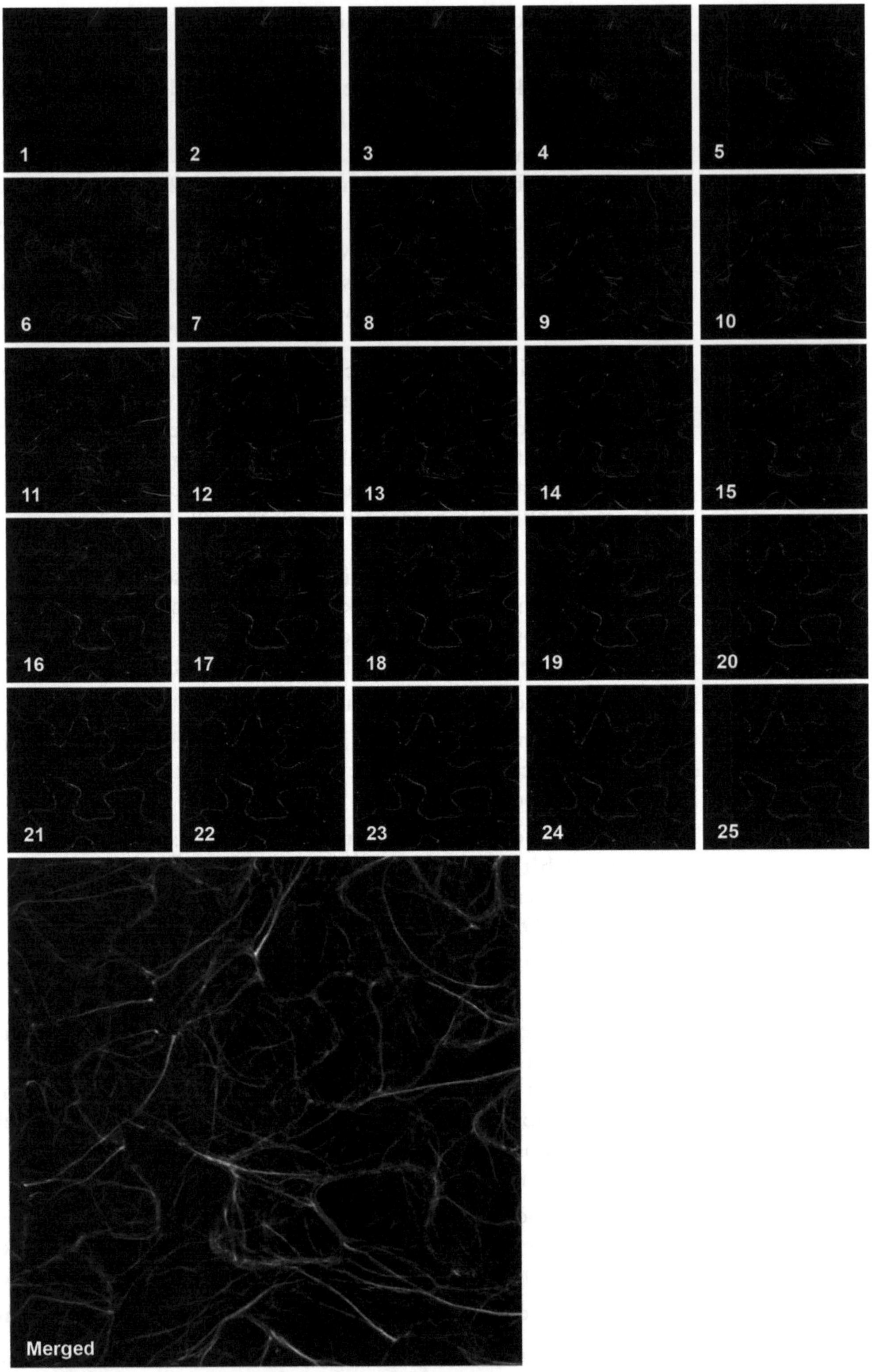

Fig. 2 Twenty-five Z-series image stacks collected using confocal laser scanning microscopy. The image stacks are collected in 0.5 μm intervals. The projection, which is the sum of the merged image stacks, is shown at the bottom left

9. Highlight and copy the output (i.e., '**sigNum**' and '**skewness**') and paste into a new Excel file sheet. In the output file, the 1st column is simply the image number for each image, the 2nd column is the 'sigNum' and the 3rd column is 'Skewness'. Save the Excel file.

3.5.2 Density Analysis: Image Processing

1. Select multiple (i.e., ca. 10) original files (.oib files), right click on the selected files, and choose '**Open With**', and select **Fiji** to open files.
2. A new '**Bio-Formats Import Options**' window will open. In the '**Bio-Formats Import Options**' window, at the '**Stack viewing**' panel, in '**View stack with:**', select '**Hyperstack**'. Remaining options (i.e., '**Dataset organization**', etc.) can be left at the default settings. From '**Color options**' panel, select '**Grayscale**'. From '**Split into separate windows**', check the box for '**Split channels**'. Select '**OK**' to complete format settings.

At this point, 2 windows will open, one that represents the confocal image, the other representing the brightfield image (Fig. 3a, b).

The remaining stack images from the experiment should be processed in the same fashion as above **steps 1–3**. As above, additional images (i.e., confocal and brightfield) will be created for each of the image stack .oib files. Repeat **steps 1–2** for each stack image. Each additional input stack image will yield a confocal and brightfield image.

3. Close all brightfield image windows (i.e., Fig. 3b). Leave the windows containing fluorescent actin filament images open (Fig. 3c).

The next step should be performed using the confocal microscopy-generated .oib images, processed as described above (i.e., those that are each contained within a separate window).

4. To apply the **Gaussian Blur** filter, choose 1 image, and from the Fiji 'Menu' bar select **Process ► Filters ► Gaussian Blur** (*see* **Note 8**).
5. In the '**Gaussian Blur**' dialog box, set '**Sigma (Radius)**' **1.0**, select '**OK**' (Fig. 3d).
6. A window '**Process Stack?**' will appear, select '**Yes**' to process all 25 images (25 series stacks).
7. Select the next image, click on it and repeat **step 4–6** until the remaining 9 images have been processed.
8. To apply the **HBP** filter, choose 1 image (.oib file generated from **steps 5–7**) and select **Process ► FFT ► Bandpass Filter** (*see* **Note 9**).
9. From the '**FFT Bandpass Filter**' window, select '**Filter large structures down to**' **40** pixels, '**Filter small structures up**

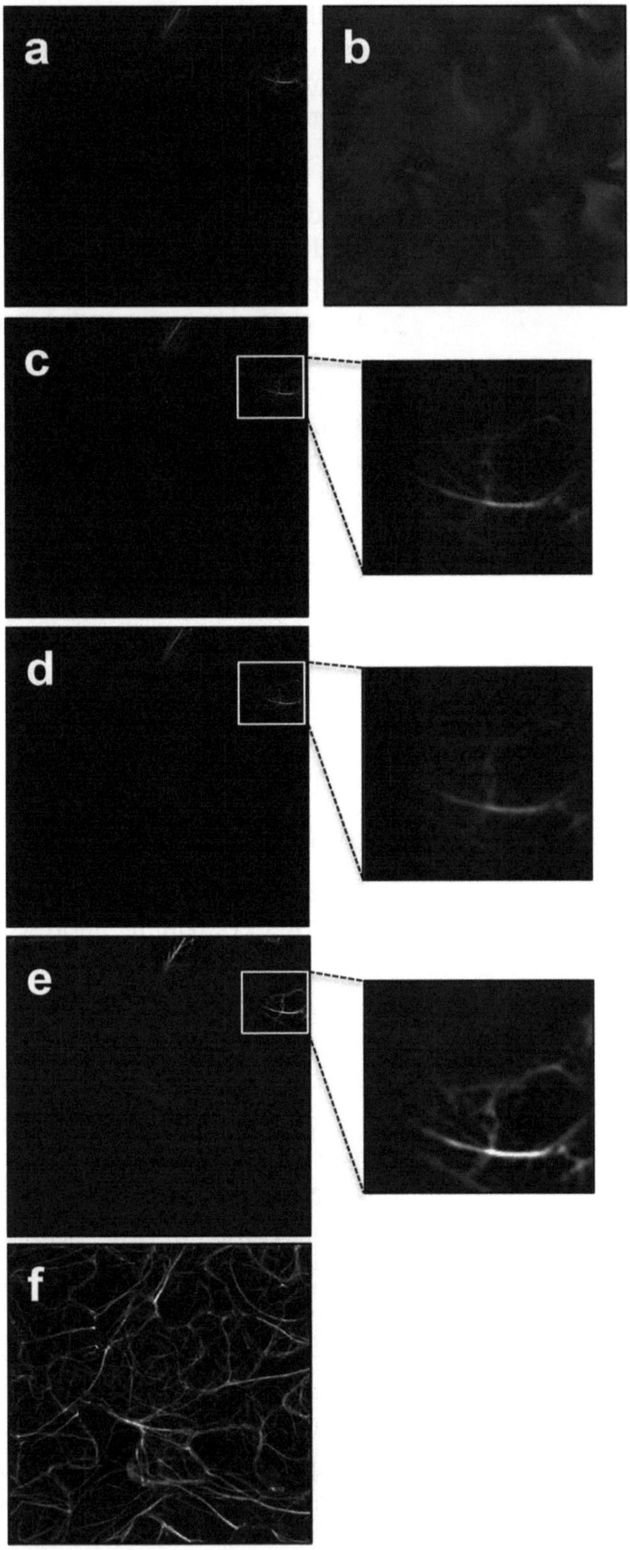
a
b
c
d
e
f

to' 0 pixels, **'Tolerance of direction'** to **0**%. Ensure that all boxes are unchecked, then select **'OK'**.

10. Repeat **step 9** for the remaining images.
11. To project serial stacks into 3D images: select 1 image processed as above (i.e., **step 10**). From the Fiji menu, select **Image ► Stacks ► Z Project**.
12. From the **'ZProjection'** window, set **'Start slice'** to **1**, **'Stop slice'** to **25**, **'Projection type'** to **'Max Intensity'**, then select **'OK'**. Repeat for each image.

Completion of step 12 *will result in the generation of a new .oib file, representing the merged image with a maximum projection. The new file will be automatically named with the prefix 'MAX', preceding the original file name. At this point, the original .oib file can be closed, and* *changes should not be saved*.

13. Using the **'MAX'** image, at the Fiji menu, select **File ► Save As ► Tiff**. Create a 'New folder' and save **'MAX'** 3D images in this folder. ► Close windows.
14. Repeat **step 13** for the remaining **'MAX'** 3D images.

*Completion of this step will result in the generation of maximum project, 3D, image files, named '*xxxx.oib.tiff*'. The resultant TIFF files will be used for 'Density Analysis: Quantitative Evaluation', immediately below.*

Density Analysis: Quantitative Evaluation

1. From the Fiji menu, select **File ► Import ► Image Sequence**, and choose the new folder containing images (in **.oib.tiff** format) generated above (**step 15**) to be analyzed. Highlight the folder, and select **'Choose'**.
2. A new dialog window, **'Sequence Options'**, will open. At this step, leave all settings at the default options, with the exception of **'Convert to 8-bit Grayscale'**, which should be selected. Click **'OK'**.

Fig. 3 A sample of four images stacks analyzed for quantitative evaluation of the Density metric. (**a**, **b**) are confocal and bright field images, respectively, generated by the confocal laser scanning microscopy. The bright field image, which is also created during the Density analysis (Subheading 3.5.2) is shown, yet is not used for quantitative evaluation (Subheading 3.5.2, **step 3**). (**c**) The original, unprocessed, confocal image. (**d**) The same image as (**c**) following processing using the Gaussian Blur filter. (**e**) The same image as in (**d**) following processing with the high bandpass (HBP) filter. (**f**) The HBP filter processed stacks for each of the 25 collected images, merged, to yield a processed 3D image. This processed image is used as the input file for quantitative Density analysis

A new window will open that contains all of the processed images. The individual images can be viewed by scrolling the sliding tab at the bottom of the window, left-to-right.

3. For image analysis, begin at the 1st image (sliding tab to far left). From the Fiji menu, select **Image ► Adjust ► Threshold**. In 'Threshold' window, check the box for **'Dark Background'**. Select **'Set'**, which will open a new dialog window **'Set Threshold Levels'.** Set **'Lower Threshold Level'** to **25**, and the **'Upper Threshold Level'** to **255**. Click **'OK'**. In the **'Threshold'** dialog, click **'Apply'**, a **'Convert Stack to Binary'** window will appear, check **'Black background (of binary masks)'**, click **'OK'**. Leave the **'Threshold'** dialog window open.
4. From the Fiji menu, select **Plugin ► Macros ► Install ►** locate the 'Density' script (.txt file), downloaded as described above. Once located, highlight the file, and select **'Open'**.
5. Return to the main Fiji menu and select **Plugin ► Macros ► Density**.

At this point, a new dialog window will open, 'Results'*, and the actin filament density analysis will initiate. The* 'Area' *will be calculated for each image processed, with the corresponding image number located to the immediate left of the* 'Area'.

6. Copy these data from the dialog window and paste into Excel. Save the resultant Excel file.

3.5.3 Quantification and Presentation of Skewness and Density Parameters

For each cotyledon, five different areas of the epidermal pavement cells are imaged to ensure a non-biased selection for analysis. Additionally, for each technical repeat, at least 25 images of cotyledon epidermal cells per genotype with each treatment are collected. This results in a total of approximately 75 images per treatment (i.e., 3 biological repeats × 25 images per experiment). Once all images are collected and analyzed using the *Skewness* and *Density* parameters, mean values, standard errors, and the standard deviation is calculated using Microsoft Excel. A one-tailed Student's *t*-test should be employed to determine statistical significance. Data can be plotted using the mean values, with error bars indicating standard deviations (Fig. 4).

4 Notes

1. Most potting soil mixtures designed for Arabidopsis growth work well for this procedure. We typically use Redi-earth (Hummert International, Earth City, MO, USA), which we have observed maintains soil moisture and does require frequent watering.

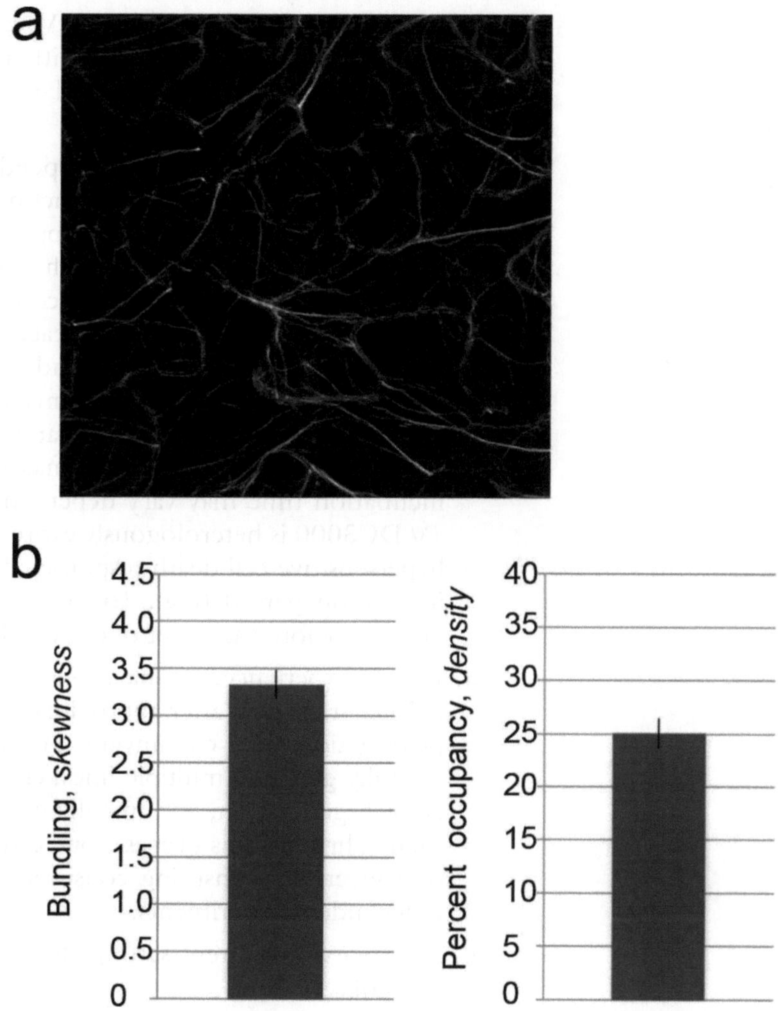

Fig. 4 Quantitative analysis of skewness and density metrics following pathogen elicitation of 12–14-day-old Arabidopsis seedlings. (**a**) Representative picture of finalized image. (**b**) Actin skewness and density are analyzed quantitatively. The data presented in the left graft (Bundling, *skewness*) represents the raw output values generated in Subheading 3.5.1. For presentation of the data shown in the graph on the right (% occupancy, *density*), the final output values (Subheading 3.5.2) are calculated whereby "% occupancy" is represented as: [100 × (Area ÷ 640,000)]. This calculation was originally described by Higaki and colleagues [8]

2. There are several companies that synthesize high quality peptides. In our experience, we have found that EZ-Biolabs (Carmel, Indiana, USA) provides high-quality peptides at a level of synthesis sufficient to last several years.

3. To ensure minimal physical injury to plant seedlings, as well as to keep the plants and soil undisturbed during the dip-inoculation procedure, we cover the soil with a fiberglass

mesh insect screen (i.e., BetterVue; Phifer Inc., Tuscaloosa, Alabama, USA), held in place with a rubber band. The screen should be cut to approximately 1-in. larger (on each side) than the dimensions of the pot.

4. We have observed a density-dependent (i.e., dose response) change in the Arabidopsis actin cytoskeleton following pathogen inoculation. To compensate for the differential growth of bacterial strains which may show significant differences in growth at 24 hpi, we recommend evaluating, side-by-side, the *in planta* growth of each strain to be used, and if direct comparisons are to be made between strains that show differential growth at 24 hpi, to make adjustments in the initial inoculum concentration such that each strain reaches a similar CFU at the desired time of imaging [6]. Additionally, the incubation time may vary depending on whether or not the *Pst* DC3000 is heterologously expressing a T3E that elicits the hypersensitive cell death response. If this is the case, a shorter incubation period (e.g., 16 hpi) may be required to image samples before the onset of cell death.
5. In our experience, we have observed silencing of the GFP-fABD2 transgene after approximately 6–8 generations of self-propagation. To circumvent problems with silencing, we typically generate multiple lines, collect an abundance of seed at each generation, and initially characterize multiple lines of each. Through this process, one can effective prolong the life of a given line, ensuring consistency in the interpretation of independent experiments.
6. Always use consistent settings for exposure time and gain for actin observation.
7. The best results are obtained when you select a relatively flat area of the cotyledon for imaging. This will ensure that the plane of vision is in focus. Also, avoid stomatal guard cells when imaging. The guard cells contribute an abundance of background fluorescence to the sample, and thus, can skew the data in favor of increased signal intensity.
8. Gaussian blur, also referred to Gaussian smoothing, is the result of blurring an image to effectively reduce image noise and detail, enabling the user to evaluate the image as a function of the individual pixels within the image. This "low pass" filtering technique for the analysis of high-resolution images results in the removal of both high and low spatial frequencies, resulting in an image that retains the middle range of the image qualities, enhances the image signals (i.e., pixels) at the edge, while reducing image noise [8].
9. The Bandpass filter is applied to minimize background autofluorescence. This is particularly important when visualizing

plant cells using laser scanning confocal microscopy, as green tissues typically exhibit high levels of chloroplast-derived autofluorescence.

Acknowledgments

Special thanks to Amy Baetsen-Young for critical review and comments. Work in the Day Lab is supported by a National Science Foundation grant (IOS-1557437).

References

1. Day B, Henty JL, Porter KJ, Staiger CJ (2011) The pathogen-actin connection: a platform for defense signaling in plants. Annu Rev Phytopathol 49:483–506
2. Henty-Ridilla JL, Shimono M, Li J, Chang JH, Day B, Staiger CJ (2013) The plant actin cytoskeleton responds to signals from microbe-associated molecular patterns. PLoS Pathog 9:e1003290
3. Porter K, Shimono M, Tian M, Day B (2012) Arabidopsis actin-depolymerizing factor-4 links pathogen perception, defense activation and transcription to cytoskeletal dynamics. PLoS Pathog 8:e1003006
4. Tian M, Chaudhry F, Ruzicka DR, Meagher RB, Staiger CJ, Day B (2009) Arabidopsis actin-depolymerizing factor AtADF4 mediates defense signal transduction triggered by the *Pseudomonas syringae* effector AvrPphB. Plant Physiol 150:815–824
5. Shimono M, Higaki T, Kaku H, Shibuya N, Hasezawa S, Day B (2016) Quantitative evaluation of stomatal cytoskeletal patterns during the activation of immune signaling in *Arabidopsis thaliana*. PLoS One 11:e0159291
6. Shimono M, Lu YJ, Porter K, Kvitko BH, Henty-Ridilla J, Creason A, He SY, Chang JH, Staiger CJ, Day B (2016) The *Pseudomonas syringae* type III effector HopG1 induces actin remodeling to promote symptom development and susceptibility during Infection. Plant Physiol 171:2239–2255
7. Dong B, Yang X, Zhu S, Bassham DC, Fang N (2015) Stochastic optical reconstruction microscopy imaging of microtubule arrays in intact *Arabidopsis thaliana* seedling roots. Sci Rep 5:15694
8. Higaki T, Kutsuna N, Sano T, Kondo N, Hasezawa S (2010) Quantification and cluster analysis of actin cytoskeletal structures in plant cells: role of actin bundling in stomatal movement during diurnal cycles in Arabidopsis guard cells. Plant J 61:156–165
9. Shaw SL, Ehrhardt DW (2013) Smaller, faster, brighter: advances in optical imaging of living plant cells. Annu Rev Plant Biol 64:351–375
10. Dyachok J, Paez-Garcia A, Yoo CM, Palanichelvam K, Blancaflor EB (2016) Fluorescence imaging of the cytoskeleton in plant roots. Methods Mol Biol 1365:139–153
11. Daniels MJ, Barber CE, Turner PC, Sawczyc MK, Byrde RJ, Fielding AH (1984) Cloning of genes involved in pathogenicity of *Xanthomonas campestris* pv. *campestris* using the broad host range cosmid pLAFR1. EMBO J 3:3323–3328
12. Sheahan MB, Staiger CJ, Rose RJ, McCurdy DW (2004) A green fluorescent protein fusion to actin-binding domain 2 of Arabidopsis fimbrin highlights new features of a dynamic actin cytoskeleton in live plant cells. Plant Physiol 136:3968–3978
13. Zipfel C (2014) Plant pattern-recognition receptors. Trends Immunol 35:345–351
14. Kaku H, Nishizawa Y, Ishii-Minami N, Akimoto-Tomiyama C, Dohmae N, Takio K, Minami E, Shibuya N (2006) Plant cells recognize chitin fragments for defense signaling through a plasma membrane receptor. Proc Natl Acad Sci U S A 103:11086–11091
15. Felix G, Duran JD, Volko S, Boller T (1999) Plants have a sensitive perception system for the most conserved domain of bacterial flagellin. Plant J 18:265–276
16. Kunze G, Zipfel C, Robatzek S, Niehaus K, Boller T, Felix G (2004) The N terminus of bacterial elongation factor Tu elicits innate immunity in Arabidopsis plants. Plant Cell 16:3496–3507

Chapter 18

Network Reconstitution for Quantitative Subnetwork Interaction Analysis

Fumiaki Katagiri

Abstract

A fundamental task in systems biology is to quantify the contributions of the systems' parts and their interactions. Here I describe a powerful concept and tool for this purpose: network reconstitution. Genotypes of an organism that represent all possible combinations of the subnetworks in question will be quantitatively phenotyped. The quantitative phenotype data is analyzed using an R script to obtain estimates for single subnetwork contributions and their interactions.

Key words Subnetwork interactions, Network reconstitution, Signaling allocation, Combinatorial genotypes, Quantitative phenotype

1 Introduction

"The whole is greater than the sum of its parts."—Aristotle. This statement is often cited as a slogan of systems biology. It literally means that interactions among a system's parts are significant: it states that the additive effects of the individual parts cannot explain the whole. In other words, both the effects of individual parts and their interactions must be quantitatively assessed in systems biology. For this purpose, we developed a network reconstitution strategy, which we formerly called signaling allocation analysis [1], and applied to analysis of the plant immune signaling network. In the plant immune signaling network, components are highly interconnected (i.e., it is a web-like network) [2]. In such a network with significant interactions among its components, classical genetic approaches, which are based on the concept that signaling is mediated by pathways, cannot be effectively used to study the quantitative relationships among the network's components. For example, in a network consisting of subnetworks A and B, deletion of subnetwork A removes the interaction between subnetworks A and B (which we denote as A:B) as well as the individual effect of subnetwork A. Thus, comparing the phenotype of the subnetwork A

Libo Shan and Ping He (eds.), *Plant Pattern Recognition Receptors: Methods and Protocols*, Methods in Molecular Biology, vol. 1578, DOI 10.1007/978-1-4939-6859-6_18,

deletion to that of wild type, which is a standard approach in classical genetics, cannot determine the individual effect of subnetwork A. Instead, in network reconstitution, the phenotype of the subnetwork B deletion (in which subnetwork A is intact) is compared to that of the subnetworks A and B double deletion (i.e., no network function), and in this way the individual effect of subnetwork A is determined. Similarly, the individual effect of subnetwork B is determined by comparing the phenotype of the subnetwork A deletion to that of the double deletion. Once the individual effects of subnetworks A and B are determined, the interaction between the subnetworks can be calculated from the comparison of the wild-type phenotype to that of the double deletion as this comparison gives the value A + B + A:B. Thus, the fundamental idea of network reconstitution is to compare all the combinatorial subnetwork deletion states to the no network function state, which allows simple mechanistic interpretations of the results.

Network reconstitution is extendable to networks with more subnetworks, as we applied it to a network comprising four subnetworks [1] (although we called them network sectors in the report, here for the sake of generalization we call them subnetworks). However, when the number of subnetworks exceeds two, the contributions from each order of the interactions (two-subnetwork interactions, three-subnetwork interactions, etc.) need to be averaged [1].

To generalize the approach, here the subnetworks were generically named subnetworks A, B, C, ... One case with four subnetworks is used as an example in the following although the script is written to be able to accommodate any subnetwork number from 2 to 26. (Since the number of genotypes needed to achieve all combinatorial subnetwork deletions increases exponentially with the number of subnetworks, e.g., 1024 genotypes for 10 subnetworks, 26 is a more than high enough upper limit. It is also likely to hit practical computational limit before hitting the highest subunit number.) The example sample data included here ("input4y.dat.txt") are based on four subnetworks: the subnetworks are denoted A, B, C, and D. The corresponding genotypes are named with the capital letters of the subnetwork names when the subnetworks are intact and named with the lowercase letters of the subnetwork names when the subnetworks are deleted. For example: the quadruple deletion is denoted abcd; the deletion of subnetworks A, B, and D is denoted abCd; the deletion of subnetworks B and C is denoted AbcD; and the intact network (wild type) is denoted ABCD. There are 16 genotypes for all combinations of the four subnetworks: one quadruple deletion; four triple deletions; six double deletions; four single deletions; one intact network.

The next step is to collect quantitative phenotype data for all 16 genotypes. Although the bacterial titer in Arabidopsis leaves after inoculation of a bacterial pathogen was used as the quantitative phenotype in our case [1], any quantitative measure that can be applied to each genotype can be used: e.g., the mRNA level of a marker gene, the area size of necrosis, and the fresh weight of a tissue. Ideally the network function should be completely abolished in the genotype with all subnetworks deleted (genotype abcd). However, in our case, the genotype still had residual network function: approximately 20% of the function remained when we measured the level of pattern-triggered immunity triggered by flg22 treatment or the level of effector-triggered immunity triggered by the effector AvrRpt2 [1]. Therefore, we used control treatment (mock treatment or inoculation with the congenic strain without AvrRpt2) to define the null network function state and to enable quantification of the remaining network function. On the other hand, we did not have an appropriate control for measurement of the immunity level against the necrotrophic pathogen *Alternaria brassicicola* to define the 0 immunity level, and the remaining network function level was not able to be estimated. In the script attached to this protocol, an option of including data with such a control treatment is available.

It is necessary to have biological replicates for phenotype measurements to be able to estimate the significances of the contributions calculated. It is highly recommended that such biological replicates are made in multiple independent experiments. Random factors can be included in the mixed-effect linear model for network reconstitution to correct the effects arising from experimental differences. For example, we planted eight individual Arabidopsis plants of different genotypes in a 6″ × 6″ box, and we had eight boxes in a 1′ × 2′ flat and multiple flats were used in a single experiment. We expected effects from different boxes, flats, and experiments on the measured bacterial titer and thus included random factors for box, flat, and experiment in the model. As our experimental design was a nested design (i.e., a particular box is in a particular flat, which is in a particular experiment), the random factors were written in the model formula as experiment/flat/pot. A well-balanced experimental design is statistically ideal, but there could be practical constraints: for example, in our case, plants for flg22 and control treatments had to be in separate flats to avoid flg22 contamination on control plants. Ill effects arising from imbalance in the experimental design can be moderated by having a higher number of independent replicates. According to the experimental design used, the random factors need to be decided, and the data file columns and the mixed-effect linear model for network reconstitution need to be modified.

Although it is outside the scope of this chapter, we demonstrated that combining network reconstitution with time-series data of each subnetwork's activity is a powerful approach to dynamically model the plant immune signaling network [3].

2 Materials

1. A computer with R, language and environment for statistical computing, installed. The computer should be running on Windows, Mac OSX, or Linux as the operating system. R version of 3.2 or later is required for the script. R can be downloaded freely (https://www.r-project.org/). Install it according to the instructions. The R script (Material 3) was tested with R version 3.2.0 on Windows. In principle, this R script should work on Mac OSX or Linux, but this functionality has not been tested.
2. The R packages nlme, RColorBrewer, and stringr installed. Install the nlme package by opening the RGui (if R is installed according to the default setting, a shortcut with "R" icon should be on the desktop), going to Packages (in the menu bar) > Install packages, selecting a CRAN mirror (somewhere in or near your country), selecting the package name (nlme), and clicking OK. Follow the same procedure to install the RColorBrewer and stringr packages.
3. The R script "network.reconst.r". The script can be downloaded from https://sites.google.com/a/umn.edu/katagiri_software/home/mimb2016. It should be in the same folder as the data file.
4. The organism of interest with the genotype for every subnetwork combination. For example, 16 genotypes for four subnetworks.
5. An experimental setup for measuring a quantitative phenotype of interest. For example, a setup for measuring bacterial titer after inoculation of a bacterial pathogen in Arabidopsis leaves.

3 Methods

1. Measure the quantitative phenotype in every genotype. As discussed in the Introduction, biological replicates should be used along with good experimental design.
2. Organize the collected data in a long table format in Microsoft Excel (or some other spreadsheet) and save it as a tab-delimited text file (select this format from the dropdown for "Save as": the file extension should be ".txt"). The long table format has one measured value per row, and additional columns contain attributes of the measured values. The top row should contain the column names. Make sure that the column names are exactly the same as the factor names used in the mixed-effect linear model formula—note that they are case sensitive. You may put the columns in any order. See the format in the sample data file ("input4y.dat.txt"), which can be downloaded

from https://sites.google.com/a/umn.edu/katagiri_software/home/mimb2016.

(a) The table must include the columns for "genotype" and "value". Make sure that the genotype uses the ABC … notation as explained in the Introduction. Note that there should be no space or anything else between the letters.

(b) If the experimental setup includes a control treatment, include a "treatment" column: the column values must be either "c" (control) or "t" (treatment).

(c) The table should also include a column for each of the random factors that are to be included in the random effects model according to the experimental design used.

3. Open RGui.

4. In the "R Console" window inside RGui, set the working directory to the folder that contains the data file from **step 2** and the R script "network.reconst.r". The output from the software will be saved in the same folder. To set the working directory, go to File (in the menu bar) > Change dir … Then browse to the folder in the dialog box.

5. Modify the script. Open "network.reconst.r" with a text editor (e.g., WordPad) or an R editor. Lines 169–171 (with treatment) or lines 184–186 (without treatment) in "network.reconst.r" are the formula for the mixed-effect linear model. The original models are:

```
## the model with treatment.
lme.rec = lme( value ~ m. -1 + genotype,
   random =~ 1|experiment/flat/box,
   data = input.dat)

## the model without treatment.
lme.rec = lme( value ~ m.,
   random =~ 1|experiment/flat/box,
   data = input.dat)
```

Modify the random factors, which follow "random =~ 1|", in the model you are using (with or without treatment). The original random factors are "experiment/flat/box". Save the script (make sure that the extension did not change from ".r").

6. Run the script. In the R Console window in RGui. Type at the prompt (">"), "source('network.reconst.r')" (type it without the double quotations) and hit Return.

7. Answer questions that show up in R Console. If the script is aborted with an error message, fix the problem and resume from **step 6**. The questions are:

(a) "Data file name?" A Windows dialog box will open for browsing to the file. Open the data file made in **step 2**. If the data file does not have "genotype" or "value" columns, the script will get aborted with an error message. If the script gives a "Warning message: NAs introduced…", it means that the value column contained non-numeric values that were made into NAs or that missing values were made into NAs.

(b) "Is control data included? (y/n)" If the data include data from both treatment and control in the treatment column as discussed in the Introduction, type "y" and hit return. Otherwise, type "n" and hit return. If "y" is given but the data file does not have the treatment column, it will be aborted with an error message, "No treatment column". If the treatment column is included and the column contains character values other than "c" or "t", it will be aborted with an error message, "treatment level must be c or t".

(c) "The number of the subnetworks? (integer, 2 to 26)" Type the integer value and hit return. If any of the genotype values are not consistent with the ABC … notation for the specified subnetwork number, "The genotype value is not correct." will be shown, and the script will be aborted. If the values in the genotype column do not cover all possible genotypes, "Warning: missing genotypes" will be shown. If the data for a small number of genotypes are missing, in most cases fitting of the mixed-effect linear model still works. However, some of the contributions will not be able to be calculated, and the mean estimates for these will be indicated as "NA" in the output file. In this case, the mixed-effect linear model automatically assumes the value of 0 for the contributions with "NA" to calculate the other contributions, which may lead to inaccurate estimation of the other contributions.

(d) "Interactions between up to how many subnetworks? (integer from 1 to the subnetwork number)" Type the integer value and hit return. If you find higher orders of interactions insignificant, removing these higher orders of interactions may produce higher statistical power in the estimates of the lower order interactions. If the value of 1 is chosen, no interactions will be estimated but single sector contributions will be.

8. Check the data. The script generates a box plot for each genotype (without treatment included) or each genotype with each treatment (with treatment included). The file is named data file name + (the highest order for the interactions estimated) + ".boxplot.pdf". This plot will allow checking of whether the data for the genotype (and treatment) factors was imported to R properly. Note that this plot ignores the random factors, so it cannot be used to check whether random factors are imported properly.

9. Diagnose if use of the linear model platform was justified. If you are familiar with linear models, view the diagnostic plots: a Q-Q plot of the residuals against the standard normal distribution and a residual-vs-fitted plot are included in the file named data file name + (the highest order for the interactions estimated) + ".diagnostic.pdf". These diagnostics assess the assumptions of linear models: (1) the residual normally distributes; (2) the residual distribution is independent of the explanatory variables. Both of these assumptions should be met. If the Q-Q plot is close to the red dashed diagonal line, assumption (1) is met. If the residual-vs-fitted plot is approximately symmetric with respect to the red dashed horizontal line for residual = 0 and if the residual distribution is pretty much the same along the entire fitted value range, assumption (2) is met. You may want to consider transformation of the measured values, such as log-transformation or, more generally, Box-Cox power transformation, if the residual distribution is affected by the fitted value: e.g., if the residual distribution has a funnel-like shape that is narrower for smaller fitted values.
10. Understand the output file. The output file will be generated in the working directory with the file name data file name + (the highest order for the interactions estimated) + ".output.txt". It is a tab-delimited text file, which can be opened in Excel. For each contribution listed in the first column ("Contribution"), the mean estimate ("Value"), its standard error ("Std.Error"), the associated P-value (p.value), and other parameters used for the *t*-test of the contribution's significance are listed for each row. The names of the contributions following "m." are the results of the network reconstitution. The names with a single letter following "m.", such as "m.A" and "m.B", are the single subnetwork contributions. The names with two letters separated by a colon following "m.", such as "m.A:B" and "m.A:C", are the interactions between the two indicated subnetworks. Higher orders of interactions are represented in a similar way: more letters separated by colons. If the treatment is included, "m.remainder" shows the remaining network activity in the all-subnetwork-deleted state (e.g., in the case of four subnetworks, the remaining activity in the quadruple deletion). If the treatment is included, the output also has the contributions whose names are "genotype" followed by the ABC ... notation of the genotypes, such as genotypeabcd and genotypeAbCd. These contributions are the mean estimates for the control treatment in the corresponding genotypes. These mean estimates were subtracted from the mean estimates for the treated in each genotype to calculate the treatment-specific effects that were used for network reconstitution calculation. If the treatment is not included, there is a contribution named "(Intercept)". This is the mean estimate for the value for the

all-subnetwork-deleted state. Since the network reconstitution is to compare the values of other network states to the all-subnetwork-deleted state, the (Intercept) value was subtracted as the baseline activity.

11. Visualize the network reconstitution results. The mean estimates and their standard errors for the network reconstitution (the contributions starting with "m.") are plotted in a bar plot in the file named data file name + (the highest order for the interactions estimated) + ".reconstitution.pdf". Different orders of interactions are colored differently. At the left side of the plot inside the box, the p-value for each contribution is shown. Note that these p-values were adjusted using the Holm multiple tests correction, and thus they are larger than the p-value in the output file. Adjusted p-values smaller than 0.05 are shown in red, and otherwise shown in blue. The plot makes it easy to see which contributions have positive or negative values and how significant they are (i.e., likely different from 0).

12. Interpret the network reconstitution results. The interpretation is straightforward if the single subnetwork contributions are all positive or all negative. If they are all positive, any subnetwork interactions that are positive show synergistic interactions (i.e., the network activity is more than additive) and any subnetwork interactions that are negative show compensatory interactions (i.e., the subnetworks compensate each other with respect to network function). Similarly, if the single subnetwork contributions are all negative, negative interactions are synergistic and positive interactions are compensatory. Even if not all single subnetwork contributions are positive, as long as the single subnetwork contributions corresponding to the elements of the interactions being considered have the same sign, the interpretation is similar. For example, in a four-subnetwork case, if A, B, and C are positive and D is negative and if A:B:C is negative, the A:B:C interaction is compensatory. However, we cannot give a simple interpretation, for example, to the A:D interaction, as the sign for the single subnetwork contributions A and D are opposite. This shortcoming that mechanistic interpretation of the results is limited to particular cases is due to the fact that network reconstitution is a quantitative description of the network function but not a mechanistic model. However, it is a far better description than comparing a subnetwork-deleted state to the intact network state (i.e., comparing a mutant phenotype to the wild-type phenotype) as the latter may not give any reliable mechanistic interpretations when interactions among subnetworks are not negligible: this issue is discussed in [4]. In fact, we have demonstrated that network reconstitution is an important concept and tool when we built a mechanistic model of the plant immune signaling network [3].

Acknowledgments

I thank Rachel Hillmer for critical reading of the manuscript. This work was supported by grants from National Science Foundation, MCB-0918908, IOS-1121425, and MCB-1518058.

References

1. Tsuda K, Sato M, Stoddard T, Glazebrook J, Katagiri F (2009) Network properties of robust immunity in plants. PLoS Genet 5(12):e1000772
2. Sato M, Tsuda K, Wang L, Coller J, Watanabe Y, Glazebrook J, Katagiri F (2010) Network modeling reveals prevalent negative regulatory relationships between signaling sectors in Arabidopsis immune signaling. PLoS Pathog 6(7):e1001011
3. Kim Y, Tsuda K, Igarashi D, Hillmer RA, Sakakibara H, Myers CL, Katagiri F (2014) Mechanisms underlying robustness and tunability in a plant immune signaling network. Cell Host Microbe 15(1):84–94
4. Katagiri F, Tsuda K (2010) Understanding the plant immune system. Mol Plant-Microbe Interact 23(12):1531–1536

Chapter 19

Stomatal Bioassay to Characterize Bacterial-Stimulated PTI at the Pre-Invasion Phase of Infection

Jeanine Montano and Maeli Melotto

Abstract

Bacterium-triggered stomatal closure is a functional output of plant immunity, also known as stomatal defense. This is an early response mediated by the recognition of pathogen-associated molecular patterns (PAMPs) by the plant's pathogen recognition receptors (PRRs). Several approaches to analyzing stomatal movement in response to bacteria have been described, but difficulties in fine-tuning the experimental procedures still exist. Here we provide a detailed method for assessing stomatal defense via high-quality microscopic imaging and explain trouble-shooting steps to obtaining robust data. Although this procedure requires minimal manipulation of the leaf sample, it is crucial to control all environmental conditions and extrinsic variables that could interfere with stomatal movement. The method described here is also suitable for in vivo characterization of stomatal response in new pathosystems and can be used in conjunction with other profiling assays to gain a detailed understanding of early PAMP-triggered immunity (PTI).

Key words Stomatal movement regulation, Guard cell, Plant–microbe interaction, Bacterial pathogens, PTI, Stomatal defense

1 Introduction

Bacteria are a natural component of the leaf epiphytic community, surviving harsh conditions such as intense UV radiation, desiccation, and poor nutrient availability [1–3]. Internalization into leaves allows phytopathogenic bacteria to escape these pressures and it is often the initiation of disease [1, 4]. Plants can sense pathogens by pathogen recognition receptors (PRRs) that bind to conserved pathogen-associated molecular patterns (PAMPs) [1, 5, 6]. Recognition of PAMPs, such as bacterial flagellin and lipopolysaccharide [7, 8], by PRRs initiates intercellular signaling that leads to the first line of plant defense against invaders, often mediated by hormonal signaling [7, 8]. This complex interaction is part of PAMP-triggered immunity (PTI) [1, 5–8]. Interestingly, the PTI-stimulated salicylic acid hormone contributes to the reduction of stomatal aperture [7, 8]. Given that stomatal closure can limit

Libo Shan and Ping He (eds.), *Plant Pattern Recognition Receptors: Methods and Protocols*, Methods in Molecular Biology, vol. 1578, DOI 10.1007/978-1-4939-6859-6_19, © Springer Science+Business Media LLC 2017

internalization of disease-causing organisms or other microbial contaminants, this output of PTI is vital for maintaining plant health and productivity.

Here we provide a detailed method for assessing stomatal response to bacteria under controlled environmental conditions, thus evaluating the initial steps of PTI (*i.e.*, <6 h of exposure to the microbe). We use surface inoculation (either whole plant dipping into or detached leaf floating onto inoculum) and inoculation by vacuum infiltration to assess stomatal defense. Subsequently, we measure stomatal aperture width under a light microscope over a period of time to determine whether and how the bacterium being tested modulates stomatal movement. Characterization of stomatal response over time allows for understanding the pathogen's ability to activate, disable, or fail to inducing PTI. Stomatal defense and its contribution to disease progression have been well described in the model pathosystem Arabidopsis–*Pseudomonas syringae* [1, 4, 8, 9]. For any new plant–microbe interaction being analyzed, it is imperative that a complete characterization of this early PTI response be multifactorial that includes other assays to provide well-rounded conclusions. Assessing stomatal aperture alone is not sufficient to determine the contribution of stomatal defense (or lack thereof) to microbial invasion of the apoplast and the outcome of disease or contamination. For instance, *Salmonella enterica* serovar Typhimurium seems to be unable to efficiently penetrate parsley leaves even though stomatal pores are open [10]. It is important to use additional methods such as determining endophytic population size after surface inoculation and vacuum infiltration as described by Katagiri et al. [9], or direct microscopic observation of bacterial behavior in the phyllosphere. These types of assays are specifically important if new pathosystems or mutants are under investigation.

2 Materials

2.1 Medium and General Laboratory Materials

1. Low Salt Luria Bertani (LS-LB) medium: 10 g tryptone, 5 g NaCl, 5 g yeast extract. Fill to 1 L with distilled water and autoclave the medium. If needed, supplement the medium with appropriate antibiotics after autoclaving. For solid medium, add 8 g/L of agar prior to autoclaving.
2. Sterile, distilled water.
3. Vac-In-Stuff (Silwet L-77®).
4. Erlenmeyer flask of sufficient size to support bacterial culture incubation.
5. 50 mL conical tubes (for centrifugation).
6. Plastic reservoir (e.g., Tupperware®).
7. Glass beaker.

8. Microscope slides.
9. Razor blade.
10. Tweezers.
11. Petri dishes.

2.2 Equipment and Software

1. Nikon ECLIPSE Ni-U upright microscope equipped with Nikon Plan Fluor 20× objective lens and NIS Elements Imaging Software Version 4.13.04. Alternate microscope models and software with the appropriate magnification and measurement capabilities will suffice.
2. Spectrophotometer and cuvettes.
3. Plant growth chamber (*see* **Note 1**) with Philips Silhouette High Output F24T5/841 HO Alto Collection 24 W bulb.
4. Laminar flow hood.
5. Centrifuge and rotor to spin 50 mL conical tubes.
6. Orbital shaker incubator at 28 °C.

2.3 Plant Tissue and Bacterial Culture

1. Young, fully extended leaves of soil-grown plants (*see* **Note 2**). Leaves should remain attached to the plant for whole surface inoculation [5], but may be detached for floating on inoculum [11].
2. Bacterial culture on agar plates freshly streaked from frozen glycerol stock (*see* **Note 3**).

3 Methods

3.1 Plant Inoculation and Sampling

1. Grow a culture of *P. syringae* to late log phase (0.8–1 OD_{600}) in liquid LS-LB in an orbital shaker at 28 °C. It may take 12 h.
2. Prepare bacterial inoculum: harvest cells by centrifugation at 1360 × *g* and 20 °C for 20 min. Discard supernatant and suspend cells in sterile, distilled water to a volume to reach 0.2 OD_{600} that corresponds to 1 × 10^8 CFU/mL of *P. syringae.*
3. At 3 h after light turns on in the morning, excise a leaf from one plant, cut a flat portion, and place it on a microscope slide (*see* **Note 4**). Observe the abaxial surface of the leaf piece under the microscope to ensure that stomata are wide open (*see* **Note 5**) and proceed with inoculation.
4. Inoculate detached leaves or whole plants and return them to the growth chamber (*see* **Notes 6 and** 7).
   (a) For detached leaf inoculation: float a leaf, abaxial side down, on an aliquot of the bacterial inoculum on a microscope slide. Add inoculum to the slide as needed to prevent leaf drying. Float leaves on water as a control. Mount one slide

per time point of the experiment as leaves must be discarded after imaging.

(b) For whole plant inoculation: dip the plant in the inoculum for 15 s as described by Katagiri et al. [9]. Mock inoculation must also be performed to serve as a control. At the time of sampling, excise one leaf to observe under the microscope (*see* **Note 4**) and keep the plants in the growth chamber for subsequent sampling times.

3.2 Leaf Surface Image Capture

1. Randomly take live microscopic images of the abaxial leaf surface at 2 and 4 h post-inoculation (hpi) (*see* **Note 8**).
   (a) Detached leaves inoculated on microscope slides can be directly imaged by simply taking one slide out of the growth chamber at each time point and placing it under the microscope.
   (b) If whole plants were inoculated, excise one leaf for each condition by holding the petiole with the tweezers and cutting with the razor blade. Float a flat section of the leaf on water (just enough to avoid leaf drying) on a microscope slide with the abaxial surface upward (*see* **Note 4**).
2. For each treatment, record images of at least 60 stomata at 200× magnification under the microscope (*see* **Note 9**).

3.3 Stomatal Aperture Data Analysis

1. Using the NIS Elements Imaging Software, zoom in on a stomate and use the measuring tool to calculate the stomatal aperture width, which lies between the guard cells (Fig. 1).

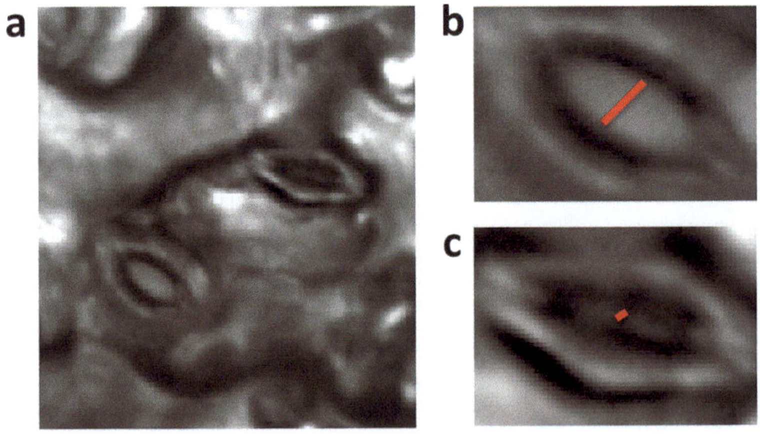

Fig. 1 Microscopic images of Arabidopsis abaxial leaf surface showing an open and a closed stomatal pore and measuring zone. The images were recorded using a Nikon Eclipse Ni-U upright microscope with Nikon Plan Fluor 20× objective lens. Images were analyzed using the NIS Elements Imaging Software Version 4.13.04. (**a**) Open and closed stomata on left and right, respectively. (**b**, **c**) Close-up images of an open (**b**) or a closed (**c**) stomate. The *red bar* indicates the measured zone

2. Record all measurements ($n > 60$ stomata per treatment) and calculate the mean and standard error (*see* **Note 10**).
3. Determine statistical significance between means by using Student's *t*-test or ANOVA (*see* **Note 11**).

4 Notes

1. Controlled environmental conditions are highly important for the success of the experiment. Optimum growth conditions must be used for the plant being tested.
2. We have used several plant species in this stomatal bioassay. Well-irrigated and healthy plants are essential to assess stomatal defense to rule out the possibility that stomatal closure is due to drought and other stresses. As an example, *Arabidopsis* seeds can be sown in hydrated professional growing mix substrate mixed with vermiculite or a 1:1:1 v:v:v ratio of substrate–perlite–vermiculite [5, 9]. Plants are maintained in growth chamber at 22 °C with 12 h of daylight with a photosynthetically active radiation of 100 μmol/m²/s (Philips Silhouette High Output F24T5/841 HO Alto Collection 24 W bulb) and 65–75% air relative humidity for 4–5 weeks. Pots should be placed on flats for underpot irrigation. Optimal Arabidopsis growth conditions and procedures are well documented [9].
3. Multiple subculturing of bacterial pathogens is not recommended as undesired genetic and physiological changes can occur. As an example, we keep glycerol stocks of *P. syringae* in the −80 °C freezer and streak new agar plates for every experiment. Culture plates are maintained at 4 °C for up to 7 days. Additionally, we re-isolate the bacterium from infected plants every 2 years to make new glycerol stocks.
4. When cutting the leaf, remove curved sections, the petiole, and the mid-vein to ensure tissue is completely flat. Any amount of curvature will reduce image quality and prevent proper analysis of the entire leaf surface. When placing the leaf piece on a microscope slide, make sure to add water in a sufficient quantity under the leaf piece to avoid dehydration of the leaf while imaging, but do not flood the upward-facing, abaxial surface. We do not use a coverslip as doing so air bubbles may form on the top of stomatal pores due to continuous transpiration of the leaf making it impossible to measure the stomatal aperture size. Furthermore, coverslip may promote buildup of humidity on the leaf surface during imaging, which could interfere with guard cell response to the bacterium being tested and lead to incorrect data.
5. It is imperative that stomata are widely open when initiating the experiment so that bacterium-triggered stomatal closure (or lack of) can be quantitatively measured with accuracy.

6. *Pseudomonas syringae* inoculum preparation and Arabidopsis inoculation procedures have been published extensively [9]. However, Silwet L-77 or other surfactants must not be used to inoculate detached leaves on slides. If whole plants are to be dip-inoculated, Silwet L-77 should be added to the inoculum at 0.03%.
7. The detached leaf floating procedure is quick and highly reproducible and is the preferred method for analyzing stomatal defense, but it only allows for stomatal analysis. Whole plant inoculation can be useful when correlating stomatal defense to apoplastic immunity, disease development, or other characteristics that require intact plants. However, whole plant inoculation requires more care than for detached leaf floating (*e.g.*, non-inoculated leaves must be marked or removed to prevent analysis of untreated leaves; additional space is required to maintain at least one plant per treatment; and timing of supplemental assays alongside stomatal analysis may be cumbersome).
8. Under normal physiological conditions including the aforementioned growth conditions, stomata of non-inoculated or mock-inoculated Arabidopsis remain open from morning until mid-afternoon. It is important to determine the circadian rhythm of stomatal movement of the plant being tested to choose the total experimental time and time points for sampling. Control leaves must have very similar stomatal aperture throughout the experiment. Some plant pathogens and human pathogens such as *P. syringae* pv. *tomato* (*Pst*) strain DC3000 [1, 3, 4] and *S. enterica* serovar Typhimurium SL1344 [5], promote initial stomatal closure due to PTI as indicated by a maximum decrease in stomatal aperture width at 2 hpi. *Pst* DC3000 can reopen stomata by producing coronatine, promoting a maximum increase in stomatal aperture at 4 hpi [1, 3, 4]. SL1344 can also completely reopen stomata at 4 hpi by an uncharacterized mechanism [5]. Other bacteria, such as the human pathogen *Escherichia coli* O157:H7, induce a strong and lasting stomatal closure completely disrupting the circadian stomatal movement [1, 4, 5]. Some bacterial species (*e.g.*, *P. syringae* pv. *syringae* B728a) lie on the other end of the spectrum and do not induce any stomatal closure [12] (Fig. 2), indicating that stomatal closure is not a general response to all bacteria. Due to the broad range of possible stomatal responses, it is highly important to perform a time course experiment to see progression of stomatal response and to include the appropriate controls to distinguish stomatal defense from circadian rhythm-driven stomatal movement.
9. Capturing images of at least 60 stomata will require multiple images and possibly more than one leaf section. Image capture should be done quickly. Do not spend time measuring stomata

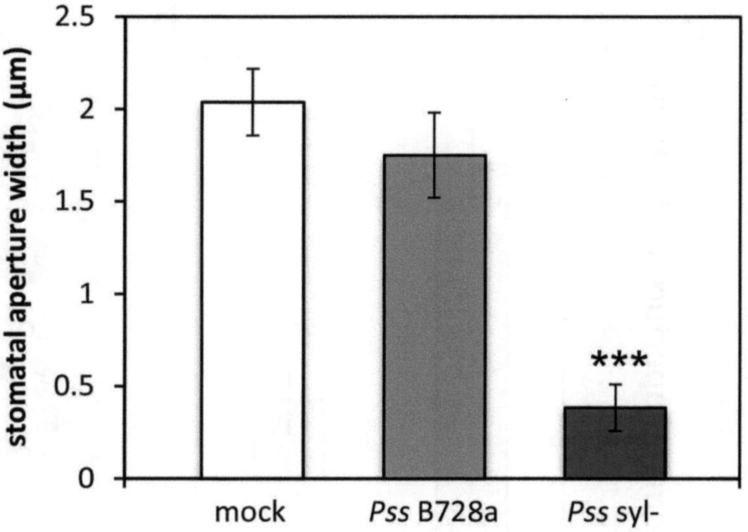

Fig. 2 Stomatal aperture width of bean plants at 2 h post dip-inoculation with *Pseudomonas syringae* pv. *syringae* (*Pss*) B728a, its syringolin A mutant (*Pss* syl-), or a water control (mock). Bacterial suspensions were prepared at 1×10^8 CFU/mL. Results are shown as the mean ($n = 60$) ± SE and statistical significance between treatments was calculated using a Student's *t*-test (***$p < 0.001$). Note that the wild type bacterium does not close the stomatal pore

and analyzing data at the same time; this practice will create inconsistency in timing and may induce stomatal closure due to temporal factors rather than PTI. Ensure that all images are in crisp focus as measurements may not be accurate if an image is blurred.

10. Record all stomata within a field of view. Our measuring tool does not allow a measurement of 0 µm. Therefore, we record the smallest possible region (typically 0.07 µm) by zooming in all the way and manually change the value to 0 µm when transferring data to a spreadsheet. Zooming-in applies to all stomata to accurately select the internal border of the guard cell and stomatal pore width. Blurry stomata are difficult to measure accurately, so these may be omitted if their aperture is not distinguishable. Recording crisply focused images of completely flat leaf sections is the best way to mitigate this problem.
11. Stomatal aperture analysis can be correlated with thermal imaging analysis due to the evaporative cooling effect of open stomata [13]. Thermal imaging can be used as a screening technique for analyzing large populations of mutant plants or bacteria. Nondestructive, time course experiments can also be performed with thermal imaging to assess the dynamic of stomatal response on the same plant. Due to the physical properties of light, all pictures should be taken in the same location

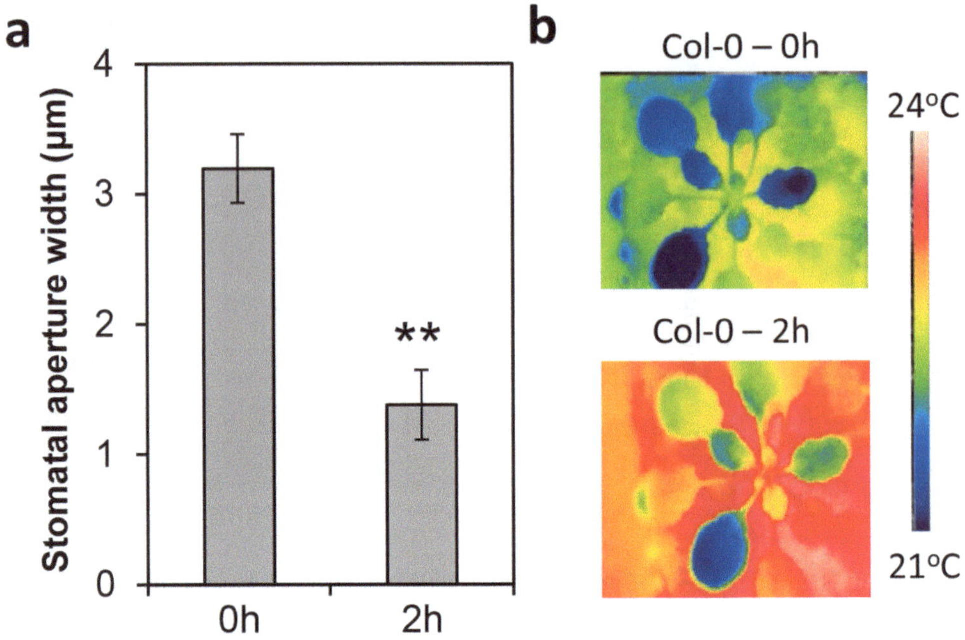

Fig. 3 Stomatal aperture width and thermal imaging correlation at 0 and 2 h post inoculation (hpi). (**a**) Stomatal aperture width of Arabidopsis leaves at 0 and 2 h after dip-inoculation with 1 × 10^8 CFU/mL *Salmonella enterica* Typhimurium SL1344. Results are shown as a mean ($n = 60$) ± SE. Statistical analysis was performed using a Student's *t*-test (**$p < 0.01$). (**b**) Thermal images of the same plants used for direct stomatal aperture measurements at 0 and 2 hpi. Temperature range on the right indicates the temperature/color correlation. Cooler plants (top, 0 hpi) indicate that stomata are mostly open, while warmer plants (bottom, 2 hpi) indicate that stomata are mostly closed

and from the same distance to allow for comparisons among treatments and time points. Figure 3 demonstrates the correlation between direct measurement of stomatal aperture width and leaf surface temperature measured with an infrared camera and associated software (FLIR T400).

Acknowledgments

Research on stomatal defense in the Melotto laboratory is supported by NIH grant 5R01AI068718, USDA-NIFA grant CA-D-PLS-2291-CG, and the Center for Produce Safety.

References

1. Melotto M, Underwood W, He SY (2008) Role of stomata in plant innate immunity and foliar bacterial diseases. Annu Rev Phytopathol 46:101–122
2. Dees M, Lysoe E, Nordskog B, Brurberg M (2015) Bacterial communities associated with surfaces of leafy greens: shift in composition and decrease in richness over time. Appl Environ Microbiol 81:1530–1539

3. Lee J, Teitzel GM, Munkvold K, de Pozo O, Martin GB, Michelmore RW, Greenberg JT (2012) Type III secretion and effectors shape the survival and growth pattern of *Pseudomonas syringae* on leaf surfaces. Plant Physiol 158:1803–1818
4. Underwood W, Melotto M, He SY (2007) Role of plant stomata in bacterial invasion. Cell Microbiol 9:1621–1629
5. Roy D, Panchal S, Rosa B, Melotto M (2013) *Escherichia coli* O157:H7 induces stronger plant immunity than *Salmonella enterica* Typhimurium SL1344. Phytopathology 103:326–332
6. Zhang W, He SY, Assmann S (2008) The plant innate immunity response in stomatal guard cells invokes G-protein-dependent ion channel regulation. Plant J 56:984–996
7. Nicaise V, Roux M, Zipfel C (2009) Recent advances in PAMP-triggered immunity against bacteria: pathogen recognition receptors watch over and raise the alarm. Plant Physiol 150:1638–1647
8. Melotto M, Underwood W, Koczan J, Nomura K, He SY (2006) Plant stomata function in innate immunity against bacterial invasion. Cell 126:969–980
9. Katagiri F, Thilmony R, He SY (2002) The *Arabidopsis thaliana-Pseudomonas syringae* interaction. In: Somerville CR, Meyerowitz EM (eds) The Arabidopsis book. American Society of Plant Biologists, Rockville, MD. doi:10.1199/tab.0039 http://www.aspb.org/publications/arabidopsis
10. Golberg D, Kroupitski Y, Belausov E, Pinto R, Sela S (2011) *Salmonella* Typhimurium internalization is variable in leafy vegetables and fresh herbs. Int J Food Microbiol 145: 250–257
11. Chitrakar R, Melotto M (2010) Assessing stomatal response to live bacterial cells using whole leaf imaging. J Vis Exp (44). doi:10.3791/2185
12. Schellenberg B, Ramel C, Dudler R (2010) *Pseudomonas syringae* virulence factor syringolin A counteracts stomatal immunity by proteasome inhibition. Mol Plant Microbe Interact 23:1287–1293
13. Mustilli AC, Merlot S, Vavasseur A, Fenzi F, Giraudat J (2002) Arabidopsis OST1 protein kinase mediates the regulation of stomatal aperture by abscisic acid and acts upstream of reactive oxygen species production. Plant Cell 14:3089–3099

Chapter 20

Using Clear Nail Polish to Make *Arabidopsis* Epidermal Impressions for Measuring the Change of Stomatal Aperture Size in Immune Response

Shuchi Wu and Bingyu Zhao

Abstract

Plant stomata are an essential route for bacterial pathogens to entry inside host tissue and cause diseases. As an important defense mechanism, plant stomata can actively restrict bacterial invasion by dynamically regulating the opening, closing, and reopening of stomatal guard cells. Therefore, accurately measuring the stomatal aperture size during the bacterial pathogenesis is an important approach to study the stomata related immunity. Several methods have been developed for stomatal aperture measurement. Here, we described a detailed protocol of using clear nail polish to make *Arabidopsis* epidermal impressions for investigate the change of stomatal aperture size in plant immune response. The application of this approach can instantly fix the status of stomatal guard cells, and provides clear, stable, and almost permanent slides of epidermal impressions for measurement of stomatal aperture size.

Key words Nail polish, *Arabidopsis thaliana*, Epidermal impressions, Stomatal guard cell

1 Introduction

Healthy plants can effectively prevent most microbes to get inside of plant tissue, and limit the colonization of the microbes on leaf surface as epiphytes [1, 2]. While the plant surface, poor with nutrient, is not a "hospitable" growth environment for most microbes [2]. The epiphytes on leaf surface can be easily washed off by heavy rain or killed by intensive UV light [3]. And the dramatic fluctuation of temperature and humidity on leaf surface also cast severe stress to most microbial cells [2]. Comparing to the "barren" surface, the inside plant mesophyll tissue is a much "fertile" land for microbial pathogens [4]. Therefore, microbial pathogens developed a number of distinct strategies to entry host plant tissue. For example, fungal pathogens (e.g., *Hyaloperonospora parasitica*) evolved a set of elaborate piercing structures such as penetration peg and haustorium to break through plant surface, and form infection structures inside of mesophyll tissues [5].

Libo Shan and Ping He (eds.), *Plant Pattern Recognition Receptors: Methods and Protocols*, Methods in Molecular Biology, vol. 1578, DOI 10.1007/978-1-4939-6859-6_20, © Springer Science+Business Media LLC 2017

Unlike fungal pathogens, bacterial pathogens do not equip such elaborate piercing structures. Therefore, they usually entry through plant surface wounds and natural openings, such as stomata, lenticels and hydathode [6, 7]. The stomata are the most common natural openings on leaf surface and therefore become one of the most important routes for bacterial invasion.

For most plant species, stomatal aperture size is controlled by two guard cells [8, 9]. It can reduce the aperture size to inhibit water loss, and increase the aperture size to promote gas exchange [8, 9]. During microbe pathogen invasion, some pathogen-associated molecular patterns (PAMPs) such as flg22, a conserved domain of bacterial flagellin can be recognized by host plant immune receptor FLS2, and trigger defense responses called PAMP-triggered immunity (PTI). The defense signal involves in salicylic acid (SA) and abscisic acid (ABA) signaling pathways that regulate guard cells to close the stomatal aperture [10]. On the other hand, some plant bacterial pathogens evolved ability to induce the reopening of stomata [10–12]. For example, phytotoxin coronatine produced by *Psuedomonas syringae* (*P. syringae*) strains can reopen stomata closed by the PTI [10]. More recently, *Xanthomonas campestris* was reported to secrete a small unidentified molecular that can enforce stomatal reopening [13]. Fungal toxins, fusicoccin, produced by *Fusicoccum amygdali* has the ability of stomatal regulation [14]. Oxalate produced by fungal pathogen *Sclerotinia sclerotiorum* can also enhance stomatal opening [15]. Therefore, measurement of stomatal aperture size is an essential and necessary indicator for studying the stomata related immunity.

Several methods have been developed to prepare microscope slides of leaf epidermal impressions for measuring the stomatal aperture size, including directly peeling the epidermis of leaf [10, 16, 17], and using nail polish, silicone adhesive, or stamp pad to make leaf surface imprint [18, 19]. Because *Arabidopsis* leaves are fragile and not easy to make leaf surface imprint, the epidermal peeling method becomes the most popular approach in the study of *Arabidopsis* stomata related immunity. However, in the process of bacterial invasion, as well as the influence of humidity and circadian, the stomatal guard cells are always in a dynamic status of opening, closing and reopening [17, 20, 21]. Thus the epidermal peeling method has a significant deficient that it cannot fix the stomata opening status, and makes the stomata counting and stomatal aperture size measurement unrepeatable.

Here, we are described an optimized protocol of using clear nail polish to make *Arabidopsis* epidermal impressions, and monitor the change of stomatal aperture size in plant immune response. Suitable clear nail polish can instantly fix the status of stomatal guard cells and provides clear, stable, and almost permanent slides of epidermal impressions for measurement of stomatal aperture size.

2 Materials

1. Plant material: Four-week-old *Arabidopsis* Co-0 plants grown in growth chamber (50% relative humidity, 100 μE/m^2/s light, 10 h light/22 °C and 14 h dark/20 °C).
2. Stomata opening buffer: 10 mM KCl, 10 mM MES, pH 5.7.
3. Stomata closure elicitors (PAMP): 5 μM flg22 peptide (GenScript, NJ).
4. Stomata closure elicitors (plant hormone): 10 μM Abscisic acid (ABA) (Sigma-Aldrich, MO).
5. Nail polish: Sally Hansen® Hard As Nails® Xtreme wear® Nail Color Invisible (*see* **Note 1**).
6. Scissors, forceps, slides, and scotch tape.
7. Light microscope: Zeiss Axio Observer A1 (Carl Zeiss MicroImaging, Inc., NY).
8. Software: AxioVision LE (Carl Zeiss MicroImaging, Inc., NY).

3 Methods

3.1 Prepare Slides of Epidermal Impressions

1. Move *Arabidopsis* plants out of growth chamber, cover the whole flat with deep dome-shape lid (Fig. 1a), and incubate in dark at room temperature for 3 h.
2. Cut off similar sized matured leaves from *Arabidopsis* plants (Fig. 1b) (*see* **Note 2**).
3. Paint a thin layer of nail polish (*see* **Note 1**) on the lower epidermis (Fig. 1c).
4. Incubate the leaves at room temperature to air dry the nail polish for about 10 min (*see* **Notes 3** and **4**).
5. Use the tip of forceps to punch a small wound at the edge of the nail polished area (Fig. 1d) (*see* **Note 5**).
6. Carefully start to peel off dried nail polish film with forceps from the wound site.
7. Stop peeling when about 0.5 cm nail polish film has been peeled off, and use scissors to cut off the peeled off nail polish film.
8. Transfer the nail polish film onto the microscope object slide and covered it with another object slide (Fig. 1e) (*see* **Note 6**).
9. Carefully press tight the slide and use scotch tape to seal the edge of the slides.
10. Keep the slides in a box container in the room temperature (*see* **Note 7**).

Fig. 1 The process of making Arabidopsis epidermal impressions. (*1*) Four-week-old *Arabidopsis* plants were maintained in higher humidity. (*2*) Detached leaf used for making the leaf reprint. (*3*) Apply the finger nail polish. (*4*) Peel the thin layer of film with a sharp forceps. (*5*) Mount the imprinted film on slide for microscope observation

3.2 Induce the Stomata Opening and Closure

1. Prepare stomata opening buffer with or without 5 μM flg22 peptide or 10 μM ABA.
2. Cut off similar sized matured leaves from *Arabidopsis* plants that have been treated for 3 h at dark, and float the leaves in stomata opening solutions.
3. Incubate the leaves floating in stomata opening buffer at room temperature under light for 3 h.
4. Take out the leaves and wipe off the buffer residue on the leaf surface.
5. Follow the protocol from **steps 3** to **10** above to make slides of epidermal impressions.
6. Take pictures of prepared slides under 400× amplification light microscope (Fig. 2).
7. Use the "ruler tools" to measure the width of 30 stomatal apertures for each treatment.

4 Notes

1. Be aware of other "clear nail polish" like makeup, such as nail strengthener or nail base. These functional "nail polishes" are not easy to be peeled off.

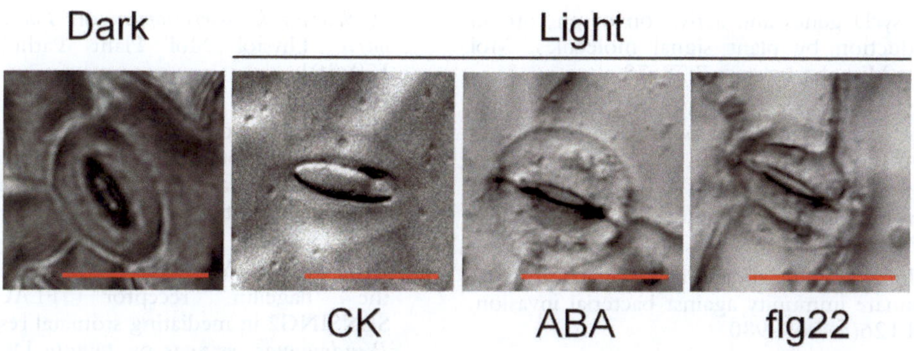

Fig. 2 The stomata on the nail polish films under 400× amplification. The left one is under 3 h dark treatment, and the right three are under another 3 h light treatment, with or without 5 μM flg22 peptide or 10 μM ABA. The legend bar represents 20 μm

2. Do not use very young leaves, which are too fragile.
3. Do not use chemical hood to air dry the nail polish paint, otherwise the leaves will also become dried and stick with the nail polish film.
4. Use the tips of forceps to test if the nail polish is dried.
5. Do not try to peel off the entire nail polish film, otherwise the film will get stretched. Use scissors to cut off 0.5 cm nail polish film, which is enough for stomata observation and measurement.
6. To use another object slide instead of a cover slide that can better protect the nail polish films.
7. The nail polish films will shrink slowly in low humidity.

Acknowledgments

This work was supported by the National Science Foundation.

References

1. Jones JD, Dangl JL (2006) The plant immune system. Nature 444(7117):323–329
2. Lindow SE, Brandl MT (2003) Microbiology of the phyllosphere. Appl Environ Microbiol 69(4):1875–1883
3. Andrews JH, Harris RF (2000) The ecology and biogeography of microorganisms on plant surfaces. Annu Rev Phytopathol 38(1):145–180
4. Beattie GA, Lindow SE (1995) The secret life of foliar bacterial pathogens on leaves. Annu Rev Phytopathol 33(1):145–172
5. Wilson M, Hirano S, Lindow S (1999) Location and survival of leaf-associated bacteria in relation to pathogenicity and potential for growth within the leaf. Appl Environ Microbiol 65(4):1435–1443
6. Bunster L, Fokkema NJ, Schippers B (1989) Effect of surface-active *Pseudomonas* spp. on leaf wettability. Appl Environ Microbiol 55(6):1340–1345
7. Quigley NB, Gross D (1994) Syringomycin production among strains of *Pseudomonas syringae* pv. *syringae*: conservation of the syrB

and syrD genes and activation of phytotoxin production by plant signal molecules. Mol Plant Microbe Interact 7:78–78

8. Fan LM, Zhao Z, Assmann SM (2004) Guard cells: a dynamic signaling model. Curr Opin Plant Biol 7(5):537–546
9. Schroeder JI et al (2001) Guard cell signal transduction. Annu Rev Plant Biol 52(1): 627–658
10. Melotto M et al (2006) Plant stomata function in innate immunity against bacterial invasion. Cell 126(5):969–980
11. Mittal S, Davis KR (1995) Role of the phytotoxin coronatine in the infection of *Arabidopsis thaliana* by *Pseudomonas syringae* pv. *tomato*. Mol Plant Microbe Interactions 8(1):165–171
12. Schulze-Lefert P, Robatzek S (2006) Plant pathogens trick guard cells into opening the gates. Cell 126(5):831–834
13. Gudesblat GE, Torres PS, Vojnov AA (2009) *Xanthomonas campestris* overcomes Arabidopsis stomatal innate immunity through a DSF cell-to-cell signal-regulated virulence factor. Plant Physiol 149(2):1017–1027
14. Turner NC, Graniti A (1969) Fusicoccin: a fungal toxin that opens stomata. Nature 223(5210):1070–1071
15. Godoy G et al (1990) Use of mutants to demonstrate the role of oxalic acid in pathogenicity of *Sclerotinia sclerotiorum* on *Phaseolus vulgaris*. Physiol Mol Plant Pathol 37(3): 179–191
16. Desikan R et al (2002) A new role for an old enzyme: nitrate reductase-mediated nitric oxide generation is required for abscisic acid-induced stomatal closure in *Arabidopsis thaliana*. Proc Natl Acad Sci U S A 99(25): 16314–16318
17. Zeng W, He SY (2010) A prominent role of the flagellin receptor FLAGELLIN-SENSING2 in mediating stomatal response to *Pseudomonas syringae* pv *tomato* DC3000 in *Arabidopsis*. Plant Physiol 153(3):1188–1198
18. Hilu KW, Randall JL (1984) Convenient method for studying grass leaf epidermis. Taxon 33:413–415
19. Behera S, Kudla J (2013) Live cell imaging of cytoplasmic Ca2+ dynamics in Arabidopsis guard cells. Cold Spring Harb Protoc 2013(7):pdb.prot072983
20. Mao J et al (2005) A role for Arabidopsis cryptochromes and COP1 in the regulation of stomatal opening. Proc Natl Acad Sci U S A 102(34):12270–12275
21. Mustilli A-C et al (2002) Arabidopsis OST1 protein kinase mediates the regulation of stomatal aperture by abscisic acid and acts upstream of reactive oxygen species production. Plant Cell 14(12):3089–3099

Chapter 21

Characterizing the Immune-Eliciting Activity of Putative Microbe-Associated Molecular Patterns in Tomato

Christopher R. Clarke and Boris A. Vinatzer

Abstract

Detection of conserved microbe-associated molecular patterns (MAMPs), such as bacterial flagellin, is the first line of active defense in plants against pathogenic invaders. Successful pathogens must subvert this immune response to grow to high population density and cause disease. Flagellin from the bacterial pathogen *Pseudomonas* was the first identified bacterial MAMP and many species across the plant kingdom have sensitive perception systems for detecting the 22-amino acid epitope known as flg22. Tomato and several other solanaceous plants are also able to independently detect a second epitope of flagellin known as flgII-28. This chapter details four experimental protocols to identify and confirm the immune response-eliciting activity of flagellin and putative MAMPs with focus on the *Pseudomonas*–tomato pathosystem.

Key words Microbe-associated molecular patterns (MAMPs), Flagellin, flg22, flgII-28, *Pseudomonas syringae*, Pattern-triggered immunity (PTI)

1 Introduction

All plants harbor a multilayered immune system capable of preventing the majority of plant-associated microbes from causing disease. One branch of this immune system monitors for the presence of microbe-associated molecular patterns (MAMPs, also known as pathogen-associated molecular patterns, PAMPs) through cell surface-localized pattern recognition receptors. Recognition of a MAMP by a pattern recognition receptor elicits myriad defense responses that a successful pathogen must subvert to be able to grow to high population density and cause disease in the plant. The first identified MAMPs included flg22, an epitope of bacterial flagellin [1], elf18, an epitope of bacterial elongation factor-tu [2], and chitin from plant-associated fungi [3]. These MAMPs were all identified based on their relative sequence conservation among related plant-associated microbes. Relative evolutionary conservation of MAMPs was originally considered a key attribute of MAMPs [4, 5] but is not essential [6, 7].

Libo Shan and Ping He (eds.), *Plant Pattern Recognition Receptors: Methods and Protocols*, Methods in Molecular Biology, vol. 1578, DOI 10.1007/978-1-4939-6859-6_21, © Springer Science+Business Media LLC 2017

The first bacterial MAMP, flg22, was identified 17 years ago and over the next decade less than a dozen more MAMPs were identified in plant-associated microbes [8]. More recently, the rate of MAMP discovery has accelerated [9]. Novel MAMPs can then be used to identify cognate plant pattern recognition receptors that will hopefully be useful in resistance breeding programs [10]. Simple, robust assays to accurately confirm the activity of putative MAMPs are therefore useful toward the identification of novel plant disease resistance genes.

We focus on MAMPs found in the bacterial plant pathogen *Pseudomonans syringae*. For work in tomato, we focus on the two MAMPs of flagellin flg22 and flgII-28 because these are the most important MAMPs in this pathosystem [11]. Unlike flg22, perception of flgII-28 is not broadly conserved across the plant kingdom but rather limited to a subset of *Solanaceae* including tomato. flgII-28 was originally identified by analyzing the genome sequence of several extremely closely related strains of *P. syringae* pv. tomato and looking for gene regions under presumably diversifying selection [12]. We and others have recently suggested that identifying protein-coding genes under diversifying selection at the subspecies level but relatively conserved at the species and genus levels is an effective starting point for the identification of novel proteinaceous MAMPs [13, 14].

MAMPs can be proteinaceous or non-proteinaceous. This set of methods is focused on characterizing the immune-eliciting activity of proteinaceous MAMPs. Robust, quantifiable assays are necessary to confirm whether or not putative MAMPs actually have immune response-eliciting activity or not and the magnitude of the response they trigger. The four immune responses we routinely measure to confirm the activity of putative MAMPs in tomato and to measure the strength thereof are: (1) production of reactive oxygen species, (2) stomatal closure, (3) callose deposition, and (4) suppression of bacterial plant pathogen growth. Our protocols described below were adapted from the works of [15–18].

The four assays we detail below were, unless otherwise noted, performed on 4- to 5-week-old tomato plants grown with 16-h day cycles using both flg22 and flgII-28 as positive control MAMPs. MAMPs are tested for their immune response-eliciting activity as purified synthetic peptides, which can be ordered from several sources (e.g., EZBioLab (Carmel, IN, USA)). To dissolve peptides, one must consider the acidity and hydrophobicity of the peptide to decide the best solvent (*see* **Note 1**). For all experiments it is critical to include a mock treatment (ddH_2O) or a known inactive peptide (e.g., elf18 in tomato ([10])) as a negative control. For all described assays, flg22 or flgII-28 [6] can be used as a positive control (*see* **Note 2**).

2 Materials

1. Plant seeds.
2. Soil.
3. Fertilizer.
4. Plant growth chamber or light shelf.
5. 96-well LUMITRAC 200 clear flat bottom white plate (Greiner Bio-One, Monroe, NC, USA).
6. Cork borer.
7. Synthesized peptide of candidate MAMP at 1 mM (EZ BioLab, Carmel, IN, USA). Store at −20 °C for short term or −80 °C for long term.
8. Horse radish peroxidase at 20 mg/ml dissolved in ddH2O. Store at −20 °C.
9. Luminol at 34 mg/ml dissolved in DMSO. Store at −20 °C.
10. ROS elicitation medium (per 1 ml): 1 μL 20 mg/ml horse radish peroxidase, 1 μL 34 mg/ml luminol, 1 μL 1 mM candidate MAMP peptide, 997 μL dH2O.
11. Biotek HD plate reader (Winooski, VT, USA).
12. Microscope slides and coverslip.
13. Axio Imager M1 upright microscope and Axiocam MRM camera (Zeiss, Oberkochen, Germany).
14. 50 ml disposable centrifuge tubes.
15. Alcoholic lactophenol: 50 ml phenol, 50 ml lactic acid, 50 ml glycerin, 50 ml ddH2O. Store in brown bottle in fume hood.
16. 95% EtOH.
17. Analine Blue: Into 100 ml ddH2O add 2.61 g K2HPO4 and 0.01 g Analine Blue. Store in a brown bottle at room temperature.
18. 1 ml TB syringes without needles.
19. KB media: Per 1 L ddH2O add 10 g proteose peptone, 1.5 g KH2PO4, 25 ml 20% glycerol, 3.2 ml 1 M MgSO4; for making plates add 15 g Agar per liter and autoclave with at least 30 min sterilization time. Add antibiotics after autoclaving.
20. 28 °C incubator.
21. 10 mM MgSO4.
22. Cuvettes.
23. Spectrophotometer.
24. Mechanical hand sprayer.
25. 2 mm glass beads.

26. 1.2 ml library tubes, library tube caps, and library tube rack (VWR, Radnor, PA, USA).
27. Software: Zeiss Axiovision, MATLAB, Excel, ImageJ.

3 Methods

3.1 Reactive Oxygen Species Production

1. Place 200 μL of ddH_2O into wells of a 96-well clear bottom plate. Fill one row of 6–12 wells—one well for each replicate plant–peptide combination in each row (*see* **Note 3**).
2. Use a 1 mm cork borer to punch a smooth leaf punch from the tomato leave avoiding the central vein and transfer to the 96-well plate floating adaxial side up. Multiple candidate MAMPs, or alleles of MAMPs, should be tested side-by-side in the same assay by taking a second punch from the same tomato leaf to put in a separate row (*see* **Note 4**).
3. Cover the plate with plastic cling wrap or plate cover to prevent water from evaporation and let leaves sit overnight (*see* **Note 5**).
4. The next day, prepare the ROS elicitation solution. Prepare ~110 μL of medium for each leaf punch to be elicited (*see* **Note 6**).
5. Setup the Biotek plate reader.
   (a) Turn plate reader on.
   (b) Start KC4 program on computer after plate reader warms up, which takes approximately 2 min, the plate reader door will open.
   (c) Click "Protocol" → new protocol.
   (d) Click "Systems" → Readers → Filters wavelengths → fluorescence/luminescence.
   (e) Confirm that at least 1 "plug" is selected in the correct wheel slot for the excitation filter and 1 "hole" is selected in the correct wheel slot for the emission filter (*see* **Note 7**).
   (f) Click "Send filters".
   (g) Click "Close" after filters are set on the plate reader (which takes approximately 10 s and will be accompanied by several clicking noises).
   (h) Select "Settings".
   (i) Click "Kinetic".
   (j) Set the plate reader to read for 50 min sampling every 3 min (*see* **Note 8**).
   (k) Make sure the "kinetic" radio button is selected.

(l) Type in the first sample (A1) and last sample (dependent on how you set up the plate—*see* **Note 9**).

(m) Click "Ok" to save and close settings.

(n) Click "Read".

6. Remove the 200 μL of ddH_2O from all of the wells preferably using a multichannel pipette (*see* **Note 10**).
7. Add 100 μL of the ROS elicitation medium to each leaf punch.
8. Place the plate in the plate holder slot with the first sample to be read in the top left corner.
9. Click "Start Reading".
10. After 50 min the plate reader will automatically eject the plate when the measurements are complete.
11. Export the data as a text file.
    (a) Click "File" → Export Data.
    (b) Type in the first and last well (as in **step 5-1** above).
    (c) Click Add.
    (d) Select M/Lum/E and click Export.
12. Open the data as a comma delimited text file in Excel or Matlab for analysis (*see* **Note 11**).

3.2 Stomatal Closure

1. Spray tomato plants with water and place in transparent plastic bag.
2. Seal bag and place in 28 °C incubator with light for 1–2 h (*see* **Note 12**).
3. Detach whole leaves and cut a square excluding the central vein of ~1–2 cm^2 to create a relatively flat leaf section and place on microscope slide (*see* **Note 13**).
4. Immerse leaf in 1 μM peptide solution and seal with coverslip. Use 4–8 leaves for each plant–peptide combination to be tested.
5. Incubate leaf in 1 μM peptide solution for 1 h.
6. Image leaf samples with a compound light microscope (we have used a Zeiss Axio Imager M1).
    (a) Focus on random section of leaf at 200× magnification (*see* **Note 14**).
    (b) Take a picture of the leaf sample and count the number of stomata present in the image and save the Zeiss image file.
    (c) Repeat the previous two steps until at least 20 stomata have been imaged for each plant sample.
7. Load the Zeiss image files into Zeiss Axiovision software and use the Measurement tool to measure the aperture of 20 stomata per leaf sample (*see* **Note 15**).

3.3 Callose Deposition

1. Detach leaf from plant and submerge in 50 ml corning tubes with 5 ml of alcoholic lactophenol solution (2:1 95% EtOH: alcoholic lactophenol).
2. Incubate leaves for 15 min at room temperature.
3. Preheat a water bath to 65 °C.
4. Place tubes in the water bath until the leaves are completely cleared of chlorophyll (*see* **Note 16**).
5. Remove the alcoholic lactophenol solution and replace with fresh alcoholic lactophenol solution.
6. Continue incubation at room temperature for 24 h.
7. Remove the alcoholic lactophenol solution and submerge leaf in 50% ethanol for 1 min.
8. Replace the ethanol and submerge leaf in ddH_2O for 2 min.
9. Replace the ddH_2O with aniline blue stain and soak for 30 min at room temperature.
10. Rinse leaf with ddH_2O.
11. Mount leaf on microscope slide with 50% glycerol and seal with coverslip.
12. Image the leaf using the fluorescent light source of a Zeiss Axio Imager M1 through the DAPI filter (*see* **Note 17**).
13. Take ten random images at 50× (check) magnification of each leaf and save as tif files.
14. Quantify the magnitude of callose deposition using Image J.
    (a) Open tif file in ImageJ.
    (b) Click “Process” → Subtract background (see **Note 18**).
    (c) Click “Image” → Adjust → Threshold.
    (d) Move lower slider (the upper threshold) to maximum.
    (e) Move upper slider (the lower threshold) unil only callose spots and not the vascular system of the leaf is red (see **Note 19**).
    (f) Click “Apply”.
    (g) Click “Analyze” → Analyze Particles.
    (h) Adjust the threshold values for area and circularity (see **Note 20**).
    (i) Click “Ok” and copy output of count and total area to Excel or other program to check for statistical differences across samples.

3.4 Plant Protection Assay

1. Streak *Pseudomonas syringae* pv. tomato strain on KB media plate with appropriate antibiotics and grow overnight in a 28 °C incubator (*see* **Note 21**).

2. Infiltrate tomato leaves with 1 μM putative MAMP or positive (flgII-28) or negative (ddH_2O) controls by pressing the axial side of the leaf with a 1 ml blunt-end syringe into the tip of a finger. Infiltrate 6–12 individual leaves across multiple tomato plants for each peptide treatment and label the leaves with a marker circling the area that was infiltrated.
3. Spray plants to be infect with water and cover with plastic dome for high humidity and keep covered on light shelf for 20 h.
4. Prepare bacterial inoculum.
   (a) In 10 mM $MgSO_4$, aseptically resuspend *P. syringae* grown for 24 h on KB media plates with antibiotics.
   (b) Measure the optical density of the bacteria at 600 nm in a spectrophotometer (*see* **Note 22**).
   (c) Make a 0.01 OD_{600} bacterial suspension for inoculation in 10 mM $MgSO_4$ (*see* **Note 23**).
5. Spray the bacterial inoculum slowly and consistently across all tomato plants that were infiltrated the day before. Ensure that all leaves that were infiltrated are fully saturated on the leaf surface with inoculum (*see* **Note 24**).
6. Cover the tomato plants again for 24 h and place back on light shelf.
7. Uncover the tomato plants and place back on light shelf for 3 days keeping the tomato plants watered.
8. Four days after inoculation of bacteria, sample leaf disks using a 3 mm cork borer for all leaves that were originally infiltrated by the putative MAMP or controls.
9. Place leaf disk in a 1.2 ml library tube containing 2 mm glass beads and 200 μL 10 mM $MgSO_4$ in a 96-well library tube rack. Repeat for all treated leaves placing in separate library tubes. If a bead beater is not available, leaf tissue can be homogenized by placing the leaf punch in a 1.5 ml centrifuge tube with 200 μL of 10 mM $MgSO_4$ and grinding manually with a centrifuge tube pestle.
10. Seal the library tubes with caps and place lid on library tube rack (*see* **Note 25**).
11. Lock the library tube rack in a bead beater and set sample to shake for 100 s to homogenize the leaf disks and release the bacteria (*see* **Notes 26** and **27**).
12. Add 180 μL of 10 mM $MgSO_4$ to five wells of a 96-well plate for each leaf disk sampled (*see* **Note 28**).
13. Add 20 μL of the homogenized leaf disk to the first well of 10 mM $MgSO_4$ and mix to dilute the sample 10×.

14. Perform a serial dilution in subsequent wells of 10 mM $MgSO_4$ in the 96-well plate until a 100,000× dilution factor (five wells, *see* **Notes 29** and **30**).
15. Divide a KB media plate with appropriate antibiotics into six sections with a marker on the bottom of the plate. Label the six sections with the sample name and on each half the three subsequent dilution factors 1000×, 10,000×, and 100,000×.
16. Plate 10 μL of the 100,000× dilution sample on the appropriately labeled section of the plate and then the 10,000× and 1000× samples on the respective correct sections of the same plate (*see* **Note 31**).
17. Repeat for all other leaf disk samples and treatments.
18. Place the plates in a 28 °C incubator for 2 days.
19. Count the colonies at one dilution factor for each leaf disk sample. Count the dilution factor that led to 10–100 colonies and record the counts and dilution factors for calculating the CFU/cm^2 (*see* **Note 32**).

4 Notes

1. Many guides are available online for choosing a solvent based on protein sequence of a peptide (e.g., http://www.lifetein.com/handling_and_storage_of_synthetic_peptides.html#q2). We recommend dissolving the peptide to 1–5 mM concentration in the best solvent and then diluting in molecular biology grade water for working stocks (typically 1 μM).
2. It is important to note that false positives are a major concern for many of these experiments because some MAMPs, such as flg22, elicit an immune response at sub nanomolar concentrations. Therefore, contamination of your working stocks of your putative MAMPs with trace amounts of a very active elicitor can lead to a false positive result [19, 20].
3. We determined white, clear bottom plates to be the best plates for the assay. The BioTek HD plate reader we used has both a top and bottom optical sensor but the bottom optical sensor is more sensitive which is ideal for this assay and thus necessitated the clear-bottom plates.
4. There is often very high variability from plant to plant and even leaf to leaf in this assay. To make comparisons between different MAMPs, or MAMP alleles, it is best to take multiple leaf punches from the same leaf to test with the different MAMPs. For convenience, put leaf punches from the same leaf in the same column of the 96-well plate and put replicate leaves in the subsequent columns. Then, in **step 7**, apply the peptide

treatment by rows. This setup allows for the data to be exported as a convenient spreadsheet where each treatment is a block of data across the time course for doing analyses and making graphs.

5. The overnight incubation is important because many components of PTI response pathways are also activated by mechanical damage [8] and are thus elicited by the leaf punching. The overnight incubation avoids interference from ROS elicitation due to the mechanical damage.
6. The ROS elicitation medium can be kept at room temperature for at least 1 h but reagents should go back in the −20 °C freezer as soon as the medium is made. Make one master mix of the ROS elicitation medium for each peptide to be tested.
7. Refer to the Biotek manual for safely removing and installing filters in the plate reader. Make note of which position you install the plug and hole. A plug is necessary for the excitation filter in a luminescence assays so light is not sent to the sample that will saturate the light produced by the sample. A hole is necessary in the emission filter to ensure all produced light is sent to the optical detector for maximum sensitivity.
8. We found 50 min to be sufficient in all plant–peptide combinations we have tested, but this can be adjusted. If a larger number of leaf samples are being tested on the plate, the plate reader may not be able to measure all wells within 3 min and will give an error message when you close out of the setting menu. Click "minimum" in the Kinetic options section and the plate reader will automatically calculate the minimum sampling time based on the number of samples you input.
9. Again, you must ensure that the leaf punches are placed in a rectangular setup with each plant–peptide combination being tested as one row. If you have three rows of eight samples each, then the first sample is A1 and the last sample is C8.
10. Tilting the plate and the pipette so that they form a "V" and then aspirating the water is the best way to remove the water without getting the leaf punches stuck to the side of the well or pipette tip. It is still necessary to visually confirm that no leaf punches get stuck to the pipette tip.
11. If you plan on comparing different plant–peptide combinations in terms of magnitude of ROS response, then accurate statistical analysis is more complex than typically considered. Please see the supplementary methods of [6] for guidelines and link to MATLAB programs for performing a robust analysis of the data using Markov-chain monte carlo modeling and Bayesian analysis.
12. These steps ensure mostly open stomata.

13. Only using a section of the leaf is necessary to have a relatively flat surface to view under the microscope so that the stomata stay in focus while moving across the leaf surface during image sampling.
14. Ensure that the focus is optimized for the epidermal layer of cells including the guard cells.
15. Exact measurement of the stomatal aperture is challenging and can be somewhat arbitrary. We therefore strongly recommend having a different individual with no knowledge of the samples perform the actual image analysis measurement of the samples.
16. This step usually takes 30 min. The leaves should turn very pale yellow to white in color.
17. Autofluorescence is also partially visible in these conditions. It is therefore critical to include ddH_2O-treated leaves as a negative control.
18. We found the default rolling ball radius to be optimal but this choice can be optimized.
19. Some callose spots are likely to be lost, but consistently adjusting the threshold to remove all vascular autofluroescence of the leaf consistently is important for the quantification. The vascular system will be bright red and then diffuse into small red dots as you increase the lower threshold, keep increasing until the diffuse red dots have just disappeared and only the large red spots from the callose remain.
20. We typically use 0.1 to infinity for area and 0.05–1 for circularity for the threshold values but this can be adjusted if concerned about capturing non callose in the analysis. It is essential to apply the same thresholds to all images though.
21. Any plant pathogen with which you are comfortable working is acceptable to use in this assay. The goal of this assay is to check for immunity activation by a putative MAMP which should reduce the growth of the plant pathogen when subsequently infected. For plant protection assays on tomato, we found that the *avrpto1-avrptoB* double mutant of *P. syringae* strain DC3000 is the best pathogen option. Pretreament of leaves with flg22 and flgII-28 did not consistently lead to a reduction in growth of wildtype DC3000, presumably because of the pathogen's ability to effectively suppress the tomato PTI response. The *avrpto1-avrptoB* double mutant is limited in its ability to suppress PTI, especially flagellin-triggered immunity [21]. This strain is an effective pathogen also on Arabidopsis and other solanaceous plants such as *N. benthamiana* [21] and *Solanum melongena* [22]. Other pathogens should be considered for different MAMPs and/or plant hosts.

22. Be sure to check the linear range of your spectrophotometer (often between 0.1 and 1) and make sure your OD_{600} of the sample is in that range. If your sample has an OD_{600} greater than 1, simply dilute it in the cuvette and keep track of the dilution factor to multiply the spectrophotometer reading to calculate the true OD_{600} of the sample.
23. The amount of bacterial suspension necessary depends on the number of plants to be inoculated. We typically make 10 ml of bacterial inoculum for each 5-week-old tomato plant to be inoculated to ensure saturation of bacterial inoculum on the leaf surface. The surfactant silwet can be added to the bacterial suspension at 3 μL/10 ml for more consistent wetting of the leaf surface.
24. We found spray inoculation to work well for the plant protection assay. Another option is infiltration inoculation. For this inoculation procedure, a lower OD_{600} of bacterial inoculum (0.0001 OD_{600} for tomato infection with *P. syringae* pv. tomato) is prepared and infiltrated into the same leaves by a blunt-end syringe 20 h after the peptides were infiltrated.
25. If the lid does not sit smoothly on top of the library tube caps, insert folded paper towels to fill the gap to prevent the caps from coming loose and the tubes leaking.
26. The amount of time the samples should be shaken in the bead beater will vary depending on the type of leaves and the speed of the bead beater. Ensure homogenous grinding of the leaf samples and equal treatment of all leaf disks.
27. Some protocols call for only measuring endophytic bacteria by killing the epiphytic bacteria by dipping the leaf punches in 70% EtOH for 10 s before placing in the library tubes. We observed more substantial reduction of growth of *P. syringae* in tomato leaves infiltrated with flg22 or flgII-28 compared to water by sampling both epiphytic and endophytic bacteria however.
28. This 96-well plate is to perform a serial dilution of the homogenized leaf samples before plating samples on a media plate. Be sure to setup the plate in such a way that the samples can be diluted using a multichannel pipette without mixing up samples.
29. The best approach is to dilute all samples for each treatment group in parallel using a multichannel pipette and continue through the rest of the plating before going back to do the dilution on the next treatment group.
30. We typically use three dilutions for each leaf disk for plating to count colonies. For *P. syringae* pv. tomato on tomato, the dilutions 1000×, 10,000×, and 100,000× are best 4 days following

spray inoculation but this will need to be optimized based on inoculation conditions, time, host, and pathogen. By dividing the media plate into six sections, two leaf disks and three dilutions per leaf disk can be plated on one plate.

31. Long 10 μL pipette tips work best for this step. Aspirate 10 μL of the sample and hold the pipette nearly horizontal to the plate and expel the sample onto the correct section of the plate. Continuing to hold the pipette nearly horizontal to the plate, use the tip to spread the droplet of sample around the section of the plate. By starting with the highest dilution and then moving to the lowest dilution, the need for changing pipette tips is obviated until moving to a new sample.

32. To calculate the CFU/cm^2, use the equation:

    (colony count × dilution factor × 20) ÷ (area of leaf punch in cm^2)

Acknowledgments

Christopher Clarke is funded by a postdoctoral research fellowship from USDA-NIFA (2015-67012-22821) Work in the Vinatzer lab is funded by the NSF (IOS-1354215 and DEB-1241068). Funding to Vinatzer was also provided in part by the Virginia Agricultural Experiment Station and the Hatch Program of the National Institute of Food and Agriculture, US Department of Agriculture.

References

1. Felix G, Duran JD, Volko S, Boller T (1999) Plants have a sensitive perception system for the most conserved domain of bacterial flagellin. Plant J 18(3):265–276
2. Kunze G et al (2004) The N terminus of bacterial elongation factor Tu elicits innate immunity in Arabidopsis plants. Plant Cell 16(12):3496–3507
3. Felix G, Regenass M, Boller T (1993) Specific perception of subnanomolar concentrations of chitin fragments by tomato cells: induction of extracellular alkalinization, changes in protein phosphorylation, and establishment of a refractory state. Plant J 4(2):307–316
4. Bittel P, Robatzek S (2007) Microbe-associated molecular patterns (MAMPs) probe plant immunity. Curr Opin Plant Biol 10(4): 335–341
5. Jones JDG, Dangl JL (2006) The plant immune system. Nature 444(7117):323–329
6. Clarke CR et al (2013) Allelic variation in two distinct Pseudomonas syringae flagellin epitopes modulates the strength of plant immune responses but not bacterial motility. New Phytol 200(3):847–860
7. Sun W, Dunning FM, Pfund C, Weingarten R, Bent AF (2006) Within-species flagellin polymorphism in *Xanthomonas campestris* pv *campestris* and its impact on elicitation of Arabidopsis FLAGELLIN SENSING2-dependent defenses. Plant Cell 18(3):764–779
8. Boller T, Felix G (2009) A renaissance of elicitors: perception of microbe-associated molecular patterns and danger signals by pattern-recognition receptors. Annu Rev Plant Biol 60(1):379–406
9. Zipfel C (2014) Plant pattern-recognition receptors. Trends Immunol 35(7):345–351
10. Lacombe S et al (2010) Interfamily transfer of a plant pattern-recognition receptor confers broad-spectrum bacterial resistance. Nat Biotechnol 28(4):365–369
11. Rosli HG et al (2013) Transcriptomics-based screen for genes induced by flagellin and

repressed by pathogen effectors identifies a cell wall-associated kinase involved in plant immunity. Genome Biol 14(12):R139

12. Cai R et al (2011) The plant pathogen *Pseudomonas syringae* pv. tomato is genetically monomorphic and under strong selection to evade tomato immunity. PLoS Pathog 7(8): e1002130
13. McCann HC, Nahal H, Thakur S, Guttman DS (2012) Identification of innate immunity elicitors using molecular signatures of natural selection. Proc Natl Acad Sci 109(11):4215–4220
14. Vinatzer BA, Monteil CL, Clarke CR (2014) Harnessing population genomics to understand how bacterial pathogens emerge, adapt to crop hosts, and disseminate. Annu Rev Phytopathol 52(1):19–43
15. Adam L, Somerville SC (1996) Genetic characterization of five powdery mildew disease resistance loci in *Arabidopsis thaliana*. Plant J 9(3):341–356
16. Chakravarthy S, Velásquez AC, Ekengren SK, Collmer A, Martin GB (2010) Identification of *Nicotiana benthamiana* genes involved in pathogen-associated molecular pattern–triggered immunity. Mol Plant Microbe Interact 23(6):715–726
17. Melotto M, Underwood W, Koczan J, Nomura K, He SY (2006) Plant stomata function in innate immunity against bacterial invasion. Cell 126(5):969–980
18. Zipfel C et al (2004) Bacterial disease resistance in Arabidopsis through flagellin perception. Nature 428(6984):764–767
19. Mueller K et al (2012) Contamination risks in work with synthetic peptides: flg22 as an example of a pirate in commercial peptide preparations. Plant Cell 24(8):3193–3197
20. Segonzac C et al (2012) The shoot apical meristem regulatory peptide CLV3 does not activate innate immunity. Plant Cell 24(8):3186–3192
21. Kvitko BH et al (2009) Deletions in the repertoire of *Pseudomonas syringae* pv. tomato DC3000 type III secretion effector genes reveal functional overlap among effectors. PLoS Pathog 5(4):e1000388
22. Clarke CR, Hayes BW, Runde BJ, Wicker E, Vinatzer BA (2014) Eggplant and related species are promising genetic resources to dissect the plant immune response to *Pseudomonas syringae* and *Xanthomonas euvesicatoria* and to identify new resistance determinants. Mol Plant Pathol 15(8):814–822

Chapter 22

Genome-Wide Analysis of Chromatin Accessibility in Arabidopsis Infected with *Pseudomonas syringae*

Yogendra Bordiya and Hong-Gu Kang

Abstract

Changes in chromatin accessibility are an important aspect of the molecular changes that occur in eukaryotic cells responding to stress, and they appear to play a critical role in stress-induced transcriptional activation/reprogramming and epigenetic changes. In plants, pathogen infection has been shown to induce rapid and drastic transcriptional reprogramming; growing evidence suggests that chromatin remodeling plays an essential role in this phenomenon. The recent development of genomic tools to assess chromatin accessibility presents a significant opportunity to investigate the relationship between chromatin dynamicity and gene expression. In this protocol, we have adopted a popular chromatin accessibility assay, DNase-seq, to measure chromatin accessibility in Arabidopsis infected with the bacterial pathogen *Pseudomonas syringae* pv. *tomato* (*Pst*). DNase-seq provides information on chromatin accessibility through the sequencing of DNA fragments generated by DNase I digestion of open chromatin, followed by mapping these sequences on a reference genome. Of the two popular DNase-seq approaches, we based our method on the Stamatoyannopoulos protocol, which involves two DNase cleavages rather than a single cleavage, followed by size fractionation. Please note that this two-cleavage approach is widely accepted and has been used extensively by ENCODE (Encyclopedia of DNA Elements) project, a public research consortium investigating *cis*- and *trans*-elements in the transcriptional regulation in animal cells. To enhance the quality of the chromatin accessibility assay, we modified this protocol by including two centrifugation steps for nuclear enrichment and size fractionation and an extra washing step for removal of chloroplasts and *Pst*. The outcomes obtained by this approach are also discussed.

Key words Biotic stress, *Pseudomonas syringae*, Arabidopsis, Chromatin accessibility, DNA footprinting, DNase-seq, DNase I hypersensitive site (DHS)

1 Introduction

The ability of living organisms to adapt to an ever-changing environment is critical for their survival. At the molecular level, this adaptability often involves substantial transcriptional reprogramming [1, 2], which is mediated by dynamic chromatin alterations triggered in part by the interaction between *cis*-regulatory DNA sequences and transcription factors [2]. To study these dynamic chromatin changes, a wide range of molecular tools have

Libo Shan and Ping He (eds.), *Plant Pattern Recognition Receptors: Methods and Protocols*, Methods in Molecular Biology, vol. 1578, DOI 10.1007/978-1-4939-6859-6_22,

been utilized. Starting in the late-1970s, David Galas and Albert Schmitz used DNA footprinting [3], which relies on the fact that DNA bound by protein(s) is protected from DNase cleavage, to identify *cis*-regulatory DNA sequences, such as promoters and enhancers, that are bound by proteins [4]. Over the last decade, the development of a wide range of next-generation sequencing platforms [5] has revolutionized chromatin accessibility analyses, as they allow stress-induced changes to be monitored on a genomic scale. Most of current genomic tools for assessing chromatin accessibility use chemicals or enzymes, such as MNase-seq, DNase-seq, FAIRE (Formaldehyde-Assisted Isolation of Regulatory Elements)-seq and ATAC (Assay for transposase-accessible chromatin)-seq, to identify accessible or protected regions; the pros and cons of these tools have been recently reviewed in detail [6].

To date, DNase-seq has been widely used to identify open and closed regions of chromatin [7]. Indeed, this technique was extensively used by the ENCODE project [8], a public research consortium investigating *cis*- and *trans*-elements in transcriptional regulation in animal cells. DNase-seq was originally developed in animal systems by two independent groups led by Drs. Crawford and Stamatoyannopoulos [9, 10]. The main difference between their approaches rests on whether single or double DNase I cleavage events are analyzed. The Stamatoyannopoulos protocol is based on a double cleavage using limiting concentrations of DNase I, followed by selection of DNA fragments between 100 and 500 bp in size [9].

The first plant genome accessibility study was published in 2012 by Dr. Jiang's group, who used the Stamatoyannopoulos protocol on Arabidopsis and rice [11, 12]. Subsequent studies have investigated genome accessibility in Arabidopsis undergoing flower development or photomorphogenesis, and in response to heat stress [13, 14]. Here, we extend these studies by providing the first analysis of genome accessibility during a plant-pathogen interaction using Arabidopsis infected with *Pseudomonas syringae* pv. *tomato* (*Pst*). Although we essentially followed the Stamatoyannopoulos protocol [9], some modifications were necessary to minimize contamination by chloroplasts and *Pst*, thereby enhancing the quality of DNase-seq. In particular, two additional centrifugation steps were added; the first was performed to remove contaminating chloroplasts and some *Pst* from the enriched nuclei, and the second was to size fractionate the digested DNA. Also, an extra washing was necessary to further remove chloroplasts and *Pst*. An overview of our DNase-seq protocol is shown in Fig. 1. Following nuclei enrichment, DNase I was used to cleave open chromatin, which is more accessible to enzymes than closed chromatin, into small fragments. These DNA fragments were then size-fractionated, purified and sequenced, followed by mapping on the reference genome to create a chromatin accessibility map. The open chromatin regions are

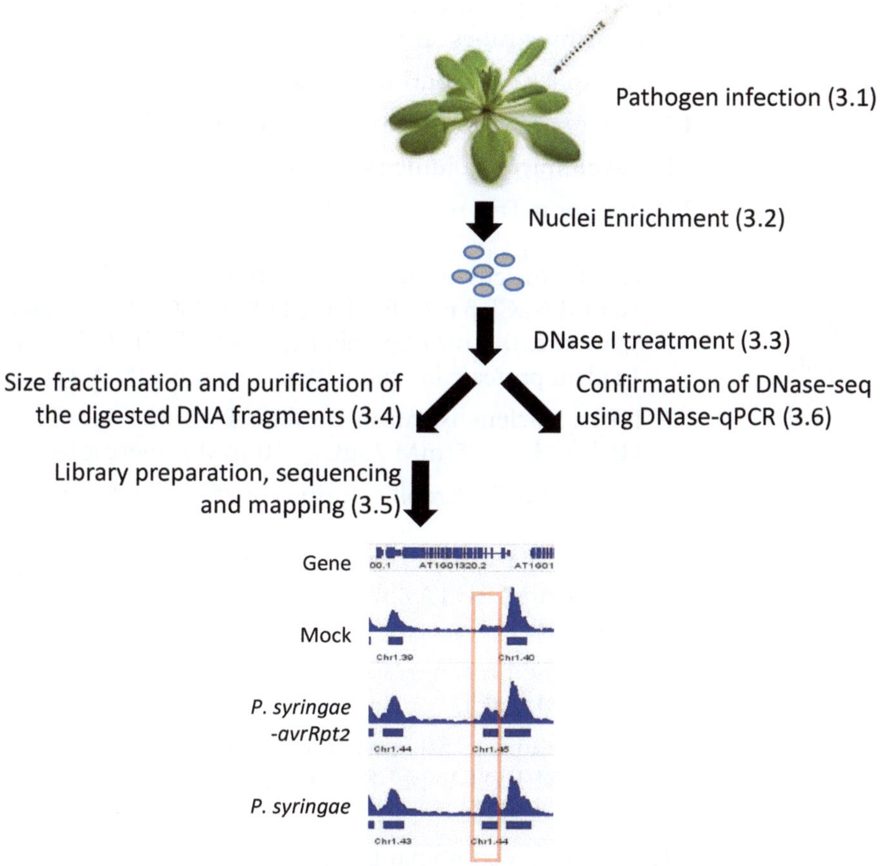

Fig. 1 Overview of DNase-seq procedure. The method section relevant to each step is noted in parenthesis. An example of DNase-seq read densities associated with At1g01320 displaying differential accessibility in response to *P. syringae* is presented. A chromatin region showing differential accessibility is highlighted in a *red box*

termed DHSs (DNase Hypersensitive Sites). A manuscript using the described protocol, which limited contamination by *Pst* to under 5%, is currently under review.

2 Materials

2.1 Materials and Equipment

1. 50 ml Falcon conical tubes (Fisher).
2. Tissue homogenizer (IKA, T10).
3. Miracloth (EMD-Millipore).
4. Hemocytometer (Hausser Scientific).
5. Compound microscope.
6. 37 and 55 °C water bath.
7. Centrifuge (Sorvall, ST 40R) and ultracentrifuge (Sorvall, WX Ultra 80).

8. 5 ml polypropylene round bottom and open top high-speed centrifuge tubes (Nalgene).
9. QIAquick PCR purification kit (Qiagen).
10. Synergy HTX multi-mode reader (BioTek).
11. Take3 micro-volume plate (BioTek).
12. Typhoon Trio scanner (GE).

2.2 Reagents

2.2.1 Reagents for Nuclei Enrichment

1. NHB (Nucleus homogenization buffer): 25 mM PIPES/pH 7.0, 10 mM NaCl, 5 mM EDTA/pH 8.0, 250 mM Sucrose, 0.15 mM Spermine, 0.5 mM Spemidine, 5 mM DTT, 0.5% TritonX-100, 1x plant protease inhibitor cocktail (Sigma; P9599).
2. NIB (Nucleus isolation buffer): 1 M Sucrose, 10 mM Tris–HCl/pH 7.2, 5 mM $MgCl_2$, 10 mM β-mercaptoethanol.
3. 80% and 35% Percoll solutions: 80% and 35% Percoll in NIB.

2.2.2 Reagents for DNase I Treatment

1. Buffer A: 15 mM Tris–HCl/pH 8.0, 15 mM NaCl, 60 mM KCl, 1 mM EDTA/pH 8.0, 0.5 mM EGTA, 0.5 mM spermidine, and 11% sucrose
2. 2× DNase I reaction solution: Buffer A with 12 mM $CaCl_2$ and 150 mM NaCl.
3. Stop solution: 50 mM Tris–HCl/pH 8.0, 100 mM NaCl, 100 mM EDTA/pH 8.0, 1 mM spermidine, 0.3 mM spermine, 0.1% SDS.
4. DNase I, recombinant (Roche).
5. RNase A (ThermoFisher).
6. Proteinase K (NEB).

2.2.3 Reagents for DNA Size Fractionation, Purification, and Quantification

1. 40%, 30%, 20%, and 10% Sucrose solutions: 40%, 30%, 20%, and 10% Sucrose (w/v) in water.
2. 5 M NaCl.
3. PB buffer (Qiagen).
4. EB buffer (Qiagen).
5. PicoGreen (Thermo Fisher).
6. SYBR Green I (Thermo Fisher).

3 Methods

3.1 Pathogen Infection

Arabidopsis plants are grown at 22 °C in growth chambers (Percival) programmed for a 16-h-light (8000–10,000 lux) and 8-h-dark cycle. 3.5-week-old Arabidopsis plants are syringe-infiltrated with *Pst* at 5×10^5 cfu/ml as previously described [15] and incubated for 24 h. Mock infection is done with the infiltration buffer (10 mM $MgCl_2$).

3.2 Nuclei Enrichment

4.5 g of leaf tissue from Arabidopsis plants is harvested in a 50 ml conical tube and incubated in prechilled diethyl ether at 4 C° for 3 min, followed by washing with cold water three times at 4 °C. The tissue is homogenized using a tissue homogenizer (IKA, T10) at the highest speed for 45 s in 15 ml NHB, filtered via double-layered Miracloth, and centrifuged at 1000 × *g* for 5 min at 4 °C. The pellet is resuspended in 1 ml NHB, which was then transferred to an Eppendorf tube. To remove excess chloroplasts and *Pst*, the pellet is washed with NHB five times by centrifugation at 1000 × *g* for 1 min at 4 °C. After resuspending the pellet in 1 ml NIB, the nuclei are further purified and enriched by performing Percoll density gradient centrifugation. The nuclei are layered on a step gradient consisting of 35% and 80% Percoll in a high-speed centrifuge tube, as shown in Fig. 2, and centrifuged at 635 × *g* for 5 min, followed by 10,154 × *g* for 15 min at 4 °C. Enriched nuclei, generally visible as white material at the interface between the 35% and 80% layers, are collected by pipetting. The collected nuclei are then diluted fivefold using NIB and pelleted by centrifugation at 4000 × *g* at 4 °C for 10 min.

3.3 DNase I Treatment

Purified nuclei are resuspended and washed two times in 200 μL of Buffer A. An aliquot of purified nuclei is counted using a hemocytometer under a microscope to determine the appropriate amount of the DNase I enzyme. As per the Stamatoyannopoulos protocol, the DNase I concentration at which "a very small portion of genomic DNA was digested and a smear under the major high molecular weight band just began to appear" is optimal [9]; it is recommended to find the optimum DNase I concentration by using a small scale pilot experiment. Thus, two Eppendorf tubes containing (in equal volumes) the nuclei in Buffer A and the 2×

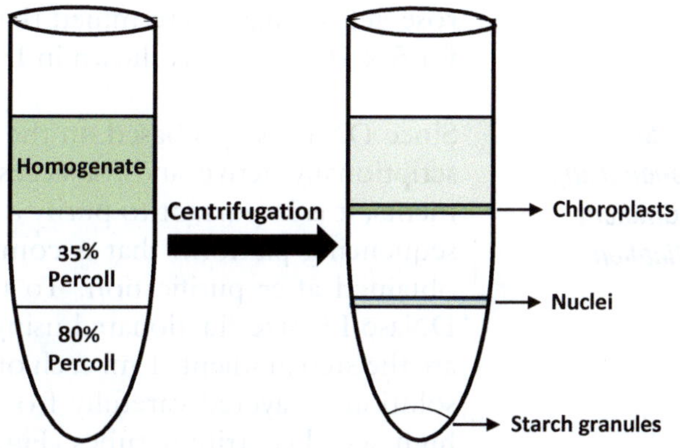

Fig. 2 Percoll density gradient centrifugation for nucleus enrichment. Homogenate was gently overlayed onto the Percoll gradient. After centrifugation, a layer of nuclei was observed at the 35/80% interface, which was then recovered by pipetting. This step plays an important role in removing chloroplasts and some *P. syringae*

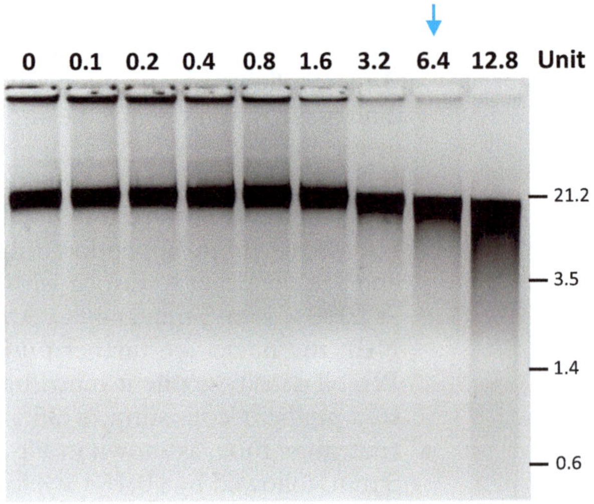

Fig. 3 Optimization of DNase I treatment. Enriched nuclei were treated with DNase I at the indicated enzyme activity levels. Digested DNA was run on a 1% agarose gel and stained with ethidium bromide for a conventional UV imaging. Size markers are shown on the right of the panel in kB. An *arrow* indicates an optimum DNase I concentration used in the subsequent procedures

DNase I reaction solution are incubated at 37 °C for 1 min to pre-warm the tubes. After pre-warming, different units of DNase I are added to the 2× DNase I reaction solutions, which were then combined with the nuclei in Buffer A and incubated at 37 °C for 10 min. 400 μL of stop solution is added to terminate the DNase I reaction. RNase A is added at the final concentration of 20 μg/μL and incubated at 55 °C for 15 min to remove RNA. Proteinase K is then added at the final concentration of 25 μg/μL and incubated at 55 °C overnight. Following electrophoresis on a 1% agarose gel, we have determined that 6.4 U of DNase I are adequate for 5 × 10^6 nuclei, as shown in Fig. 3.

3.4 DNA Size Fractionation Using Sucrose Gradient and Purification

Since DNase-seq is based on the principle that DNase I cuts transcriptionally active and/or accessible chromatin into small fragments, it is important to purify a proper size fraction and choose a sequencing platform that is compatible with the DNA fragments obtained after purification. To this end, chromatin treated with DNase I is size-fractionated using a sucrose step gradient. To create the step gradient, 1 ml each of 40%, 30%, 20%, and 10% sucrose solution is layered carefully from bottom to top, respectively, in high-speed centrifuge tubes (Fig. 4a). After 1 h at room temperature, the step gradients are placed at 4 °C for 30 min. After adding 128 μL of 5 M NaCl to the DNA solution that has been digested with RNase A and proteinase K, the sample is loaded onto the sucrose gradient and centrifuged at 77,000 × *g* for 24 h at room temperature. Several fractions containing DNA in different size are

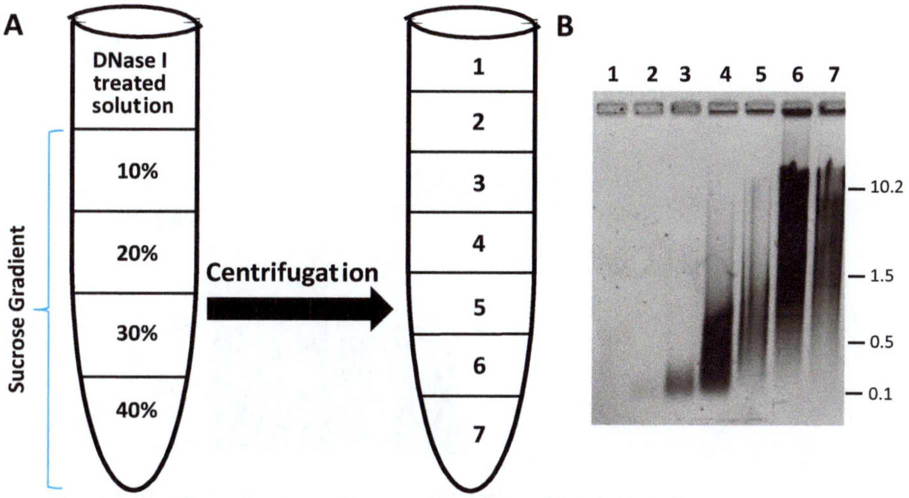

Fig. 4 Sucrose density gradient centrifugation for DNA size fractionation. (**a**) The sucrose density step gradient was generated as shown on the left. After centrifugation, seven fractions in equally diving volume were collected by pipetting from the top. (**b**) The fractionated DNA was purified using a QIAquick PCR purification kit and run on a 1% agarose gel. The DNA was stained with SYBR green I and imaged in a Typhoon Trio scanner at 488 nm. Size markers are shown on the right of the panel in kB

collected (Fig. 4a). In the presented case, we arbitrarily divide the centrifuged solution to seven fractions of equal volume. PB buffer provided in the QIAQuick PCR purification kit is added (five times the volume of the recovered sample in each fraction) and the DNA is purified using the QIAQuick PCR purification kit as directed by the manufacturer. Following elution of purified DNA using 35 μL of EB buffer (the first elution in 25 μL and the second in 10 μL), the DNA concentration is determined using PicoGreen and BioTek's Take3 multi-volume plate. Purified DNA is run on a 1% agarose gel containing SYBR Green I, which is imaged by a Typhoon Trio scanner using a 488 nm laser to assess the DNA quality and quantity (Fig. 5).

3.5 Library Preparation, Sequencing, and Mapping

Once DNA quality and quantity are assessed, DNase-seq libraries are prepared using a Genomic DNA Sample Prep kit (Illumina) as suggested by the manufacturer. This library preparation procedure has also been described in an earlier report in which we did not find any major difference from ours [16]. An Illumina system in 100 bp mode is used for the sequencing.

To identify uniquely mapped reads for DHS analysis, we first discard the sequencing reads that aligned with the Arabidopsis chloroplast genome via Bowtie [17] (allowing up to two mismatches). The remaining reads are then aligned to the Arabidopsis genome via Bowtie, allowing up to two mismatches, and the uniquely mapped reads are used for F-seq [18] to identify DHSs. Once the number of reads mapped to each DHS in mock and pathogen-inoculated plants

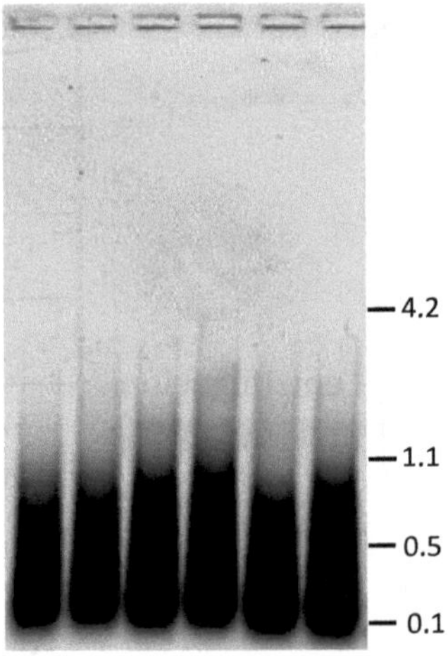

Fig. 5 Agarose gel electrophoresis was used to assess DNA quantity and quality before library preparation. DNA from six independent samples was analyzed for length and concentration using a 1% agarose gel. Size markers are shown on the right of the panel in kB. DNA was stained with SYBR Green I and visualized by a Typhoon Trio scanner at 488 nm

is counted and normalized to reads per million mapped reads, this information is fed into DESeq [19] to identify DHSs that differ between mock- and pathogen-infected plants.

3.6 Confirmation of DNase-seq Data Using DNase-qPCR

When the DNase-seq analysis is completed, we use real-time quantitative PCR (qPCR) to verify that the DNase-seq data are reproducible. Note that DNase-seq provides information from the ends of DNA fragments generated by DNase I digestion, whereas DNase-qPCR amplifies DNA that has not been disrupted by DNase I. Since DNase I digestion of open chromatin yields less undigested DNA than that of closed chromatin, genomic regions with greater accessibility will display reduced levels of amplification via qPCR, but generate higher read counts from DNase-seq. To perform DNase-qPCR, nuclei preparation and DNase I treatment are performed as described above for DNase-seq. However, unlike DNase-seq, in which small fragments are fractionated and then purified, total DNA purification is performed for DNase-qPCR without size fractionation using conventional CTAB genomic DNA preparation as described previously [20]. Real-time qPCR is then performed using primers selected from different parts of the genome based on a DNase-seq data set.

4 Notes

Since chloroplasts are highly abundant in plant cells and are of a similar size as nuclei, they are a major source of contamination in preparations of enriched plant nuclei. In addition, since our analyses involve monitoring chromatin accessibility in *Pst*-infected plants, bacterial DNA is another major source of contamination that must be minimized. Vigorously washing the pelleted nuclei with NHB in Subheading 3.2 substantially reduced *Pst* DNA contamination. However, we noted that washing more than five times often led to poor recovery of the nuclei. Since *Pst* is a gram-negative bacterium with an outer lipopolysaccharide layer surrounding its peptidoglycan layer, it is fairly resistant to bacterial cell wall degrading enzymes like lysozyme [21]. It has been reported that EDTA helps disrupt the lipopolysaccharide layer, and thus its combination with lysozyme helps dissolve the *Pst* cell wall [22]. However, we found that adding EDTA and lysozyme while homogenizing the tissue in NHB not only reduced *Pst* contamination, but also significantly reduced both the quality and quantity of the nuclei, likely because the EDTA/lysozyme treatment also compromises the nuclei and/or the bacterial cells are tightly trapped in the plant tissues.

Acknowledgments

We thank D'Maris Dempsey and Angela H. Kang for critical comments on the manuscript. This work is supported by Texas State University-Faculty Startup Program, Multi-disciplinary Internal Research Grant and National Science Foundation Grant (IOS-1553613) to H.G.K.

References

1. Tao Y et al (2003) Quantitative nature of Arabidopsis responses during compatible and incompatible interactions with the bacterial pathogen Pseudomonas syringae. Plant Cell 15(2):317–330
2. Rivas S (2011) Nuclear dynamics during plant innate immunity. Plant Physiol 158(1):87–94
3. Gross DS, Garrard WT (1988) Nuclease hypersensitive sites in chromatin. Annu Rev Biochem 57:159–197
4. Galas DJ (2001) The invention of footprinting. Trends Biochem Sci 26(11):690–693
5. Metzker ML (2010) Sequencing technologies—the next generation. Nat Rev Genet 11(1):31–46
6. Tsompana M, Buck MJ (2014) Chromatin accessibility: a window into the genome. Epigenetics Chromatin 7(1):33
7. Jiang J (2015) The 'dark matter' in the plant genomes: non-coding and unannotated DNA sequences associated with open chromatin. Curr Opin Plant Biol 24:17–23
8. Encode Project Consortium (2012) An integrated encyclopedia of DNA elements in the human genome. Nature 489(7414):57–74
9. Hesselberth JR et al (2009) Global mapping of protein-DNA interactions in vivo by digital genomic footprinting. Nat Methods 6(4):283–289
10. Boyle AP et al (2008) High-resolution mapping and characterization of open chromatin across the genome. Cell 132(2):311–322
11. Zhang W et al (2012) High-resolution mapping of open chromatin in the rice genome. Genome Res 22(1):151–162
12. Zhang W et al (2012) Genome-wide identification of regulatory DNA elements and protein-binding footprints using signatures of open chromatin in Arabidopsis. Plant Cell 24(7):2719–2731
13. Sullivan AM et al (2014) Mapping and dynamics of regulatory DNA and transcription factor networks in *A. thaliana*. Cell Rep 8(6):2015–2030

14. Pajoro A et al (2014) Dynamics of chromatin accessibility and gene regulation by MADS-domain transcription factors in flower development. Genome Biol 15(3):R41
15. Kang HG et al (2008) CRT1, an Arabidopsis ATPase that interacts with diverse resistance proteins and modulates disease resistance to Turnip Crinkle Virus. Cell Host Microbe 3(1):48–57
16. Zhang W, Jiang J (2015) Genome-wide mapping of DNase I hypersensitive sites in plants. Methods Mol Biol 1284:71–89
17. Langmead B et al (2009) Ultrafast and memory-efficient alignment of short DNA sequences to the human genome. Genome Biol 10(3):R25
18. Boyle AP et al (2008) F-Seq: a feature density estimator for high-throughput sequence tags. Bioinformatics 24(21):2537–2538
19. Anders S, Huber W (2010) Differential expression analysis for sequence count data. Genome Biol 11(10):R106
20. Clarke JD (2009) Cetyltrimethyl ammonium bromide (CTAB) DNA miniprep for plant DNA isolation. Cold Spring Harb Protoc 2009(3):pdb.prot5177
21. Hiroshi Nikaido MV (1985) Molecular basis of bacterial outer membrane permeability. Microbiol Rev 49(1):1–32
22. Ellison RT, Giehl TJ (1991) Killing of Gram-negative bacteria by lactofernn and lysozyme. J Clin Invest 88(4):1080–1091

Chapter 23

Small RNA and mRNA Profiling of Arabidopsis in Response to *Phytophthora* Infection and PAMP Treatment

Yingnan Hou and Wenbo Ma

Abstract

Small non-coding RNAs (smRNAs) regulate gene expression at both transcriptional and post-transcriptional levels. Well known for their roles in development, smRNAs have emerged as important regulators of plant immunity. Upon pathogen perception, accumulation of specific smRNAs are found to be altered, presumably as a host defense response. Therefore, identification of differentially accumulated smRNAs and their target genes would provide important insight into the regulation mechanism of immune responses. Here, we describe the detailed experimental procedure using Illumina sequencing to analyze the expression profiles of smRNAs and mRNAs in Arabidopsis. We focus on a newly developed pathosystem using *Phytophthora capsici* as the pathogen and include the treatment of Arabidopsis leaves with *p*athogen-*a*ssociated *m*olecular *p*atterns (PAMPs) of *Phytophthora*.

Key words Small RNA-Seq, RNA-Seq, Next-generation sequencing, Small RNA-regulated defense response, Pattern-triggered immune response, *Phytophthora* infection

1 Introduction

RNA silencing is a ubiquitous gene regulation mechanism in eukaryotes. Central players in RNA silencing are non-coding RNAs of 20–30 nucleotides in length that guide sequence-specific repression of target genes. Plants produce two major types of small RNAs (smRNAs): microRNAs (miRNAs) and small interfering RNAs (siRNAs) [1]. These smRNAs bind to target transcripts based on sequence complementarities, leading to gene silencing by inducing DNA methylation, transcript cleavage, or translational repression [2, 3].

Accumulating evidence suggests that smRNAs play important regulatory functions during plant defense upon pathogen infection. Challenged by microbial parasites in the environment, plants can defend themselves from most of the attacks through their innate immune system. A basal layer of plant immunity relies on the recognition of conserved molecular signatures called *p*athogen- or

Libo Shan and Ping He (eds.), *Plant Pattern Recognition Receptors: Methods and Protocols*, Methods in Molecular Biology, vol. 1578,
DOI 10.1007/978-1-4939-6859-6_23, © Springer Science+Business Media LLC 2017

*m*icrobe-*a*ssociated *m*olecular *p*atterns (PAMP or MAMP) [4, 5]. PAMP perception leads to activation of defense signal transduction and transcriptional reprogramming. In Arabidopsis, several miRNAs are induced after the treatment of flg22, the major PAMP of the bacterial pathogen *Pseudomonas syringae* [6–8]. Among them, miR393 acts as a positive regulator of this pattern-triggered immunity (PTI) [7]. Similarly, miR393 is also induced by the cell wall extract of *Phytophthora sojae* (equivalent to PAMP treatment) in soybean roots. Knocking down miR393 led to enhanced susceptibility to *P. sojae*, suggesting that miR393 is also required for soybean defense against this oomycete pathogen [9].

Concomitant with the maturation of next-generation sequencing technology, RNA-seq emerged as a high throughput, sensitive, and powerful tool for genome-wide profiling of gene expression [10–12]. Importantly, this technology is particularly suitable for small RNA analysis. Using the Illumina platform, smRNA-seq provided key information in the pioneering studies of smRNA characterization and will continue to be a core technology to understand the role of smRNAs in various plant processes including immunity.

Here, we describe a detailed protocol for smRNA-seq using Arabidopsis leaf tissues inoculated by *Phytophthora capsici* or treated with the culture filtrates (equivalent to a PAMP prep) of *P. capsici*. In particular, our protocol combines library constructions for both smRNA-seq and RNA-seq using the same tissue samples. Because the bioinformatic prediction of smRNA-targeted genes is relatively straightforward, this experimental strategy will detect not only changes in smRNA accumulation, but also opposite changes in the abundance of their corresponding targets. We focus on the newly established Arabidopsis-*P. capsici* pathosystem [13, 14], which is more advantageous compared to many other systems because both the host and the pathogen can be genetically manipulated. Moreover, the genus *Phytophthora* contains some of the most destructive plant pathogens, including the notorious Great Irish Famine culprit *Phytophthora infestans* [15]. Parallel analyses of smRNA and mRNA profiles will enable the identification of regulatory smRNAs and genes contributing to PTI, and in particular, plant defense against *Phytophthora* infection.

2 Materials

2.1 Making the Culture Filtrates of *Phytophthora capsici*

1. Pea medium: Autoclave 283 g frozen peas (preferably unsalted) in 700 mL of water for 30 min. Collect supernatant after straining through two layers of cheesecloth. Bring up the volume to 1 L with water and autoclave again before use.
2. Miracloth (Millipore, Cat. 475855).
3. Four-week-old plants of Arabidopsis eco Col-0. Plants are grown in a growth room at 22 °C under 12–12 light–dark regime.

4. *Phytophthora capsici* isolate LT263 [16] grown on a V8 agar plate (*see* Subheading 2.2).
5. 0.45 μm cellulose acetate syringe filters (VWR, Cat. 28145-481).

2.2 Whole Plant Inoculation of Phytophthora capsici

1. 10% V8 medium: Aliquot V8 juice into 50 mL conical centrifuge tubes. Centrifuge at 4000 × *g* for 5 min (*see* **Note 1**). Transfer the supernatant into new tubes. Dilute the "cleared" juice to 10% using ddH_2O. Add $CaCO_3$ to a final concentration of 0.02% (w/v) into the 10% V8 solution to use it as a liquid medium for *Phytophthora*. To make the solid medium, add agar to a final concentration of 1.2% (w/v) into the 10% V8 solution. Autoclave before use.
2. Autoclaved tap water.
3. Microscope (Primo Star, Zeiss).
4. Miracloth (Millipore).
5. Four-week-old plants of Arabidopsis eco Col-0.
6. *Phytophthora capsici* isolate LT263.

2.3 RNA Extraction and Separation of RNAs with High or Low Molecular Weight

1. Tris-saturated phenol.
2. RNA extraction buffer: 0.1 M LiCl, 100 mM Tris–HCl (pH 8.0), 10 mM EDTA and 1% SDS.
3. Hot phenol buffer: equal volume of the RNA extraction buffer and phenol, pre-warmed at 80 °C.
4. 4 M LiCl.
5. Chloroform–isoamyl alcohol (v/v = 24:1).
6. TE buffer: 10 mM Tris–HCl (pH 8.0), 1 mM EDTA.
7. Diethyl pyrocarbonate (DEPC)-treated TE buffer (pH 8.0).
8. DEPC-treated ddH_2O.

2.4 DNase Treatment of the High Molecular Weight RNA Extract

1. Tris-saturated phenol.
2. 3 M NaOAc.
3. Chloroform–isoamyl alcohol (v/v = 24:1).
4. 70% ethanol.
5. 100% ethanol.

2.5 mRNA Library Construction

1. AMPure XP beads (Beckman Coulter).
2. Magnetic separation stands for PCR tubes (Diagenode).
3. 5 M NaCl.
4. 40% PEG-8000 (w/v).
5. DEPC-treated ddH_2O.
6. Diluted Ampure XP beads: For a final volume of 1 mL, place 500 μL of Ampure XP beads in a 1.5 mL Eppendorf tube on a

magnetic separation stand for 5 min. Remove the supernatant and wash the beads with 500 μL of DEPC-treated ddH_2O, pipette gently to suspend the beads. Place the tube back to the magnetic separation stand and repeat the washing step. Suspend the beads in 500 μL of 5 M NaCl, then add 500 μL of 40% PEG-8000 and mix well.

2.6 Small RNA Isolation

1. 2× small RNA loading dye: 80% formamide (Sigma), 0.1% xylene FF (Sigma), 0.1% bromophenol blue (Sigma).
2. DEPC-treated 0.4 M NaCl.

3 Methods

3.1 Making Culture Filtrates using Phytophthora capsici Mycelia

1. Grow *P. capsici* on a fresh 10% V8 plate at room temperature in the dark for 3–4 days until the surface of the plate is fully covered by mycelia.
2. Cut an agar plug (approximately 0.3 cm × 0.3 cm) with growing mycelia from the plate (Fig. 1a); incubate the plug in a 250 mL flask containing 20 mL of pea medium at 25 °C in the dark without agitation for 4–5 days (Fig. 1b).
3. Filtrate the *P. capsici* culture with one layer of miracloth (Fig. 1c), and then with 0.45 μm cellulose acetate syringe filter to obtain the culture filtrates that will be used as a PAMP prep (Fig. 1d and e).

3.2 Treatment of Arabidopsis with P. capsici Culture Filtrates

1. Adult leaves of 4-week-old Arabidopsis plants are infiltrated with the culture filtrates of *P. capsici* mycelia. Leaves infiltrated with clean pea media are used as mock-treated controls (Fig. 1f). Ten plants are used for each treatment with 3–4 leaves of each plant being infiltrated.
2. Keep the infiltrated plants in a high humidity (~85% RH) growth chamber at 22 °C.
3. Collect leaf tissues during a time course at different time points (e.g., 0, 3, 6 h post treatment) for RNA extraction.

3.3 Inoculation of Arabidopsis with P. capsici

1. Grow *P. capsici* on a fresh 10% V8 plate at room temperature in the dark for 3–4 days.
2. Cut the agar with growing mycelia into small pieces (*see* **Note 2**) and subculture the strain by incubating the mycelial plugs in petri dishes containing 20 mL of 10% V8 broth at room temperature for 2 days in the dark.
3. Remove the medium, and then wash the mycelium plugs with sterile water for three times. Keep the plugs in sterile water at room temperature for 24 h in the dark.

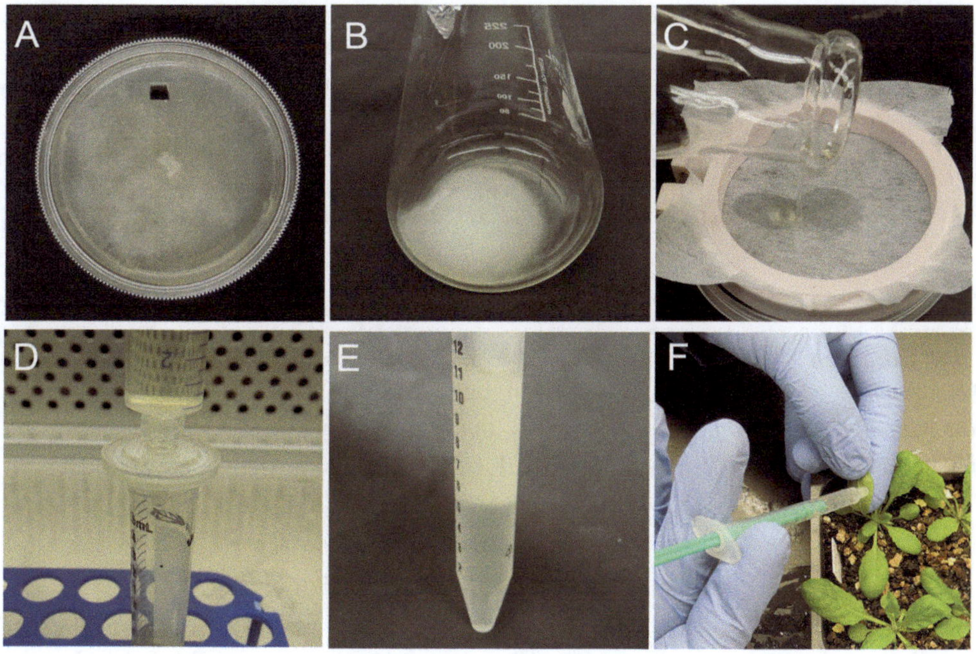

Fig. 1 Treatment of Arabidopsis leaves with culture filtrates (a PAMP prep) of *Phytophthora capsici*. Culture filtrates of *P. capsici* (or any other *Phytophthora* species) can be used as a crude PAMP extract to elicit pattern-triggered immunity in plants. (**a**) A mycelial agar plug was cut from V8 agar plate with growing *P. capsici* isolate LT263. (**b**) The agar plug was subcultured in a flask containing the liquid pea medium for 4–5 days. (**c**) *P. capsici* culture was filtrated using one layer of miracloth. (**d**) The primary filtrates were further sterilized using 0.45 μm cellulose acetate syringe filter to obtain the final *P. capsici* culture filtrates (**e**). (**f**) Sterilized culture filtrates were used to infiltrate Arabidopsis leaves using a 1 mL syringe without a needle

4. Place the petri dishes under illumination at room temperature for another 24 h.
5. Incubate the petri dishes at 4 °C for 30 min and then under illumination for 30 min at room temperature to induce the release of zoospores.
6. Check the zoospore concentration under a microscope. Ideal concentration of the zoospore suspension for plant inoculation should be around 10^2–10^3 zoospores/μL. If the concentration is too low, repeat the induction step (**step 5**) to collect more zoospores.
7. Use one layer of miracloth to remove agar and mycelia. The zoospore suspension is then ready to be used for inoculation.
8. Dip a cotton swab into the zoospore suspension and gently apply to the abaxial side of adult rosette leaves of 4-week-old Arabidopsis plants. Ten plants are needed for each treatment with 3–4 leaves of each plant being inoculated. Leaves treated with autoclaved tap water are used as mock-treated controls.

9. Place the inoculated plants in a growth chamber with high humidity (~85% RH) at 22 °C. Keep the plants in the dark during the first 24 h and then under illumination at a 12–12 light–dark regime.
10. Collect samples for RNA extraction during a time course at different time points (e.g., 0, 6, 12 h post inoculation). At least one leaf from each inoculated plant should remain on the plant to confirm the disease symptoms that will be developed at 3 days post inoculation.

3.4 RNA Extraction and Separation of RNAs with High or Low Molecular Weight

1. Grind 0.5 g of inoculated leaves (30–40 leaves) into fine powder in liquid nitrogen.
2. Transfer the powder into a 15 mL conical centrifuge tube containing 2 mL of hot phenol buffer, vortex for 20 s.
3. Add 1 mL chloroform–isoamyl alcohol (v/v = 24:1) and vortex for another 20 s.
4. Centrifuge at 15,000 × *g* for 15 min at 4 °C.
5. Transfer the aqueous phase to a new 15 mL conical centrifuge tube, mix with an equal volume of 4 M LiCl. Incubate at −20 °C overnight.
6. Separate the high molecular weight (HMW, >100 bp) and the low molecular weight (LMW, <100 bp) RNA molecules by centrifugation at 18,000 × *g* at 4 °C for 15 min. The pellet contains the HMW RNA molecules and will be used for RNA-seq; the supernatant (containing the LMW RNA molecules) will be used for small RNA purification.
7. To further purify the HMW RNA, resuspend the pellet in 350 μL of DEPC-treated TE buffer and transfer the RNA sample to a new 1.5 mL Eppendorf tube.
8. Add an equal volume of isopropanol and incubate at −20 °C for 1 h.
9. Centrifuge at 15,000 × *g* for 15 min at 4 °C to precipitate the RNA. Wash the pellets twice with 1 mL of 70% ethanol before they are dried in a chemical fume hood.
10. Resuspend the pellets in 50 μL of DEPC-treated ddH_2O. Store at −80 °C or use it immediately for DNase treatment.

3.5 DNase Treatment of the High Molecular Weight RNA

1. Incubate the reaction mix in a 1.5 mL tube at 37 °C for 30 min.

RNA sample	30 μL
10× DNase I buffer	5 μL
DNase I (RNase free)	2 μL
RNase inhibitor	1 μL
DEPC-treated ddH_2O	12 μL

2. Add 150 μL of DEPC-treated ddH_2O and 200 μL phenol: chloroform:isoamyl alcohol (25:24:1) into the reaction; vortex for 10 s.
3. Centrifuge at 15,000 × *g* for 15 min at 4 °C.
4. Transfer the supernatant (approximately 200 μL) into a new tube, add 20 μL of 3 M NaOAc and 500 μL of pre-cooled 100% ethanol to precipitate the RNA at −20 °C for 1 h.
5. Centrifuge at 15,000 × *g* for 15 min at 4 °C. Wash the pellets with 1 mL of 70% ethanol. Suspend the RNA in 25 μL of DEPC-treated ddH_2O.
6. Determine the concentration of the RNA samples and check their quality using a UV spectrophotometer (A260/A280 should be 1.7–2.0 and A260/230 >2.0). Proceed to construct libraries for RNA-seq.

3.6 mRNA Library Construction and Illumina Sequencing

1. Follow the Illumina TruSeq RNA sample prep v2 (RS-122-2001) guide for library construction (*see* **Notes 3** and **4**).
2. Determine the quality of the mRNA libraries using a Bioanalyzer (*see* **Note 5**) (Fig. 2a).
3. Pool the libraries into a mixture with equal molar concentration of each library (*see* **Note 6**).
4. Sequence with Illumina Hiseq using the 100 bp single end run.

3.7 Small RNA Purification and Library Construction

1. Precipitate small RNAs from the supernatant obtained in Subheading 3.4 by adding 1/10 volume of 3 M NaOAc (pH 5.2) and 2 volumes of 100% ethanol. Incubate the sample at −20 °C for at least 3 h.
2. Centrifuge at 15,000 × *g* for 15 min at 4 °C. Wash the pellet with 70% ethanol twice and dissolve the small RNAs in 25 μL of DEPC-treated ddH_2O.
3. Determine the concentration of the small RNA samples and check their quality using a UV spectrophotometer. Proceed to further purify small RNAs within a specific length range using urea-PAGE.
4. Make 15 mL of 15% urea PAGE gel mix (*see* **Note 7**).

Urea	6.3 g
5×TBE	1.5 mL
40% acrylamide–bis-acrylamide (29:1)	5.63 mL
10% ammonium persulfate	120 μL
TEMED	9 μL
ddH_2O	~ 2 mL

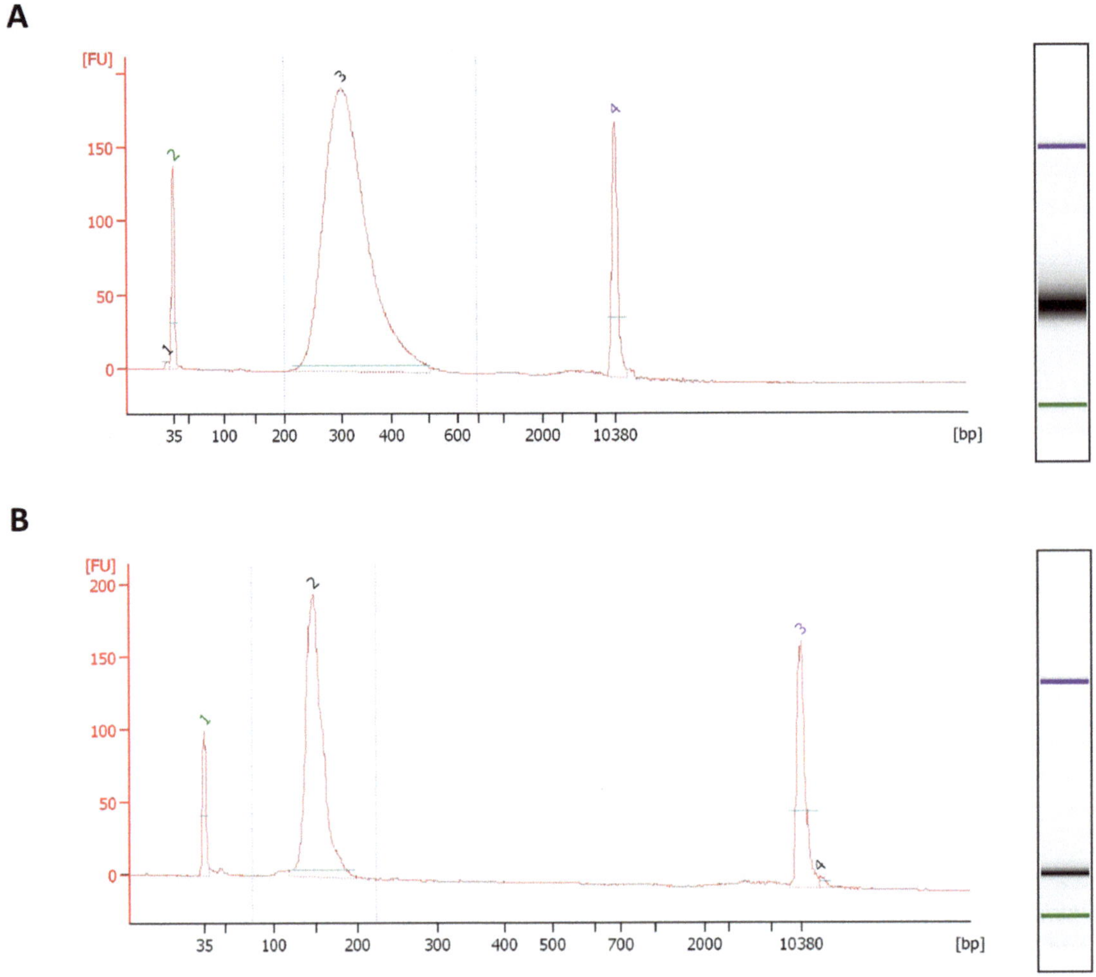

Fig. 2 Typical bioanalyzer results of libraries for Illumina sequencing. (**a**) Bioanalyzer result of an mRNA library (diluted four times). The median size of this library was approximately 300 bp. The region for concentration counting was selected between 200 to 500 bp (the black "band" shown in the *right panel*). This sample generated 9.7 million reads. (**b**) Bioanalyzer result of a smRNA library pool (diluted ten times). The median size of these multiplexed libraries is approximately 150 bp. The region for concentration counting was selected between 125 to 175 bp (shown in the *right panel*). This pooled sample produced 8–16 million reads per library

5. Pour the gel mix into a Bio-Rad mini-gel cassette with the 1.0 mm spacer, place a ten-well comb on the top, let the gel polymerize for at least 30 min.
6. Mix 25 μL of small RNA extract with an equal volume of 2× small RNA loading dye.
7. Denature the small RNA samples at 70 °C for 10 min, and then place the samples on ice immediately.
8. Load the RNA samples in the 15% urea PAGE gel with one lane containing 10 μL of a 10 bp DNA marker (Thermo, Cat.

SM1211). Run the gel in 0.5×TBE buffer at 150 V for 1.5 h (*see* **Note 8**).

9. Stain the gel in 50 mL of 0.5×TBE buffer containing 2 μL of ethidium bromide (10 mg/mL) for 10 min.
10. Using the 10 bp DNA marker as a reference for size, cut and collect the gel between the region of 15–35 bp; place the gel slices into 1.5 mL Eppendorf tubes and smash them into small pieces using a 1 mL pipette tip.
11. Add 500 μL of 0.4 M DEPC-treated NaCl into each Eppendorf tube and make sure the gel pieces are sufficiently suspended in the solution. Shake the tubes for 6 h or overnight at 4 °C.
12. Centrifuge at 18,000 ×*g* for 1 min at 4 °C to eliminate gel debris. Repeat this step 2–3 times until the samples are free of gels.
13. Add 1 μL glycogen, 50 μL 3 M NaOAc and 1 mL 100% ethanol into each Eppendorf tube, mix well, and precipitate the small RNA for 6 h at −20 °C.
14. Centrifuge at 15,000 ×*g* for 15 min at 4 °C. Wash the pellets using 1 mL of 70% ethanol. Dissolve the small RNA in 6 μL of DEPC-treated ddH_2O.
15. Follow the NEBNext Multiplex small RNA library prep (NEB #E7580) guide to construct small RNA-seq libraries.

3.8 smRNA Libraries Recovery and Illumina Sequencing

1. Make 15 mL of 12% PAGE gel.

5×TBE	1.5 mL
40% acrylamide–bis-acrylamide (29:1)	4.5 mL
10% ammonium persulfate	105 μL
TEMED	5.25 μL
ddH_2O	8.85 mL

2. Add 10 μL of 6× DNA loading dye into each of the libraries, mix well, and load the samples into the PAGE gel with one lane containing 10 μL of 10 bp DNA marker.
3. Run the gel in 0.5×TBE buffer at 150 V for about 1.5 h (*see* **Note 9**).
4. Cut the gel slices containing cDNA molecules that are approximately 140–150 bp in length (*see* **Note 10**). Recover the smRNA libraries from the gel using the same method as described in Subheading 3.7.
5. Precipitate the cDNA by centrifugation, wash the pellet with 70% ethanol and dissolve the library in 20 μL of ddH_2O. Proceed for Illumina sequencing.

3.9 Illumina smRNA Sequencing

1. Determine the quality of the smRNA libraries using a Bioanalyzer (*see* **Note 5**) (Fig. 2b).
2. Pool the libraries into a mixture with equal molar concentration of each library.
3. Sequence with Illumina Hiseq using the 50 bp single end runs.

4 Notes

1. To obtain clean supernatant of V8 juice for *Phytophthora* cultivation, filtrate the supernatant after centrifugation with one layer of miracloth.
2. For subculturing *P. capsici* in V8 broth from mycelial plugs, cut agar slices from the outermost region of each petri dish, which contains newly grown mycelia.
3. Always vortex and sufficiently mix the beads before pipetting into RNA/DNA samples. The beads settle quickly and a uniform suspension is important for optimal binding.
4. DO NOT REMOVE BEADS!! If beads are sucked up in the pipette, return them to the tube so that they can be re-collected by the magnetic stand. A way to check if there are any beads stuck in the pipette tips is to hold the tip above a piece of whitepaper; the presence of the beads would be indicated by the appearance of brown color in the tip.
5. For library quality checkup, Qubit DNA HS 1000 Assay kit (Invitrogen) or the Bioanalyzer DNA 1000 kit (Agilent Technologies) could be used instead of Nanodrop (Nanodrop loses accuracy at low concentrations). The Bioanalyzer is advantageous compared to Qubit because it provides the concentration of each fragment with a particular size, which best reflects the quality of the libraries. Qubit only gives the overall concentration, but is easier and faster.
6. The total volume of the pooled sample is 10× the number of libraries. For example, the volume for two samples is 20 μL, and the volume for 12 samples is 120 μL. Caution needs to be taken to avoid pooling two libraries with the same index.
7. Dissolve urea at 42 °C in the gel buffer without APS and TEMED. If the final volume is less than 15 mL, add ddH_2O to make it up.
8. PAGE should be stopped when the lower band of the 2× small RNA loading dye is 1 cm above the bottom of the gel. The dyes in the loading buffer run at around 10 and 40 bp, respectively.
9. It is important to run the gel for more than 1.5 h so that the cDNA bands containing molecules of 140–150 bp in length

can be clearly separated from the bands representing self-ligation of the adapter, which are approximately 120 bp in length.

10. For the final step of PCR amplification, the number of cycles can be increased (one or two more cycles) if the DNA band is faint.

Acknowledgment

This research is supported by grants from USDA-NIFA (2014-67013-21554) and NSF-PGRP (IOS-1340001) for W. Ma.

References

1. Chen X (2010) Small RNAs—secrets and surprises of the genome. Plant J 61:941–958
2. Rogers K, Chen X (2013) Biogenesis, turnover, and mode of action of plant microRNAs. Plant Cell 25(7):2383–2399
3. Matzke MA, Mosher RA (2014) RNA-directed DNA methylation: an epigenetic pathway of increasing complexity. Nat Rev Genet 15(6): 394–408
4. Jones JD, Dangl JL (2006) The plant immune system. Nature 444(7117):323–329
5. Boller T, He SY (2009) Innate immunity in plants: an arms race between pattern recognition receptors in plants and effectors in microbial pathogens. Science 324(5928):742–744
6. Zhang W et al (2011) Bacteria-responsive microRNAs regulate plant innate immunity by modulating plant hormone networks. Plant Mol Biol 75(1–2):93–105
7. Navarro L et al (2006) A plant miRNA contributes to antibacterial resistance by repressing auxin signaling. Science 312(5772):436–439
8. Fahlgren N et al (2007) High-throughput sequencing of Arabidopsis microRNAs: evidence for frequent birth and death of MIRNA genes. PLoS One 2(2):e219
9. Wong J et al (2014) Roles of small RNAs in soybean defense against *Phytophthora sojae* infection. Plant J 79(6):928–940
10. Nagalakshmi U et al (2008) The transcriptional landscape of the yeast genome defined by RNA sequencing. Science 320(5881): 1344–1349
11. Lister R et al (2008) Highly integrated single-base resolution maps of the epigenome in Arabidopsis. Cell 133(3):523–536
12. Mortazavi A, Williams BA, McCue K, Schaeffer L, Wold B (2008) Mapping and quantifying mammalian transcriptomes by RNA-Seq. Nat Methods 5(7):621–628
13. Wang Y, Bouwmeester K, van de Mortel JE, Shan W, Govers F (2013) A novel Arabidopsis-oomycete pathosystem: differential interactions with *Phytophthora capsici* reveal a role for camalexin, indole glucosinolates and salicylic acid in defence. Plant Cell Environ 36(6): 1192–1203
14. Xiong Q et al (2014) Phytophthora suppressor of RNA silencing 2 is a conserved RxLR effector that promotes infection in soybean and *Arabidopsis thaliana*. Mol Plant Microbe Interact 27(12):1379–1389
15. Kamoun S et al (2015) The Top 10 oomycete pathogens in molecular plant pathology. Mol Plant Pathol 16(4):413–434
16. Donahoo RS, Lamour KH (2008) Interspecific hybridization and apomixis between *Phytophthora capsici* and *Phytophthora tropicalis*. Mycologia 100(6):911–920

Chapter 24

Mapping and Cloning of Chemical Induced Mutations by Whole-Genome Sequencing of Bulked Segregants

Jian Hua, Shuai Wang, and Qi Sun

Abstract

Here we describe a method of gene mapping and cloning of chemical induced mutations using next-generation sequencing on their backcrossed segregants. This method utilizes polymorphisms induced by chemical mutagens for mapping and therefore does not require outcrossing to a different background or a large segregating population. It can be used to clone causal mutations generated in various accessions or mutant backgrounds for dissecting complex processes such as pattern recognition receptor pathways in plant immunity.

Key words Mapping, Cloning, Next-generation sequencing, Mutagenesis

1 Introduction

Forward genetics is a powerful approach to dissect complex processes and regulations in living organisms such as plant immunity. Mutations disrupting the process can be screened from mutagenized populations or natural populations. Map based cloning is often used to reveal the causal mutations disrupting import players in the process of interest. It has been utilized to identify many critical components in Pattern Recognition Receptor (PRR) signaling. For instances, the PRRs Xa21 in rice and FLS2 in Arabidopsis were isolated by map-based cloning, and so was the co-receptor BAK1 in Arabidopsis [1–3]. This cloning method used to be a long tedious process requiring crossing mutant to a wild type in a different accession so that polymorphisms between the two accessions can be used as molecular markers for linkage mapping [4]. Next-generation sequencing technologies can now be used to accelerate this process by combining multiple steps of mapping and cloning in one step 'Mapping by Sequencing' [5]. The first implement of such an approach in Arabidopsis is the SHOREmap where a mutant in one accession was crossed to a wild-type plant in another accession and mutant plants from F2 progenies were subject to

Libo Shan and Ping He (eds.), *Plant Pattern Recognition Receptors: Methods and Protocols*, Methods in Molecular Biology, vol. 1578, DOI 10.1007/978-1-4939-6859-6_24, © Springer Science+Business Media LLC 2017

next-generation sequencing [6]. Another implement used a backcross to the same accession rather than to a different accession to identify causal mutation induced by chemical mutagen EMS [7]. There, random mutations induced by the mutagen, rather than polymorphisms between two accessions in SHOREmap, were used as polymorphisms for linkage mapping. This method eliminates the potential modification of mutant phenotypes by different accession backgrounds and the lengthy process of generation of a mapping parent. It can be used to clone causal mutations in any chemical or physical mutagen induced mutants even when no traditional mapping parents are available. This might be particularly useful for dissecting PRR signaling because some critical components such as FLS2 are not present in all accessions. This protocol describes the design and flow of how to map and clone of chemical induced mutant by Illumina next generation sequencing of bulked segregants. It is written for *Arabidopsis thaliana*, but can be generally applied to other plants with reference genomes.

2 Materials

1. Plant Genomic DNA midi preparation kit.
2. Next generation sequencing DNA library construction kit.
3. General plant molecular biology reagents and equipment.

3 Methods

3.1 Generating Mapping Population and Preparing Pooled Genomic DNA Libraries

1. Cross chemical induced mutant (*see* **Note 1**) to its un-mutagenized starting line. Obtain F1 seeds.
2. Grow F1 plants from F1 seeds (*see* **Note 2**). Determine if the mutation is dominant or recessive (*see* **Note 3**). Collect F2 seeds from individual F1 plant.
3. Grow approximately 200 F2 plants under proper conditions. Phenotype these plants at appropriate stage. Analyze the segregation ratio of the mutant phenotype to determine if it fits the single gene segregation (*see* **Note 4**).
4. Collect tissues from approximately 50 mutants and 50 non-mutants by cutting with scissors leaf tissues approximately 1 cm × 0.5 cm in size from each plant. Pool tissues of mutants and non-mutants separately (*see* **Note 5**). Tissues are quick frozen in liquid nitrogen, either for immediate DNA extraction or storage at −80 °C till use.

5. Isolate genomic DNA from pooled samples using genomic DNA midi preparation kit following manufactory's instruction (*see* **Note 6**).
6. Quantify DNA concentration and quality by standard spectrophotometer or Nano-drop (*see* **Note 7**).

3.2 Next-Generation Sequencing Analysis

1. Construct DNA libraries for next generation sequencing according to manufacturer's instruction. In case of using NEB product, treat 1–2 μg genomic DNA with dsDNA fragmentase (NEBNext dsDNA Fragmentase, M0348) to generate the desired fragment size of 50–200 bp. Fragmented DNA is then subject to DNA end repair, Adaptor Ligation, Size Selection of Adaptor Ligated DNA, PCR Enrichment of Adaptor Ligated DNA, and Cleanup of PCR Amplification by NEBNext Ultra II DNA Library Prep Kit for Illumina (NEB #E7645S or NEB #E7645L) (*see* **Note 8**).
2. Submit samples for next generation sequencing (*see* **Note 9**).
3. Use BWA software [8] to align sequence reads from the two pools to the reference genome.
4. Use SAMtools-mpileup [9] to quantify non-reference allele frequency at each polymorphic site (*see* **Note 10**).
5. Filter SAMtools-mpileup results with a threshold of non-reference allele frequency above 90% in the mutant pool and less than 50% in the control pool (*see* **Note 11**).
6. Make a dot-plot of filtered SNPs with *x*-axis represents genome position and *y*-axis represents non-reference allele frequency (*see* **Note 12**).
7. Use software Annovar [10] to annotate SNP.

3.3 Confirmation of Causal Mutation

1. Polymorphisms at the map region will be inspected individually and potential candidate will be selected (*see* **Note 13**).
2. If candidate mutation is a loss of function, search for a second loss of function mutant allele to determine its phenotype. Such a second allele can be searched in Arabidopsis T-DNA insertion collections available from TAIR or T-DNA Express (http://signal.salk.edu/cgi-bin/tdnaexpress/).
3. Verify or isolate homozygous mutant of the second allele by PCR genotyping method (*see* **Note 13**).
4. Analyze expression of the candidate gene in the mutant by Reverse Transcription-PCR to determine whether or not the mutant line is a good loss of function allele.
5. Determine the phenotype of the second mutant allele (*see* **Note 14**).

6. Complementation test is done in parallel or in alternative to **steps 2–5**. PCR amplify the wild-type gene of the candidate and transform it into the mutant if the mutation is recessive. Amplify the mutant gene and transform it into the wild type if the mutation is dominant. Use standard techniques of plant molecular biology, transformation, and selection for this test.

4 Notes

1. EMS mutagenesis under commonly used strength (0.25% for 12 h on seeds) will generate mutation rate at 1 bp change in every 125 kb in the Arabidopsis genome. This rate provides sufficient polymorphism for mapping.
2. All F1 plants should have the same phenotype. If not, the parent mutant plant was not a homozygous or some of the putative F1 plants are selfed progenies not crossed F1 progenies.
3. If F1 shows mutant phenotype, it is a dominant mutation. If F1 shows wild-type phenotype, it is a recessive mutation. If the F1 shows intermediate phenotype, it is a semi dominant mutation.
4. For a recessive mutation, mutant versus wild type phenotypes should be 1:3. For a dominant mutation, mutant versus wild type should be 3:1. For a semi-dominant mutation, the ratio should be 1:2:1 for strong mutant, weak mutant, and wild type.
5. Same amount of similar tissues are collected from each plant to ensure their equal representation in the pool. Collect two sets of tissues with one stored in −80 °C as a backup.
6. Most commercial kits will do. Check quality after preparation. Load 1 μL DNA (from total of 50 μL) on 1% agarose gel. The genomic DNA band should be larger than 20 kb without smear or RNA contamination.
7. The concentration of the DNA should be 50–200 ng/ml. A260/A280 should be between 1.8 and 2.0, and A260/A230 should be above 2.0.
8. Multiple genomic libraries can be run on a single lane. Use different index oligos for each library. For instance, use NEBNext adaptor oligos provided in NEBNext Multiplex (NEB #E7335, #E7500, #E7600, #E7535 and #E6609).
9. Achieve 20–50 times coverage of the genome. When using Illumina Hi-Seq machine with single end reads of 51 bp, four samples can be run on a single lane.
10. About eight SNPs (Single Nucleotide Polymorphisms) or small INDELs (INsertion and DELetions) are expected to occur per

Mb window. Misalignment of the reads due to repeats could result in a large number of false positive SNPs.

11. This can be done in Excel sheet or using a PERL script available through http://cbsu.tc.cornell.edu/software/bulkseg.htm.
12. The dot plot can be made in Excel, with one plot for each chromosome.
13. Polymorphisms that can be the causal mutations are listed below largely (not strictly) to the order of likelihood. (a) Nonsense mutations which will cause a truncated protein. (b) Frameshift mutations which will cause a truncated protein. (c) Missense mutations at critical residues that will change the protein activity. (d) Mutation at intron–exon junction that will change the mRNA transcript and therefore protein sequences. Test the hypothesis by analyzing the cDNA sequence of the candidate gene from the mutant.
14. Primers can be designed using the website (http://signal.salk.edu/tdnaprimers.2.html).

Acknowledgment

This work was supported by National Science Foundation IOS–1353738 to J.H.

References

1. Gomez-Gomez L, Boller T (2000) FLS2: an LRR receptor-like kinase involved in the perception of the bacterial elicitor flagellin in Arabidopsis. Mol Cell 5(6):1003–1011
2. Song WY et al (1995) A receptor kinase-like protein encoded by the rice disease resistance gene, Xa21. Science 270(5243):1804–1806
3. Aarts N et al (1998) Different requirements for EDS1 and NDR1 by disease resistance genes define at least two R gene-mediated signaling pathways in Arabidopsis. Proc Natl Acad Sci U S A 95(17):10306–10311
4. Lukowitz W, Gillmor CS, Scheible WR (2000) Positional cloning in Arabidopsis. Why it feels good to have a genome initiative working for you. Plant Physiol 123(3):795–805
5. James GV et al (2013) User guide for mapping-by-sequencing in Arabidopsis. Genome Biol 14(6):R61
6. Schneeberger K et al (2009) SHOREmap: simultaneous mapping and mutation identification by deep sequencing. Nat Methods 6(8): 550–551
7. Zhu Y et al (2012) Gene discovery using mutagen-induced polymorphisms and deep sequencing: application to plant disease resistance. Genetics 192(1):139–146
8. Li H, Durbin R (2009) Fast and accurate short read alignment with Burrows-Wheeler transform. Bioinformatics 25(14): 1754–1760
9. Li H et al (2009) The Sequence Alignment/Map format and SAMtools. Bioinformatics 25(16):2078–2079
10. Wang K, Li M, Hakonarson H (2010) ANNOVAR: functional annotation of genetic variants from high-throughput sequencing data. Nucleic Acids Res 38(16):e164

Chapter 25

Rapid Construction of Multiplexed CRISPR-Cas9 Systems for Plant Genome Editing

Levi Lowder, Aimee Malzahn, and Yiping Qi

Abstract

Multiplex CRISPR-Cas9 nuclease mediated genome editing is an efficient method for disrupting gene function in plants. Use of CRISPR-Cas9 has escalated rapidly in recent years and is expected to become routine practice in molecular biology and related fields of research. Due to the relatively novel and widespread adoption of this technology, first-time users may not have regular access to experienced guidance or technical support from peers or mentors. Here, we offer guidance and technical support in the form of a detailed and tested protocol for simultaneous targeting of three separate loci on the TRANSPARENT TESTA 4 (TT4) gene in *Arabidopsis thaliana* using multiplex CRISPR-Cas9. Although we target multiple loci on a single gene in *Arabidopsis*, the same approach can be used to target multiple genes or alleles in other plant species as well. We recommend the use of a molecular toolkit to streamline the process and make recommendations for this type of approach. The protocol starts with an overview of the reagents and covers designing of gRNAs and assembly of components into a final T-DNA delivery molecule through Golden Gate cloning and Multisite Gateway LR recombination.

Key words CRISPR/Cas9, Plant genome editing, Multiplex, Golden gate assembly, Gateway cloning

1 Introduction

Clustered regularly interspaced short palindromic repeats (CRISPR)-Cas9 derived from Streptococcus pyogenes (SpCas9) [1] is quickly becoming a standard genome editing tool for generating mutants and investigating functional genetics in plant systems. Expediting this trend is the widespread availability of kits and reagents for streamlining multiplexed CRISPR-Cas9 use in various organisms. However, Cas9 technology has so recently appeared that many labs face the prospect of starting a CRISPR-Cas9 based experiment without having any prior functional experience using the technique. Here, we present a detailed protocol for disrupting three independent loci simultaneously using a molecular toolbox that simplifies the process. This protocol is intended to serve as a

Libo Shan and Ping He (eds.), *Plant Pattern Recognition Receptors: Methods and Protocols*, Methods in Molecular Biology, vol. 1578, DOI 10.1007/978-1-4939-6859-6_25, © Springer Science+Business Media LLC 2017

learning tool for investigators who are designing and assembling a CRISPR-Cas9 experiment in plants for the first time and may not be routinely familiar with the techniques, reagents, pitfalls and procedures involved. We first start with some background and introduction to the methodology and reagents and then move to a detailed procedure for designing a CRISPR/*Sp*Cas9 experiment and finally, the assembly of reagents into a final T-DNA vector ready for delivery to plant cells.

CRISPR-Cas9 activity in plants has been demonstrated in Arabidopsis, rice, tobacco, orange, sorghum, wheat and tomato [2, 3]. CRISPR associated protein 9 (Cas9) is an RNA guided sequence specific nuclease (SSN). Small RNA molecules, in-complex with Cas9 protein, recognize ~20 base pair (bp) target DNA sequences through Watson–Crick base-pairing and the recognition also requires a 3′ NGG protospacer adjacent motif (PAM). Cas9 nucleases function in nature as part of bacterial adaptive immune systems. i.e., Cas9 in-tandem with CRISPR genetic elements in bacteria and archaea can disable invading phage DNA. In natural systems, Cas9 nucleases form complexes with CRISPR RNA (crRNA) and transactivating CRISPR RNA (tracrRNA) for function (*see* [4] for review). To simplify genome editing applications, crRNA and tracrRNA are fused together forming a single gRNA (sgRNA) [1, 5]. This short sgRNA sequence, along with Cas9, can be expressed under heterologous promoters to form a functional Cas9-sgRNA complex. This complex scans host cell genomes for PAM sequences directly 3′ to a matching 20 nucleotide sgRNA sequence. Cas9-sgRNA complex recognizes the target DNA and generates a DNA double-strand break which is then repaired by either error prone nonhomologous end joining (NHEJ) or homologous recombination (HR) DNA repair pathways. NHEJ commonly inserts small insertion and deletion (indel) mutations about 3-bp upstream from PAM sites on target loci [1, 5, 6]. HR based repair occurs rarely and generally only in actively dividing cell tissues. By default, sister chromatid sequences act as templates for HR under natural circumstances. For genome targeting applications, whereby heterologous DNA sequences can be inserted at specific sites within host cell genomes, an exogenous DNA donor molecule is supplied to act as HR template [7].

Multiplexing, or targeting multiple genomic loci simultaneously can be achieved using any class of SSN such as zinc finger nuclease (ZFN) [8] and TAL effector nuclease (TALEN) [9]. However, CRISPR/Cas9 has significant advantages for multiplexing due to their relatively simple design for generating target specificity. To target multiple sites, only multiple sgRNAs need to be expressed along with the Cas9 protein. In contrast, ZFN and TALEN require the engineering of protein domains for DNA targeting, which makes these classes of SSNs ill-suited for multiplexing and their scalability is thus relatively limited. Although most multiplexed CRISPR studies target three to five loci simultaneously, the current upper threshold of successfully targeted separate loci appears to settle around ten individual sites [10]. However, at the time of this writing there has been no rigid determination of maximum scalability using CRISPR/Cas9. Multiplexing opens up a range of flexible options for genome editing activities: large chromosomal deletions and inversions can be generated by simultaneously targeting loci that are between 50 bp and 245,000 kb away from each other [11]. Large deletions offer simplified genotyping operations due to differences in PCR amplicon fragment sizes. So called "paired nickases" can be used to make overhanging DSBs by directing two Cas9 nickases that lack functionality on a single nuclease domain such that they only cut a single strand of DNA. Thus, when paired together they can generate DSBs if the two targets are close to each other. Pared nickases offer significant reduction, or elimination of off target effects [12]. Here we offer a protocol for polymerase chain reaction (PCR)-independent assembly of a triplex Cas9 genome editing T-DNA. As an example, we choose to target the *TRANSPARENT TESTA 4* (*TT4*) gene in *Arabidopsis thaliana* (*AtTT4*).

We recently presented a multipurpose molecular toolbox for CRISPR-Cas9 use in plants [13]. This toolbox streamlines multiplexed CRISPR-Cas9 experiments by providing a preassembled vector system utilizing Golden Gate and Multisite Gateway cloning strategies. The kit allows an end-user to plug in annealed 20-mer synthetic oligonucleotides (to function as guide RNAs) and produce a single T-DNA construct ready for delivery to target plants. A Cas9 promoter (RNA Polymerase II) specific T-DNA backbone, multiplex gRNA expression array and Cas9 specific modules are assembled in a single step using Multisite Gateway recombination (Fig. 1). Utilizing a preassembled kit, such as the one utilized in this chapter, drastically simplifies and expedites the process of building multiplexed gRNA expression vectors for effective genome editing.

The toolkit assembly workflow starts by annealing forward and reverse gRNA oligonucleotides into designated Golden Gate entry vectors (pYPQ131-138) (Fig. 1) using BsmBI Type II restriction sites. The pYPQ130 series (pYPQ130-138) of Golden Gate entry vectors contain a small RNA Polymerase III promoter in-frame with inserted gRNA, scaffold and transcriptional terminator sequence. Successful insertion of the user's chosen gRNA into this vector results in a functional and independent gRNA expression cassette under a small RNA promoter of choice (U3 or U6) and is Golden Gate compatible for assembly with other gRNA expression cassettes. The next step is to carry out a single step Golden Gate reaction and subsequent blue-white screening assays using Golden Gate recipient vectors, pYPQ142-pYPQ148 (Fig. 1). After assembly of two or more Golden Gate entry cassettes has been carried out,

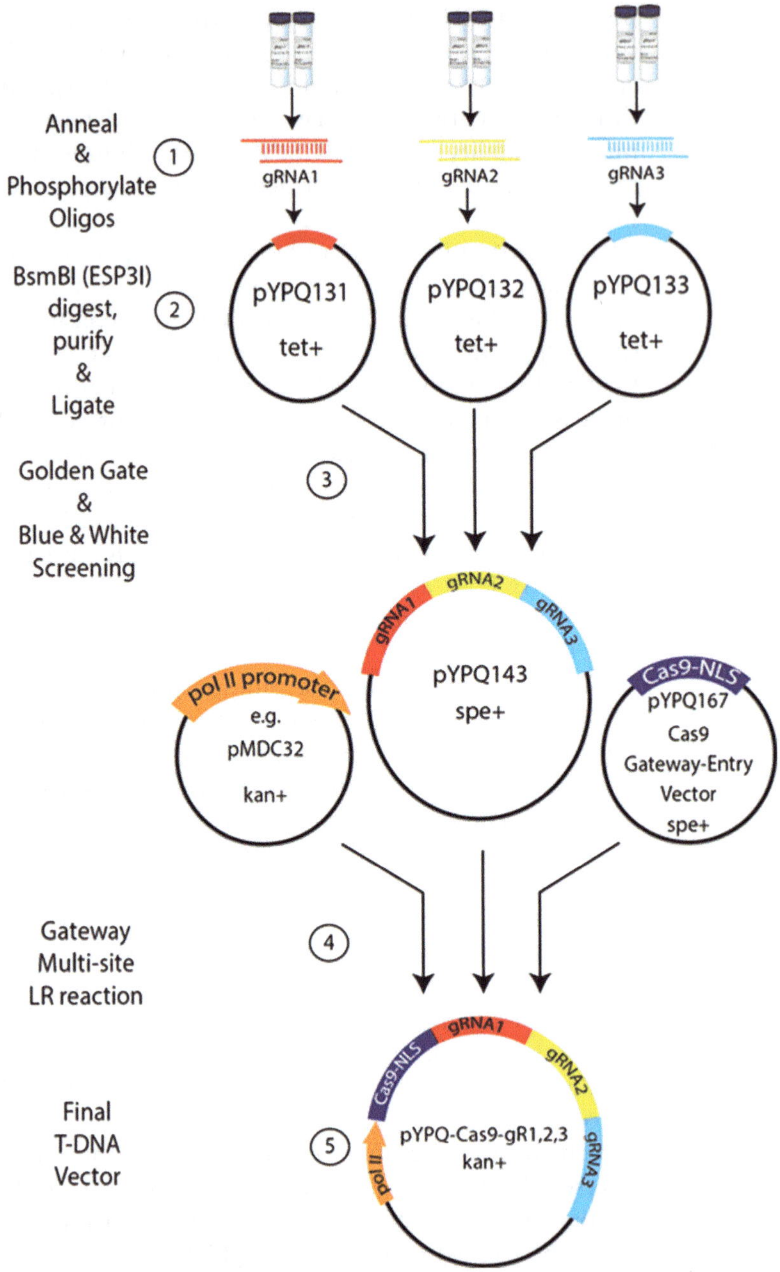

Fig. 1 Toolkit workflow. Figure outlines multiplex gRNA assembly into a single T-DNA expression vector using Golden Gate and Multisite Gateway cloning methods. (*1*) gRNAs are purchased as forward and reverse synthetic oligonucleotides. These oligonucleotides are phosphorylated and annealed. (*2*) Phosphorylated and annealed oligonucleotides are then ligated into gRNA entry vectors at BsmBI (Esp3I) sites. (*3*) gRNA expression cassettes are then assembled together using Golden Gate assembly and blue-white screening using the Multisite Gateway entry vector pYPQ143. (*4*) pYPQ143 (gRNA cassettes flanked by attL5 and attL2) are assembled with Cas9 nuclease vector of choice (with attL1 and attR5) and an attR1-attR2 T-DNA destination vector containing Polymerase II promoter (for Cas9 expression). This assembly uses Gateway Multisite LR recombination

a functional multiplex gRNA expression array module is formed. Each gRNA cassette in the array contains its own promoter, scaffold, and terminator sequences. This array module is now ready for Gateway LR recombination with T-DNA destination vectors and Cas9 entry vectors to form the final T-DNA vector ready for delivery to plants (Fig. 1) (*see* Ref. 13 for review).

2 Materials

1. DNA editing software (DNA Star, Vector NTI, ApE, etc.) and target genomic sequence of interest (*AtTT4*; Arabidopsis Resource Center accession #: AT5G13930).
2. gRNA oligonucleotides (*see* Table 1).
3. Sequencing oligonucleotides.
4. Plasmids pYPQ131A, pYPQ132B, pYPQ133A, pYPQ143, pYPQ167, and pMDC32 [13] (All except pMDC32 [14] are available from Addgene.org, Arabidopsis Resource Center or Kerafast.com.)
5. Restriction Enzymes: BglII, SalI, BsmBI (isoschizomer Eps3I), BsaI, BamHI, EcoRV, and EcoRI.
6. Restriction Enzyme buffers: New England Biolabs 10× Buffer 3.1 and CutSmart Buffer; Thermo Fisher 10× OPTIZYME Buffer 4.
7. 20 mM DTT (dithiothreitol) molecular biology grade. Sterile filtered and frozen in aliquots.
8. T4 DNA Ligase and 10× Ligase Buffer.
9. T4 Polynuclease Kinase (PNK) and PNK 10× Buffer.
10. Invitrogen Gateway LR Clonase® II enzyme mix.

Table 1
Primary gRNA oligonucleotides for synthesis

gRNA	Corresponding oligo sequence
gR1-Forward	5′-**GATT*g***TTCTTTGGATGAGATCAGAC-3′
gR1-Reverse	5′-**AAAC**GTCTGATCTCATCCAAAGAA***c***-3′
gR2-Forward	5′-**GGTC*a***GATACTCCGCCTGAAGCACA-3′
gR2-Reverse	5′-**AAAC**TGTGCTTCAGGCGGAGTATC***t***-3′
gR3-Forward	5′-**GATT**GGGCTGGCCCCACTCCTTGA-3′
gR3-Reverse	5′-**AAAC**TCAAGGAGTGGGGCCAGCCC-3′

11. Chemically Competent DH5α *E. coli* (or equivalent cloning strain) with at least 1 × 10^6 cfu/μg pUC19 DNA transformation efficiency.
12. Solid phase Silica microcentrifuge DNA "spin" column purification and miniprep kit (Qiagen QIAquick gel extraction kit and QIAprep mininprep or some other comparable brand).
13. A thermocycler (for Golden Gate reaction).
14. 37 °C shaking and static incubator.
15. LB medium (1% tryptone, 0.5% yeast extract, 1% sodium chloride, 1.5% agar for solid medium).
16. S.O.C. liquid medium (2% tryptone, 0.5% yeast extract, 10 mM sodium chloride, 2.5 mM potassium chloride, 10 mM magnesium chloride, 10 mM magnesium sulfate, 20 mM glucose).
17. Antibiotic stocks (sterile filtered and frozen in 1 ml aliquots): 10 mg/ml tetracycline in 50% ethanol (use at 10 μg/ml). 100 mg/ml spectinomycin (use at 50–100 μg/ml). 50 mg/ml kanamycin (use at 50 μg/ml).
18. 2 and 1.7 ml centrifuge tubes and 0.2 ml PCR tubes.
19. A NanoDrop spectrophotometer is preferred for DNA quantification.
20. Agarose gel equipment and supplies; DNA ladders.
21. 42 °C water bath, heat block, and magnetic heat plate.
22. 20 mg/ml X-gal in DMSO and 0.1 M IPTG stocks (sterile-filtered and frozen in 1 ml aliquots).

3 Methods

3.1 Selection of Target Sites and gRNA Design

1. Choose target gene(s) of interest (GOI) for study. In this case, we are targeting multiple loci on the *TRANSPARENT TESTA 4* (*TT4*) gene in *Arabidopsis thaliana* (*AtTT4*) (*see* **Note 1** and Fig. 3). *TT4* encodes for chalcone synthase, a key precursor enzyme involved in flavonoid biosynthesis. Arabidopsis *tt4* mutants give rise to the appearance of yellow-looking seeds as the seed coat lacks pigmentation.
2. Assemble target genomic sequence information into DNA editing software of choice. You can choose whether to select your gRNA target sites manually or by using software tools available online or for installation. While online tools are improving constantly, we find that selecting gRNA target sites manually following a systematic approach can be more effective (*see* **Note 2**). If large chromosomal deletions are not generated, as demonstrated in this chapter, we prefer to target

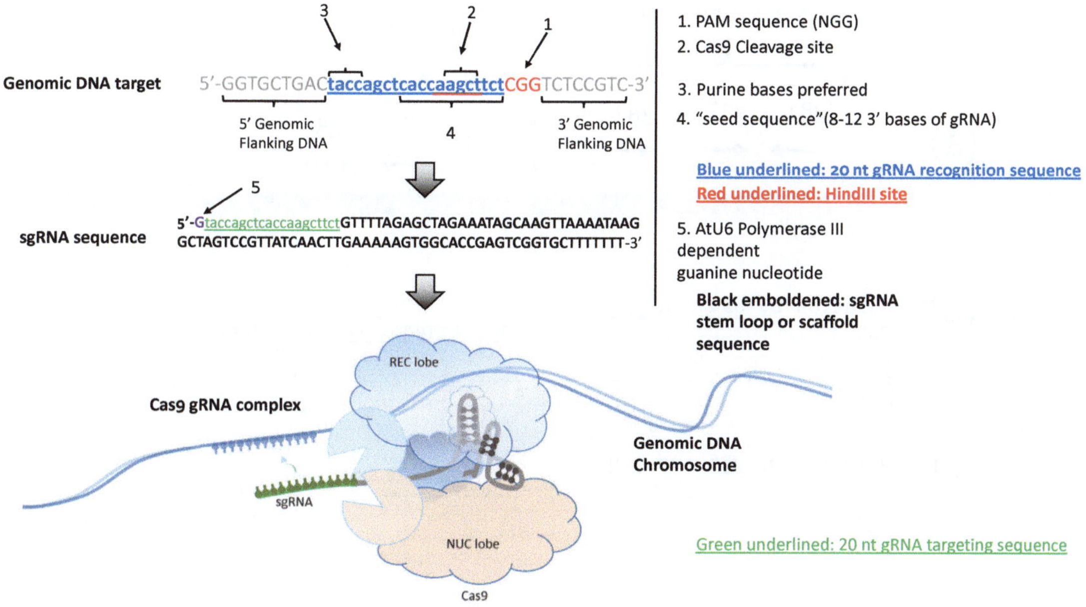

Fig. 2 Design of a CRISPR-Cas9 target. Figure shows design parameters and important features of a CRISPR-Cas9 target sequence. (*Top*) gRNA target sequence within host genome. Important parameters for design at this stage (target identification) include PAM sequence, Cas9 cleavage site and nucleotide preferences for optimal Cas9 activity and specificity. Note that a HindIII site is designed within the Cas9 cleavage motif. (*Middle*) Single guide sgRNA design for targeting genomic sequence shown at top. A 5′ guanosine residue is added for AtU6 polymerase III activity. *Green underlined sequence* in lowercase typeset represents the cognate gRNA sequence used for Watson–Crick base-paring/targeting. This targeting gRNA sequence along with the 5′ guanosine (and overhang sequences for cloning *see* Table 1) is the sequence cloned into gRNA entry vectors. *Uppercase typeset nucleotides* represent the 3′ gRNA scaffold or stem-loop sequence which mediates Cas9 recruitment, binding and activity. (*Bottom*) Illustration showing complex formation after sgRNA and Cas9 expression in plant cell nuclei. Cas9 in complex with sgRNA scans host cell genomic DNA until sgRNA recognizes cognate target site and loads the Cas9 complex onto target site for double strand break formation. *Green section* of sgRNA represents the user defined targeting sequence. *Grey* and *black regions* of the sgRNA represent how the scaffold sequence interacts with Cas9 (Scaffold interaction is shown schematically; *see* [27] for literal structure)

specific genomic cut sites so that a restriction enzyme site is abolished or mutated (Fig. 2). Mutation or excision of restriction enzyme sites greatly simplifies screening of Cas9-induced mutations following transformation if large deletions are not generated. Screening of excised restriction sites can be accomplished by digestion followed by PCR or vice versa.

3. Locate *Sp*Cas9 PAM sequences and design first gRNA. On the sense DNA strand of your GOI sequence, search for forward direction (**NGG**) PAM sequences or reverse compliment (**CCN**) PAM sequences (*see* **Note 3**). Select PAM sequence at 402 bp of TT4 genomic DNA target and highlight 20 bp upstream of this sequence. This is the 20-nt *genomic* gRNA

(a)

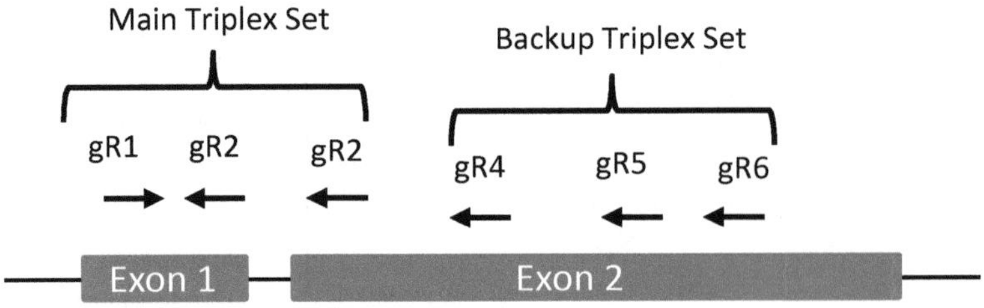

(b)

```
Deletion by gR4 and gR5
CCGGCGGTACTGTCCTCCGTATC...//...CCGCACTCATTGTGGGGTCGGAC (genomic)
CCGGCG---------------------------TCATTGTGGGGTCGGAC (-162 bp, 6X)
CCGGC----------------------------TCATTGTGGGGTCGGAC (-163 bp)
CCGG-------------------------------TTGTGGGGTCGGAC (-167 bp)
```

(c)

```
Deletion by gR5 and gR6
CCGCACTCATTGTGGGGTCGGAC...//...CCGCTCAGACCATCCTTCCAGAC (genomic)
CCGCAC---------------------------AGACCATCCTTCCAGAC (-69 bp, 5X)
CCGCAC----------------------------GACCATCCTTCCAGAC (-70 bp)
CCGC-------------------------------CCATCCTTCCAGAC (-74 bp)
```

Fig. 3 Targeted deletions at AtTT4 locus in *Arabidopsis*. Panel (**a**) shows relative positions and directions of gRNA targets against the *AtTT4* locus. Panel (**b**) shows sequence confirmation of on-target gene editing after Cas9 activity and large chromosomal deletion/NHEJ repair. Regions assayed result from Cas9 activity at gRNA4 and gRNA5 sites. Genomic regions were PCR amplified, gel-purified, extracted, cloned, and sequenced. Top genomic sequence is uncut control and the preceding three sequences show separate but successful gene editing events. (**c**) Shows the same as (**b**) but large chromosomal breaks are between gRNA5 and gRNA6

targeting sequence (TT4 gR1: 5′-TTCTTTGGATGAGATC AGAC**AGG**-3′). Repeat for TT4 gR2: 5′-**CCA**TGTGCTTCA GGCGGAGTATC-3′ and TT4 gR3: 5′-**CCA**TCAAGGAGT GGGGCCAGCCC-3′. Note that TT4 gR2 and gR3 both have PAM sequences oriented on the antisense strand (Fig. 3).

4. Cross-check gRNA sequences against the entire genome using bioinformatics tools available online (We recommend CRISPR-P or *see* [15] for more software recommendations.)
5. In the event that gRNAs 1–3 do not function as planned, show off target effects or are not robust we also target another three sites on Exon 2 of TT4 as a backup plan (TT4 gRNA 4, 5 and 6) (Fig. 3).
6. Design gRNA oligonucleotides for subcloning into gRNA entry vectors by adding proper nucleotide overhangs. For AtU6 based promoters in pYPQ131A, 132A, 133A, etc., add a five nucleotide (**GATTg**) overhang to the 5′end of the forward oligo sequence. This overhang anneals to sticky ends created by BsmBI (Esp3I) digestion (**GATT**). The 3′ **g** of the overhang is required for U6 RNA polymerase III activity (add only if the native sequence lacks a G at this site). For the Reverse oligonucleotide add a four nucleotide overhang (**AAAC**) on the 5′ end for annealing to BsmBI sticky ends.

Finally, add a single cytosine (**c**) to the 3′ end of the reverse oligonucleotide overhang. This terminal C is the reverse compliment of the 5′ terminal G needed for U6 RNA polymerase III activity. For AtU3 promoters, add (**GGTCa**). Thus genomic gRNA sequences from **step 3** should be modified as followed to form functional gRNA targeting sequences as listed in Table 1. Note that gR2 uses the overhang for the U3 promoter. Also note that the G overhang for U6 activity is included in gR3 native sequence and thus does not need to be added.

7. Design PCR primers flanking gRNA sites such that large deletions between gRNA sites can be amplified and compared to control cells not transformed with active Cas9.
8. Obtain all oligos before or while carrying out following sections. Additionally, we recommend you design and obtain sequencing primers for confirmation of proper gRNA clones (*see* **Note 4**).

3.2 Assembly of Reagents into a Final T-DNA Expression Vector

1. Prepare gRNA entry vectors for ligation by sequential digestion: Double digest 2 μg each of pYPQ131A, pYPQ132B and pYPQ133A from 2 h to overnight at 37 °C using BglII, SalI and New England Biolabs (NEB) buffer 3 (*see* **Note 5**). Prepare each vector backbone simultaneously but in separate tubes.
2. Purify BglII/SalI digestion product on silica microcentrifuge columns (*see* **Note 6**).
3. Digest entire volume from the BglII/SalI digestion using 2 μl BsmBI (also called Eps3I), 2 μl of 20 mM DTT and 10× OPTIZYME buffer 4 (Thermo Fisher) in 40 μl end volume. Digest at 37 °C from 2 h to overnight and follow by deactivation at 80 °C for 20 min. Purify using silica microcolumn and quantify concentration (*see* **Note 7**).
4. As you linearize gRNA entry vector backbones, you can prepare gRNA oligonucleotides for ligation. First, phosphorylate gRNA oligonucleotides by mixing 1 μl of each forward and reverse oligo (from a 100 μM working stock), 1 μl 10× T4 polynucleotide kinase reaction buffer, and 1 μl of polynucleotide kinase. Bring final volume to 10 μl with sterile ddH_2O and incubate for 30 min at 37 °C. Since there are six total gRNA to clone (three primary and three "backup") we start with 12 total oligos and will pair them in individual tubes for phosphorylation and annealing. After phosphorylation, there will be six total tubes with each containing a forward and reverse gRNA pair of oligos (gR1, gR2, gR3 and gR4, gR5, gR6).
5. Anneal phosphorylated oligos by immersion in boiling water for 10 s. Let the oligos and water cool to room temperature very slowly (*see* **Note 8**).

6. Dilute cooled oligos 1:200 by adding 1 μl of annealed oligos to 199 μl of ddH_2O at room temperature and mix well.
7. Ligate phosphorylated and annealed gRNA oligos into linearized vector backbones from step three (above). Make sure to ligate gRNAs into corresponding vector backbones as described in Table 1. Setup ligation reactions with an end volume of 10 μl in a PCR tube. Ligate using 1 μl of 1:200 diluted oligos, 1 μl of linearized vector backbone, 1 μl of T4 DNA ligase buffer and 0.5 μl of T4 DNA ligase. Incubate ligation at room temperature from 2 h to overnight.
8. Transform 50 μl of competent *E. coli* cells with 2 μl of ligation reaction (*see* **Note 9**). Plate transformed cells onto LB Tetracycline solid media for selection. Incubate the plates overnight at 37 °C.
9. Pick two or more colonies and culture overnight in selective media with tetracycline (at 10 μg/ml) with vigorous shaking at 37 °C (*see* **Note 10**).
10. Obtain high quality purified DNA for each culture via miniprep.
11. Sequence DNA minipreps using TC14-F2 primer or any primer that will read through the gRNA cloning site cassette (*see* **Note 11**).
12. Perform Golden Gate reactions by assembling components according to Tables 2 and 3. DNA concentration for each vector should be around 100 ng/μl.
13. Incubate Golden Gate reactions as indicated in Table 4 and use a thermocycling program as indicated in Table 5. Transform the finished reactions each with competent *E. coli* cells and plate to LB Spectinomycin with X-gal and IPTG (*see* **Note 12**). Culture overnight at 37 °C.

Table 2
gRNA vector ligation key

gRNA	Corresponding vector for ligation
gR1	pYPQ131A
gR2	pYPQ132B
gR3	pYPQ133A
gR4	pYPQ131A
gR5	pYPQ132B
gR6	pYPQ133A

Table 3
Golden Gate reaction key: primary triplex assembly

Component	Volume (μl)
ddH_2O	4
10× T4 Ligase buffer	1
T4 DNA Ligase	.5
BsaI	.5
pYPQ131A-gR1	1
pYPQ132B-gR2	1
pYPQ133A-gR3	1
pYPQ143	1
Total end volume	10

Table 4
Golden Gate reaction key: backup triplex assembly

Component	Volume (μl)
ddH_2O	4
10× T4 Ligase buffer	1
T4 DNA Ligase	.5
BsaI	.5
pYPQ131A-gR4	1
pYPQ132B-gR5	1
pYPQ133A-gR6	1
pYPQ143	1
Total end volume	10

14. Using blue-white screening, pick two or more white colonies. Culture overnight in selective media with vigorous shaking at 37 °C (*see* **Note 10**).
15. Miniprep *E. coli* cultures.
16. Confirm Golden Gate reaction success by restriction digest. Double Digest minipreps using BamHI, EcoRV, and NEB Cutsmart buffer (*see* **Note 13**). Successful Golden Gate reactions will produce two gRNA cassette assemblies (pYPQ143-TT4-gR1,2,3 and pYPQ143-TT4-gR4,5,6) ready for entry into MultiSite Gateway™ destination vectors.

Table 5
Golden Gate reaction: thermocycling profile

Stage number	Temperature (°C)	Time (min)	Cycles
1	37	5	10
2	16	10	
3	50	5	1
4	80	5	1

Table 6
MultiSite Gateway™ LR assembly

Component	Volume (μl)
Cas9 entry vector (40 ng/μl)	2
Guide RNA entry vector (40 ng/μl)	2
Destination vector (100 ng/μl)	2
LR Clonase II	1
Total end volume	7

17. Quantify and dilute gRNA cassette assemblies (pYPQ143-TT4-gR1,2,3 and pYPQ143-TT4-gR4,5,6) to 40 ng/μl.
18. Setup Multisite Gateway LR reactions as described in Table 6 (*see* **Note 14**). Incubate separate LR reactions for pYPQ143-TT4-gR1,2,3 and pYPQ143-TT4-gR4,5,6 respectively using pYPQ167 (Cas9p—plant codon-optimized Cas9 entry vector) for Cas9 entry vector and pMDC32 (2× 35S promoter; *see* [14]) as destination vector (*see* **Note 15**).
19. Transform competent *E. coli* and plate to LB Kanamycin. Culture overnight at 37 °C. Pick two or more colonies, culture and minprep. Verify constructs by restriction digest using EcoRI (*see* **Note 16**).
20. Move on to transforming an appropriate *Agrobacterium* strain for floral dip transformation of *Arabidopsis thaliana* (*see* **Note 17**).

4 Notes

1. Most likely, you will have predetermined a target gene based upon your own research objectives. If possible, however, choose genes which will provide a visible or at least measurable

phenotype. For example, null AtTT4 (*att*) gene mutants produce a colorless seed coat [16–19]. Alternatively, genotypic assays can detect Cas9 induced gene alterations relatively easily without the use of a reporter gene or induced phenotype. For example, by targeting three different loci along the *AtTT4* gene we are able to generate large chromosomal deletions that span the distance from two or more adjacent gRNA sites. This strategic approach for knocking out TT4 is advantageous as successful Cas9 activity can be screened at the molecular level by performing a simple PCR based experiment using primers that flank gRNA sites. Successful mutation of TT4 loci using this approach will yield a defined fragment size based upon PCR primers used while Cas9 negative control lines will yield a relatively larger PCR fragment.

A great advantage of targeting a single gene using a multiplex approach is that failure or inefficiency of any given gRNA activity can be overcome by other gRNAs. Thus there is redundancy built into this strategy which makes it less prone to gRNA specific problems.

2. Although we recommend selecting target sites manually by adhering to our systematic approach, we do recommend that online tools or other software be used to check for off targeting by cross-referencing the entire genome for similar gRNA sites (specifically the trailing 8–12 nucleotide seed sequence) that reside next to PAM sequences. Significant off targeting effects of CRISPR/Cas9 have been reported in the literature, especially within mammalian cell systems, but reports in plants have shown off-targeting to be of less concern [20] and outcrossing can alleviate these effects in many instances. However, both the prevalence and significance of off targeting effects in plants remains to be rigorously studied and the details of off-targeting in plants remains controversial. New and more effective bioinformatic tools are surfacing periodically and we recognize that software based tools for directing the placement of gRNAs will likely become standard procedure; still, understanding the intricacies of gRNA functionality and design should prove valuable. Refer to Graham and Root [15] for an extensive list of currently available software packages that aid in gRNA design and off target prediction.

3. Doench et al. show that CGG PAM motifs are more effective than the generalized NGG motif (where N $\neq$ C) and guanine is preferentially restricted from the 3′ terminus of the PAM, i.e., NGGH (where H = A, T, or C) [21]. Moreover, four of the most 5′ terminal nucleotides on the gRNA recognition sequence are preferentially purine bases (Fig. 2) [21, 22]. If trying to disrupt gene function, make sure to target more 5′ regions on exons of your GOI as 3′ frame shifts or stop codon

insertions may have a lesser effect on gene functionality. Similarly, be cautious of targeting exon regions, especially if the splicing isoform of your GOI to be knocked out is unknown. Last, target open chromatin regions if possible as Cas9 functions better when directed to open chromatin regions [15, 23]. If greater than normal specificity is desired, limit the length of your gRNA recognition sequence to 17 or 18 nucleotides as these have been shown to give lower off target activities [24].

4. We use the following sequencing primer for confirming our gRNA insertions prior to performing Golden Gate reactions: TC14-F2: 5′-CAAGCCTGATTGGGAGAAAA-3′.

5. It is essential to start with highly pure DNA preps of at least 100 ng/μl. BglII/SalI preliminary digestion reduces the risk of backbone self-ligations further downstream. Notice use of the AtU3 (B) promoter in the 2nd gRNA slot (pYPQ132B). We use this promoter buffering approach to minimize repeated inserts of AtU6 promoters along the entire gRNA expression cassette, which may lead to increased rates of transcriptional suppression [25, 26]. pYPQ131, pYPQ132, pYPQ133 and many other kit plasmids referenced in this chapter are available through Addgene, the Arabidopsis Resource Center or Kerafast.

6. Linearized gRNA entry vector using BglII/SalI yields a 3790-bp backbone and 13-bp fall-out sequence. Silica column purification removes the 13-bp fall-out fragment due to its small size. Alternatively, the 13-bp fragment can be removed by gel purification of the 3790-bp backbone. Silica column purification followed by elution with 35 μl of warm molecular biology grade sterile water (sterile ddH_2O) generally obtains the best yield of purified linear backbone with the least effort. We recommend Qiagen DNA minicolumns or any similar product.

7. BglII/SalI/BsmBI sequential digestion can be carried out far in advance and linearized backbone can be stored in the −20 °C freezer for instant use whenever needed. Quantify DNA using spectrophotometry ("NanoDrop") or agarose gel densitometry.

8. We boil our annealed oligos in PCR tubes secured to a floating incubation raft. The tubes and raft are boiled in a beaker of water. After 10 s in boiling water, the entire beaker is removed from the heat plate and is cooled to room temperature. Alternatively you can use a thermocycler: 95 °C for 5 min and then ramping down to 25 °C at 5 °C/min or slower. We have noticed higher efficiency cloning using the boiling water method.

9. We use chemically competent DH5α cells for transformation of gRNA entry vectors. These cells are prepared in our lab using common reagents and protocols. New England Biolabs

offers chemically competent *E. coli* cells for those who wish to purchase this item. Competent *E. coli* cells are transformed by heat-shock. A useful tip for plating transformed *E. coli* is to gently centrifuge cells after the outgrowth step in S.O.C. media, remove the media, resuspend in 200 μl S.O.C. and only plate one third or half of recovered cells to LB tet+ plates and save the rest for backup. We typically observe many *E. coli* colonies due to high-efficiency ligation and transformation.

10. Be sure to grow rapidly dividing cells with plenty of aeration. High rpm shaking and culture volume should not exceed 25% of culture vessel capacity.
11. It is important not to skip the sequencing confirmation step. It is not uncommon to observe self-ligations or incorrect clones up to this point. Sequencing to ensure correct gRNA insertion is critical.
12. Add 50 μl of 20 mg/ml X-gal in DMSO and 50 μl of 0.1 M IPTG per plate. Spread around and let soak in for 30 min or up to 1 h.
13. Restriction digest of pYPQ143-TT4-gR1,2,3 using EcoRV and BamHI yields a 2704-bp backbone fragment and a 2504-bp gRNA cassette fragment. These two fragments are similarly sized so be sure to analyze the fragments on a low percentage agarose gel (e.g., 0.8% agarose) with maximum separation/ running time otherwise it may be difficult to observe two distinct band fragments.
14. Precise concentration of plasmid components in the MultiSite Gateway™ LR reaction is critical. Take care to accurately quantify and dilute plasmids to concentrations recommended in Table 6.
15. The pMDC32 vector is available at ABRC. Other attLR1-attR2 destination vectors with a Pol II promoter of choice can be used.
16. Digestion with EcoRI yields a 9761-bp backbone fragment, a 4554-bp Cas9p fragment, and a 2787-bp gRNA cassette fragment.
17. Be sure to make a glycerol stock of liquid bacterial cultures. Freeze stocks at −80 °F.

Acknowledgments

This work is supported by startup funds from East Carolina University and a Collaborative Funding Grant from North Carolina Biotechnology Center and Syngenta (2016-CFG-8003) to Y.Q.

References

1. Jinek M, Chylinski K, Fonfara I, Hauer M, Doudna JA, Charpentier E (2012) A programmable dual-RNA-guided DNA endonuclease in adaptive bacterial immunity. Science 337(6096):816–821. doi:10.1126/science.1225829
2. Belhaj K, Chaparro-Garcia A, Kamoun S, Patron NJ, Nekrasov V (2015) Editing plant genomes with CRISPR/Cas9. Curr Opin Biotechnol 32:76–84
3. Baltes NJ, Voytas DF (2015) Enabling plant synthetic biology through genome engineering. Trends Biotechnol 33(2):120–131
4. Deltcheva E, Chylinski K, Sharma C, Gonzales K, Chao Y, Pirzada Z, Eckert M, Vogel J, Charpentier E (2011) CRISPR RNA maturation by trans-encoded small RNA and host factor RNase III. Nature 471:602–609
5. Mali P, Yang L, Esvelt KM, Aach J, Guell M, DiCarlo JE, Norville JE, Church GM (2013) RNA-guided human genome engineering via Cas9. Science 339(6121):823–826. doi:10.1126/science.1232033
6. Cong L, Ran FA, Cox D, Lin S, Barretto R, Habib N, Hsu PD, Wu X, Jiang W, Marraffini LA, Zhang F (2013) Multiplex genome engineering using CRISPR/Cas systems. Science 339(6121):819–823. doi:10.1126/science.1231143
7. Li J-F, Norville J, Aach J, McCormack M, Zhang D, Bush J, Chruch G, Sheen J (2013) Multiplex and homologous recombination-mediated genome editing in *Arabidopsis* and *Nicotiana benthamiana* using guide RNA and Cas9. Nat Biotechnol 31:688–691
8. Carroll D (2011) Genome engineering with zinc-finger nucleases. Genetics 188(4):773–782. doi:10.1534/genetics.111.131433
9. Christian M, Cermak T, Doyle EL, Schmidt C, Zhang F, Hummel A, Bogdanove AJ, Voytas DF (2010) Targeting DNA double-strand breaks with TAL effector nucleases. Genetics 186(2):757–761. doi:10.1534/genetics.110.120717
10. Weber J, Ollinger R, Mathias F (2015) CRISPR/Cas9 somatic multiplex-mutagenesis for high-throughput functional cancer genomics in mice. Proc Natl Acad Sci U S A 112(45):13982–13987
11. Zhou H, Liu B, Weeks D, Spalding M, Yang B (2014) Large chromosomal deletions and heritable small genetic changes induced by CRISPR/Cas9 in rice. Nucleic Acids Res. doi:10.1093/nar/gku806
12. Ran A, Hsu P, Lin C-Y, Gootenberg J, Konermann S, Trevino A, Scott D, Inoue A, Matoba S, Zhang Y, Zhang F (2013) Double nicking by RNA-guided CRISPR Cas9 for enhanced genome editing specificity. Cell 154:1380–1389
13. Lowder L, Zhang D, Baltes N, Paul J III, Tang X, Zheng X, Voytas D, Hsieh T-F, Zhang Y, Qi Y (2015) A CRISPR/Cas9 toolbox for multiplexed plant genome editing and transcriptional regulation. Plant Physiol 169:971–985
14. Curtis MD, Grossniklaus U (2003) A gateway cloning vector set for high-throughput functional analysis of genes in planta. Plant Physiol 133(2):462–469. doi:10.1104/pp.103.027979
15. Graham D, Root D (2015) Resources for the design of CRISPR gene editing experiments. Genome Biol 16:260
16. Shirley B, Kubasek W, Storz G, Bruggemann E, Koornneef M, Ausubel F, Goodman H (1995) Analysis of Arabidopsis mutants deficient in flavonoid biosynthesis. Plant J 8(5):659–671
17. Sun W, Meng X, Liang L, Jiang W, Huang Y, He J, Hu H, Almqvist J, Gao X, Wang L (2015) Molecular and biochemical analysis of chalcone synthase from Freesia hybrid in flavonoid biosynthetic pathway. PLoS One. doi:10.1371/journal.pone.0119054
18. Zhang F, Maeder M, Unger-Wallace E, Hoshaw J, Reyon D, Christian M, Li X, Pierick C, Dobbs D, Peterson T, Joung K, Voytas D (2010) High frequency targeted mutagenesis in *Arabidopsis thaliana* using zing finger nucleases. Proc Natl Acad Sci U S A 107(26):12028–12033
19. Qi Y, Zhang Y, Zhang F, Baller J, Cleland S, Ryu Y, Starker C, Voytas D (2013) Increasing frequencies of site-specific mutagenesis and gene targeting in Arabidopsis by manipulating DNA repair pathways. Genome Res 23:547–554
20. Feng Z, Mao Y, Zhang B, Wei P, Yang D-L, Wang Z, Zhang Z, Zheng R, Yang L, Zeng L, Liu X, Zhu J-K (2014) Multigeneration analysis reveals the inheritance, specificity, and patterns of CRISPR/Cas-induced gene modifications in *Arabidopsis*. Proc Natl Acad Sci U S A 111(12):4632–4637
21. Doench J, Hartenian E, Graham D, Tothova Z, Hegde M, Smith I, Sullender M, Ebert B, Xavier R, Root D (2014) Rational design of highly active sgRNAs for CRISPR-Cas9-mediated gene inactivation. Nat Biotechnol 32:1262–1267

22. Wang T, Wei J, Sabatini D, Lander E (2014) Genetic screens in human cells using the CRISPR-Cas9 system. Science 343:80–84
23. Singh R, Kuscu C, Quinlan A, Qi Y, Adli M (2015) Cas9-chromatin binding information enables more accurate CRISPR off-target prediction. Nucleic Acids Res 43(18). doi:10.1093/nar/gkv575
24. Tsai S, Zheng Z, Nguyen N, Liebers M, Topkar V, Thapar V, Wyvekens N, Khayter C, Iafrate J, Le L, Aryee M, Joung K (2015) GUIDE-seq enables genome-wide profiling of off-target cleavage by CRISPR-Cas nucleases. Nat Biotechnol 33(2):187–197
25. Vaucheret H (2006) Post-transcriptional small RNA pathways in plants: mechanisms and regulations. Genes Dev 20:759–771
26. De Wilde C, Van Houdt H, De Buck S, Angenon G, De Jaeger G (2000) Depicker A. Plants as bioreactors for protein production: avoiding the problem of transgene silencing 43:347–359
27. Nishimasu H, Ran A, Hsu P, Konermann S, Shehata S, Dohmae N, Ishitani R (2014) Crystal structure of Cas9. Cell 156:935–949

Chapter 26

Chitin-Triggered MAPK Activation and ROS Generation in Rice Suspension-Cultured Cells

Koji Yamaguchi and Tsutomu Kawasaki

Abstract

Suspension-cultured cells respond sensitively to a number of stimuli including pathogen-derived molecules. Therefore, they are used in a simple, single-cell model system in order to gain insights into immune signaling pathways in rice. Here we describe protocols for studying chitin-induced MAPK activation and ROS generation in rice suspension-cultured cells.

Key words Immunity, Chitin, Rice, MAPK, ROS

1 Introduction

Plants have evolved a sophisticated immune system to detect and defend against potential pathogens. The plant innate immune system is generally initiated via the recognition of conserved microbe-associated molecular pattern (MAMP) ligands, including bacterial flagellin or peptidoglycan and fungal chitin [1, 2]. These MAMPs are perceived by pattern recognition receptors (PRRs). The PRRs identified to date are receptor-like protein kinases (RLKs) and receptor-like proteins (RLPs), which are localized at the plasma membrane [3].

Chitin, a β-1,4-linked polymer of *N*-acetylglucosamine, is a constituent of the fungal cell wall. Chitin oligomers are typical MAMPs that elicit various defense responses [4]. In rice, chitin is recognized by two PRRs with an extracellular lysine motif (LysM) domain; LysM-RLP CEBiP directly binds to chitin, and then interacts with LysMs-RLK OsCERK1 [5, 6], leading to the rapid activation of mitogen-activated protein kinase (MAPK), OsMPK3, OsMPK4, and OsMPK6 [7] as well as the generation of reactive oxygen species (ROS) mediated by NADPH oxidase. Disruption of the OsCERK1 gene has been shown to compromise these chitin-induced immune responses, which culminates in the slowing

Libo Shan and Ping He (eds.), *Plant Pattern Recognition Receptors: Methods and Protocols*, Methods in Molecular Biology, vol. 1578, DOI 10.1007/978-1-4939-6859-6_26, © Springer Science+Business Media LLC 2017

or halting of pathogen proliferation [8]. We found that the receptor-like cytoplasmic kinase OsRLCK185 associates with and is directly phosphorylated by OsCERK1 at the plasma membrane [9]. The silencing of OsRLCK185 has been shown to suppress chitin-induced immune responses, including MAPK activation and ROS generation. These findings support a role for OsRLCK185 as an essential immediate downstream signaling partner of OsCERK1 in chitin-induced immunity in rice. Thus, MAPK activation and ROS generation constitute two main classes of early immune responses triggered by the perception of ligands by PRR.

Rice suspension-cultured cells respond with high sensitivity to MAMPs. Furthermore, a cell line is relatively easy to establish from many kinds of cultivars or transgenic plants. Therefore, this single cell model system is suitable for analyzing immune signaling pathways in rice. Here we describe methods for the production of suspension-cultured cells from rice seeds. These protocols describe the basis for studying chitin-induced MAPK activation in a simple western blot analysis and measurements of ROS production in response to chitin.

2 Materials

Unless otherwise noted, deionized water (ADVANTEC) and basic molecular biology regents and equipment are used to prepare all of the following solutions.

1. Twelve-well plates (e.g., True line).
2. Hexa-N-acetylchitohexaose (e.g., Seikagaku Biobusiness).
3. Extraction buffer: 50 mM HEPES (pH 7.4), 5 mM EDTA, 5 mM EGTA, 50 mM β-glycerophosphate, 10 mM Na_3VO_4 (*see* **Note 1**), 10 mM NaF, 2 mM DTT, Complete EDTA-free protease inhibitor cocktail (Roche), 1% SDS, or Triton 100 (*see* **Note 2**).
4. Bradford assay (e.g., Thermo Scientific).
5. SDS-PAGE running buffer: 25 mM Tris base, 192 mM glycine, 0.1% (w/v) SDS.
6. Gel running apparatus.
7. 3-mm paper (Whatman).
8. Polyvinylidenefluoride (PVDF) membrane (e.g., Millipore).
9. Tank-blotting apparatus (e.g., ORIENTAL INSTRUMENTS).
10. Transfer buffer: 25 mM Tris, 192 mM glycine, 20% (v/v) methanol.
11. Ponceau S Red staining solution: 0.1% (w/v) Ponceau S Red, 5% (v/v) glacial acetic acid.

12. Bovine serum albumin (BSA).
13. Tris-buffered saline (TBS; 10×): 1.5 M NaCl and 0.1 M Tris–HCl (pH 7.5) (*see* **Note 3**).
14. TBS containing 0.1% Tween 20 (TBS-T).
15. An anti-pMAPK antibody (e.g., Cell Signaling).
16. Can Get Signal Solution 1 (TOYOBO).
17. ECL Anti-Rabbit IgG, HRP-linked species-specific whole antibody (e.g., GE Healthcare).
18. Can Get Signal Solution 2 (TOYOBO).
19. ECL reagents (e.g., GE Healthcare).
20. Chemiluminescence imager (e.g., GE Healthcare).
21. White 96-well plates (e.g., Greiner Bio-One).
22. L-012 (WAKO Chemicals): Dilute 20 mM in deionized water. Maintain in the dark.
23. Peroxidase (e.g., TOYOBO): Dilute Peroxidase (1 mg/ml) in deionized water. Maintain at 4 °C.
24. Luminescence microplate reader (e.g., Berthold).

3 Methods

3.1 Production of Rice Suspension-Cultured Cells

3.1.1 Induction of Rice Calli

1. 2N6P media preparation (solid agar plates, 1 L): 4.0 g N6 salts, MS Vitamins (1×), 30 g Sucrose, 4 mg 2,4-D (*see* **Note 4**), 2.88 g Proline, 8 g agarose, adjust the pH to 5.7 with 1 M NaOH. Sterilize by autoclaving (*see* **Note 5**).
2. Sterilize rice seeds with 16% sodium hypochlorite (NaOCl) for 1 h with gentle shaking.
3. Wash five times with sterilized water (*see* **Note 6**).
4. Place the seeds onto a 2N6P plate (10 seeds/plate) at 30 °C for 5 weeks. Calli are generated during this period from seed embryos.
5. An example of rice calli is shown in Fig. 1.

3.1.2 Establishment of Rice Suspension-Cultured Cells

1. Sterilized flask (100 ml) with foil (*see* **Note 7**).
2. R2S media preparation (liquid media, 1 L): 20% (v/v) R2 macro, 0.1% (v/v) R2 micro, 0.2% (v/v) Fe•EDTA, MS Vitamins (1×), 30 g Sucrose, 2 mg 2,4-D (*see* **Note 4**), adjust the pH to 5.7 with 1 M NaOH. Sterilize by autoclaving (*see* **Notes 8** and **9**).
3. R2 macro media preparation (liquid media, 1 L): 80 g KNO_3, 6.7 g $(NH_4)_2SO_4$, 5.0 g $MgSO_4{\bullet}7H_2O$, 3.0 g $CaCl_2{\bullet}2H_2O$, 5.46 g $NaH_2PO_4{\bullet}2H_2O$. Maintain at 4 °C (*see* **Note 5**).

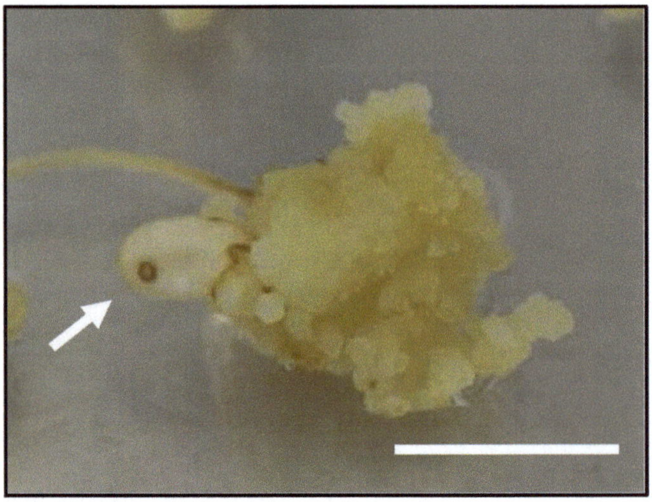

Fig. 1 Generation of rice calli from seed embryos. Rice calli were developed from rice seeds. Rice calli were photographed 5 weeks after plating. The *arrow* indicates a rice seed. Bar: 1 cm

4. R2 micro media preparation (liquid media, 100 ml): 160 mg $MnSO_4{\bullet}4H_2O$, 220 mg $ZnSO_4{\bullet}7H_2O$, 12.5 mg $CuSO_4{\bullet}5H_2O$, 300 mg H_3BO_3, 12.5 mg $NaMoO_4{\bullet}2H_2O$. Maintain at 4 °C (*see* **Note 5**).
5. Fe•EDTA media preparation (liquid media, 100 ml): 375 mg Na_2EDTA, 275 mg $FeSO_4{\bullet}7H_2O$. Maintain at 4 °C (*see* **Note 5**).
6. Transfer rice calli to a sterilized flask containing 22 ml of R2S liquid medium.
7. Maintain the flasks at 30 °C on a platform shaker set at 60 rpm.
8. Replace liquid medium with fresh R2S liquid medium (*see* **Note 6**) every week (*see* **Notes 10** and **11**).
9. Grind rice calli finely using a sterilized stainless mesh.
10. Incubate for 3 weeks to establish rice suspension-cultured cell lines.

3.2 Detection of Phosphorylated MAPKs

3.2.1 Treatment of Chitin for Rice Suspension Cells

1. Subculture rice suspension-cultured cells for 3 days in fresh R2S liquid medium, and then divide them on 12-well plates (150 mg cells, 2 ml fresh medium per well).
2. Treat rice suspension-cultured cells with 2 μg/ml hexa-N-acetylchitohexaose.
3. Remove liquid medium, flash-freeze, and maintain rice suspension-cultured cells at −80 °C until used.

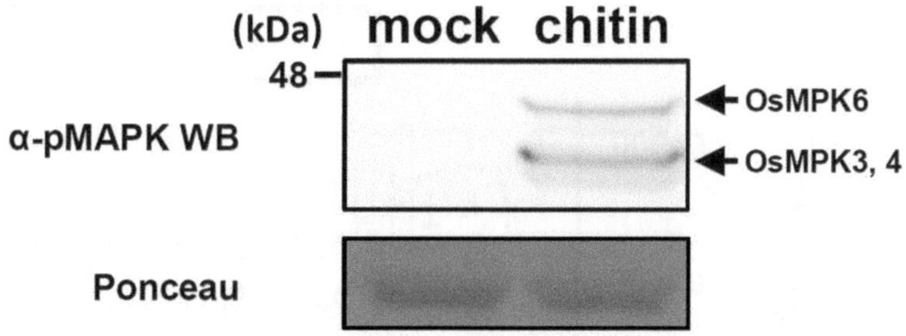

Fig. 2 In vivo activation of OsMPK3, OsMPK4, and OsMPK6 in WT suspension-cultured cells subjected to chitin and mock treatments. The activation of MAPK was analyzed by immunoblotting with α-pMAPK. The protein loading control was shown by Ponceau staining. Total proteins were extracted from WT suspension-cultured cells subjected to the chitin and mock treatments for 15 min. WT suspension-cultured cells were generated by the rice cultivar Nipponbare. The *arrow* indicates the position of OsMPK6 with a size of 45 kDa and OsMPK3, 4 with a size of 43 kDa

3.2.2 Protein Extraction for Phosphorylated MAPKs

1. Grind rice suspension-cultured cells in liquid nitrogen (*see* **Note 12**) and thaw them in chilled protein extraction buffers.
2. Let stand on ice for 1 h (*see* **Note 13**).
3. Centrifuge the mixture at 18,000 × *g* for 10 min at 4 °C.
4. Transfer supernatants into clean chilled microcentrifuge tubes.
5. Measure protein concentrations using the Bradford assay.

3.2.3 Immunoblot of Phosphorylated MAPKs

1. Separate 20 μg protein by electrophoresis on a 10% SDS polyacrylamide gel (*see* **Note 14**), and transfer to a PVDF membrane (*see* **Notes 15** and **16**).
2. Wash two times with deionized water (*see* **Note 17**).
3. Confirm transfer efficiency by Ponceau S staining.
4. Wash four times with TBS-T (5 min for each).
5. Incubate the membrane in TBS-T containing 1% (W/V) BSA (*see* **Note 18**) at room temperature for 2–12 h.
6. Wash four times with TBS-T (5 min for each).
7. Incubate the membrane in the Can Get Signal solution containing the anti-phospho p44/42 MAPK antibody at room temperature for 90 min.
8. Wash four times with TBS-T (5 min for each).
9. Incubate the membrane in the Can Get Signal solution containing the HRP-conjugated secondary antibody at room temperature for 1 h.
10. Wash four times with TBS-T (5 min for each).
11. Visualize immunoreactive signals using ECL reagents and the chemiluminescence imager.
12. An example of immunoblotting showing the phosphorylation of OsMPK3, OsMPK4, and OsMPK6 is shown in Fig. 2.

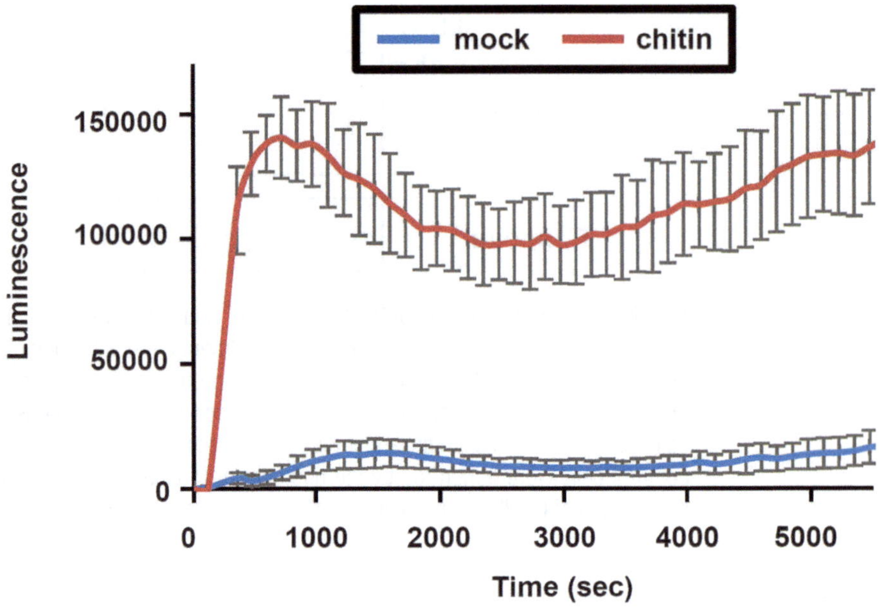

Fig. 3 Chitin-induced ROS production. WT suspension-cultured cells were treated with chitin in a solution containing 500 μM L-012 and 10 μg/ml peroxidase. Cells were subjected to the chitin and mock treatments at 0 s using injectors. ROS production was quantified using a luminescence microplate reader. Data are means ± SD

3.3 Detection of Chitin-Induced ROS Production

3.3.1 Preparation of Materials

1. Subculture suspension-cultured rice cells for 3 days in fresh R2S liquid medium.
2. Divide the cells (6.4 mg/100 μl) on white 96-well plates.
3. Incubate the plates in the dark at 30 °C for 16 h.

3.3.2 Measurement of Chitin-Induced ROS Production

1. Replace liquid medium with 100 μl of fresh medium containing 0.5 mM L-012 (*see* **Notes 19** and **20**) and 10 μg/μl peroxidase (*see* **Note 21**).
2. Add 50 μl of hexa-N-acetylchitohexaose (2 μg/ml).
3. Monitor chemiluminescence using the luminescence microplate reader (*see* **Note 22**).
4. An example of ROS production is shown in Fig. 3.

4 Notes

1. Preparation of 0.2 M Na_3VO_4: Dissolve Na_3VO_4 in water and adjust to pH 10.0 with HCl. Boil until the solution becomes colorless, then cool to room temperature and adjust to pH 10.0. Repeat the same steps until the solution remains at pH 10.0. Store in aliquots at −20 °C.

2. Protein extracts of rice suspension-cultured cells exhibit strong phosphatase activity. Do not remove the phosphatase inhibitor from the extraction buffer.
3. In order to prepare 1 M Tris–HCl pH 7.5, weigh 121.14 g of Tris-Base in 1000 ml distilled water and mix well. Adjust the pH to 7.5 by adding approximately 65 ml of HCl (35%). Sterilize by autoclaving. Maintain at room temperature.
4. Dissolve 2,4-D in 70% EtOH.
5. Ensure that no lumps remain in the solution.
6. During the handling of rice calli and suspended cells, perform all steps on a clean bench.
7. Flasks need to be sterilized by dry heating (180 °C, 3 h).
8. While preparing any solution, do not dissolve all the solute at the same time in the full volume of the solvent.
9. Prepare fresh sterilization solution every time.
10. If the growth of calli is slow, discard the sample and prepare a new cell line.
11. The normal color of calli is yellow. If the color becomes black, discard the sample and prepare a new cell line.
12. Never allow powdered rice suspension-cultured cells to thaw. Keep them continuously frozen under liquid nitrogen.
13. In order to prevent foaming of the detergent, do not vortex the solution.
14. Acrylamide is neurotoxic. The dispensing and handling of acrylamide solution prior to gel polymerization needs to be carried out under a fume hood with gloves and protective clothing.
15. All equipment needs to be cleaned with methanol prior to use.
16. The PVDF membrane needs to be pretreated with methanol.
17. Do not dry the PVDF membrane bound with proteins.
18. BSA is a better blocking reagent for phosphoprotein detection. Milk contains the phosphoprotein casein. Blocking with milk may produce a high background.
19. L-012 is a new luminol derivative. L-012 is more active than luminol.
20. L-012 may also be used. L-012 is light sensitive. Protect it from light while preparing the solution containing L-012.
21. After taking the tube containing peroxidase out of freezer, do not open it immediately. Keep it unopened at room temperature for least 3 h.
22. Do not use black plates to measure ROS because of the low intensity of luminescence.

Acknowledgments

This research was supported by Grants-in-Aid for Scientific Research (A)(15H02489), for Scientific Research on Innovative Areas (25114517 and 15H01242), and by the Strategic Project to Support the Formation of Research Bases at Private Universities: Matching Fund Subsidy from the Ministry of Education, Culture, Sports, Science and Technology, 2011-2015 (S1101035) to T.K., and Grants-in-Aid for Scientific Research (15K18649) and Basic Science Research Projects from Sumitomo Foundation to K.Y.

References

1. Jones JD, Dangl JL (2006) The plant immune system. Nature 444:323–329
2. Miya A, Albert P, Shinya T, Desaki Y, Ichimura K, Shirasu K, Narusaka Y, Kawakami N, Kaku H, Shibuya N (2007) CERK1, a LysM receptor kinase, is essential for chitin elicitor signaling in Arabidopsis. Proc Natl Acad Sci U S A 104: 19613–19618
3. Monaghan J, Zipfel C (2012) Plant pattern recognition receptor complexes at the plasma membrane. Curr Opin Plant Biol 15:349–357
4. Kombrink A, Sanchez-Vallet A, Thomma BP (2011) The role of chitin detection in plant–pathogen interactions. Microbes Infect 13: 1168–1176
5. Kaku H, Nishizawa Y, Ishii-Minami N, Akimoto-Tomiyama C, Dohmae N, Takio K, Minami E, Shibuya N (2006) Plant cells recognize chitin fragments for defense signaling through a plasma membrane receptor. Proc Natl Acad Sci U S A 103:11086–11091
6. Shinya T, Motoyama N, Ikeda A, Wada M, Kamiya K, Hayafune M, Kaku H, Shibuya N (2012) Functional characterization of CEBiP and CERK1 homologs in Arabidopsis and rice reveals the presence of different chitin receptor systems in plants. Plant Cell Physiol 53: 1696–1706
7. Kishi-Kaboshi M, Okada K, Kurimoto L, Murakami S, Umezawa T, Shibuya N, Yamane H, Miyao A, Takatsuji H, Takahashi A, Hirochika H (2010) A rice fungal MAMP-responsive MAPK cascade regulates metabolic flow to antimicrobial metabolite synthesis. Plant J 63:599–612
8. Kouzai Y, Mochizuki S, Nakajima K, Desaki Y, Hayafune M, Miyazaki H, Yokotani N, Ozawa K, Minami E, Kaku H, Shibuya N, Nishizawa Y (2014) Targeted gene disruption of OsCERK1 reveals its indispensable role in chitin perception and involvement in the peptidoglycan response and immunity in rice. Mol Plant Microbe Interact 27:975–982
9. Yamaguchi K, Yamada K, Ishikawa K, Yoshimura S, Hayashi N, Uchihashi K, Ishihama N, Kishi-Kaboshi M, Takahashi A, Tsuge S, Ochiai H, Tada Y, Shimamoto K, Yoshioka H, Kawasaki T (2013) A receptor-like cytoplasmic kinase targeted by a plant pathogen effector is directly phosphorylated by the chitin receptor and mediates rice immunity. Cell Host Microbe 13:347–357

Chapter 27

Chitin-Induced Responses in the Moss *Physcomitrella patens*

Simon Bressendorff, Magnus Wohlfahrt Rasmussen, Morten Petersen, and John Mundy

Abstract

A MAP kinase pathway below a chitin receptor in the moss *Physcomitrella patens* induces immune responses including rapid growth inhibition, a novel fluorescence burst, and cell wall depositions. The molecular mechanisms producing these three responses are currently unknown but warrant further investigation in this simple model system. Here we describe qualitative, time-lapse, and quantitative assays to monitor and measure these responses.

Key words Bryophyte, Plant innate immunity

1 Introduction

The moss *Physcomitrella patens*, which shared a last common ancestor with flowering plants at least 420 million years ago [1], is representative of early land plants. *P. patens* is amenable to reverse genetic studies as gene knockouts and knockins can be generated via homologous recombination [2]. These characteristics make *P. patens* a model for studies of plant adaptations to terrestrial environments including the evolution of immunity to pathogens. Infection of *P. patens* with bacterial, oomycete, and fungal pathogens induces responses similar to those in flowering plants such as *Arabidopsis* including production of reactive oxygen species (ROS), expression of defense related transcripts, hormone perturbations and cell wall modifications [3–8]. *P. patens* also responds to the microbe associated molecular pattern (MAMP) chitin, an oligosaccharide from fungal cell walls [9, 10].

In a recent study using targeted knockouts we identified *P. patens* components involved in signal transduction upon chitin perception [11]. These include a chitin receptor PpCERK1 and a MAP kinase cascade involving two MAP kinase kinase kinases (PpMEKK1a and b) at least two MAP kinase kinases (PpMKK1a and c) and two MAP kinases (PpMPK4a and b). We found that

Libo Shan and Ping He (eds.), *Plant Pattern Recognition Receptors: Methods and Protocols*, Methods in Molecular Biology, vol. 1578, DOI 10.1007/978-1-4939-6859-6_27, © Springer Science+Business Media LLC 2017

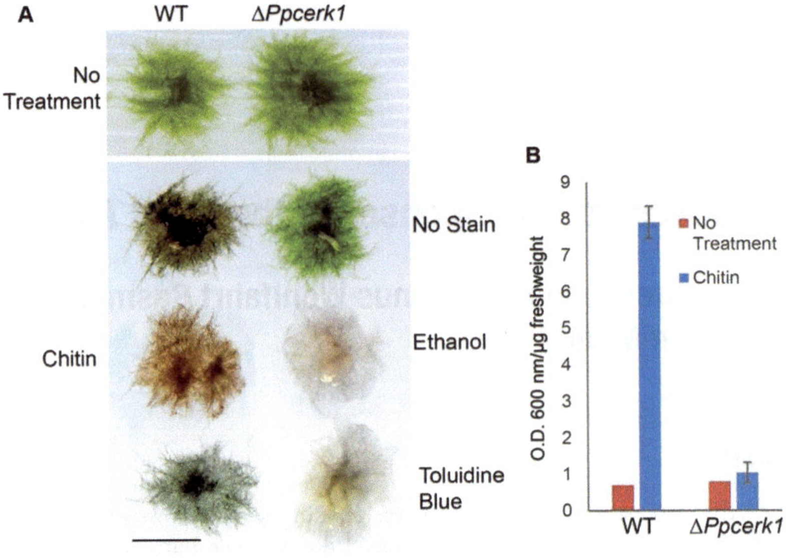

Fig. 1 Chitin induced cell wall modifications in moss. (**a**) Wild type (WT) and Δ*Ppcerk1* with no treatment and 24 h after transfer to medium supplemented with 300 μg/ml chitin. The browning of WT but not of Δ*Ppcerk1* is visible without staining and becomes pronounced when cellular content is removed by ethanol and stained with toluidine blue. The bar represents 5 mm. (**b**) Quantification of toluidine blue staining of WT and Δ*Ppcerk1* 24 h after transfer to medium supplemented with 300 μg/ml chitin. An average of three samples are shown with standard deviation as error bar

chitin induces rapid growth inhibition within 2 min after exposure in wild type but not in knockouts of the chitin receptor (Δ*Ppcerk1*). This growth inhibition is close to a primary response as it does not require the kinase cascade components. In *Arabidopsis* the bacterial MAMPs flg22 and elf18 induce growth inhibition, which is slow and usually measured after 1–2 weeks of exposure [12, 13]. Thus, although the mechanism behind the rapid chitin induced growth inhibition in *P. patens* is unknown, it may be a novel MAMP response. Protocols are provided here for visualizing chitin induced growth inhibition by time lapse photography.

We also found that chitin induced accumulation of the defense related transcripts of *Phenylalanine Ammonia Lyase 4* (*PAL4*) and *Chalcone Synthase* (*CHS*) is dependent on the chitin receptor as well as the MAP kinase signaling cascade [11]. Furthermore, we found that chitin induces reddish brown, cell wall associated depositions and that their induction depends upon the chitin receptor and the MAP kinase cascade (Fig. 1a). This reddish brown coloration is visible within 4–6 h and increases until approximately 48 h after the start of chitin treatment. By modifying an earlier protocol [14] used to quantify cotton fiber growth, we developed an assay,

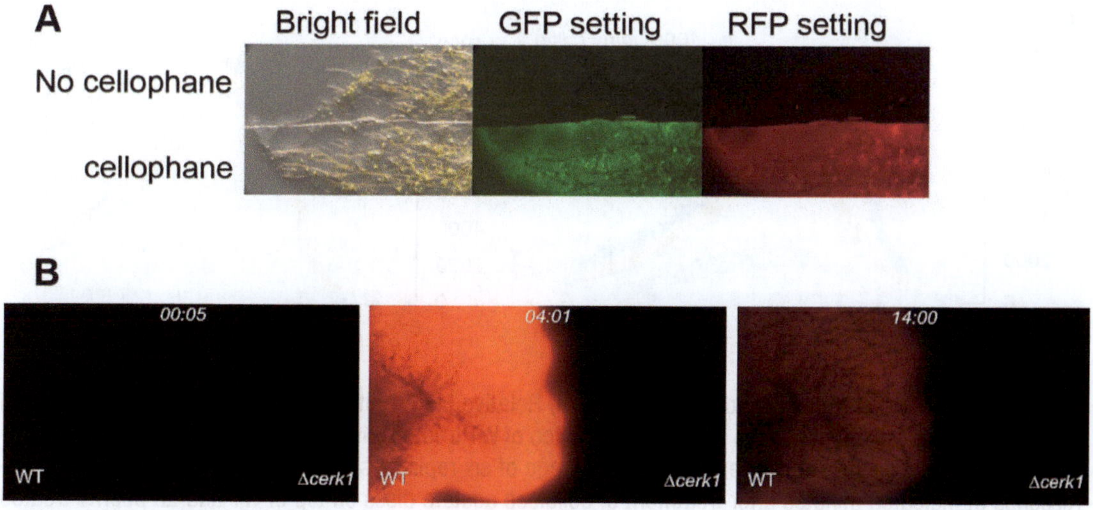

Fig. 2 Transient fluorescent burst induced by chitin. (**a**) WT moss 4 h after transfer to BCD medium supplemented with 300 μg/ml chitin and partly overlaid with cellophane. Pictures taken with bright field GFP (excitation <488 nm and detection >510 nm) and RFP settings (excitation <563 nm and detection >582 nm). Fluorescence is clearly visible in areas with the membrane. (**b**) WT and Δ*Ppcerk1* plants grown on BCD medium overlaid with cellulose acetate sheets and transferred to BCD supplemented with 300 μg/ml chitin. *Left picture* taken 5 min after transfer. *Middle picture* taken after 4 h when the fluorescent signal of WT moss peaks. *Right picture* taken after 14 h when the fluorescent signal of WT is decaying. Pictures taken with RFP detection settings

described below, for quantifying these chitin induced cell wall modifications (Fig. 1b).

Moss is often grown on cellophane or cellulose acetate membranes to facilitate transfer to fresh or other media. Interestingly, a transient fluorescent burst is observed when the moss is transferred on such membranes from normal media to media containing chitin (Fig. 2a). This fluorescent burst can be observed with settings for GFP (excitation <488 nm and detection >510 nm) or RFP (excitation <563 nm and detection >582 nm). It appears after 1–2 h, peaks at approximately 4 h, and decreases the following 10 h. This novel MAMP triggered response is not detectable in the Δ*Ppcerk1* KO line and is also dependent on the MAP kinase signaling cascade (Fig. 2b).

The fluorescence can be quantified using a plate reader. The signal is dose dependent and has the largest linear dynamic range using 300 μg/ml chitin (Fig. 3a). The fluorescence is absent in the Δ*Ppcerk1* KO line and also dependent upon the MAP kinase cascade (Fig. 3b).

In the following protocols *P. patens* is grown under standard conditions described by Ashton and Cove [15]. The moss plants are propagated weekly on minimal BCD medium or nitrogen supplemented BCDAT medium and grown at 22 °C under 16 h light (50 μE/m^2/s) and 8 h dark.

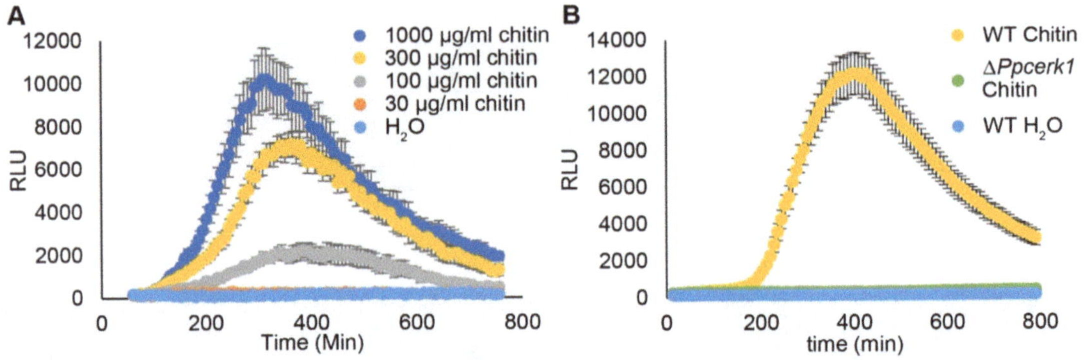

Fig. 3 Quantitation of chitin induced fluorescence. (**a**) Relative Light Units (RLU) of fluorescence at indicated minutes after treatment of cellulose acetate discs on top of WT treated with H_20 or H_20 with indicated concentrations of chitin. Each data point represents an average of six replicates with standard error. (**b**) RLU of fluorescence at indicated minutes after treatment of cellulose acetate discs on top of WT and Δ*Ppcerk1* treated with 300 µg/ml chitin and WT treated with H_20. Data points represent an average of nine samples with standard error

2 Materials

2.1 Rapid Growth Inhibition of P. patens by Chitin

1. BCDAT media or BCD (BCDAT without ammonium tartrate) [15].
2. Cellophane (325P from AA Packaging Ltd., UK) or cellulose acetate discs (pore size 0.45 µm, type WCA from Whatman).
3. Microscope with camera for time-lapse photography.

2.2 Quantitative Toluidine Staining of Chitin Induced Cell Wall Modifications

1. 50% EtOH, 0.05% toluidine blue, 50 mM citrate–citric acid buffer, pH 3.5.
2. Citrate–citric acid buffer 50 mM pH 3.5, 100 ml: Mix 39 ml 0.1 M citric acid with 11 ml 0.1 M sodium citrate and add 50 ml EtOH.

2.3 Visualization of Chitin Induced Transient Fluorescent Burst

1. Fluorescence microscope with GFP (excitation <488 nm/detection >510 nm) or RFP settings (excitation <563 nm/detection >582 nm).

2.4 Quantitative Measure of Chitin Induced Fluorescent Burst

1. Microtiter plate reader set to record fluorescence with excitation peak 485 nm and emission 535 nm.

3 Methods

3.1 Rapid Growth Inhibition of *P. patens* by Chitin

1. Grow moss on BCD media plates overlaid with cellophane for 7–10 days during which protonemal tip growth is rapid and radiates from the center (*see* **Note 1**).
2. Keep the lid on the petri dish with the moss and place under a microscope with a camera suitable for time-lapse photography. We use a Sony A7S camera attached directly to the eye piece. Zoom in until 5–10 clearly separated apical protonemal cells are in the frame. Ensure proper lighting for moss growth, ideally 20–50 $\mu E/m^2/s$. Normal table lamps will also work but should be placed on both sides to avoid directional growth towards the light. Record the moss growth for at least 4 h by taking a picture every second minute.
3. Quickly, between two pictures, carefully transfer the moss on the cellophane with forceps to a BCD plate supplemented with 300 μg/ml of chitin (Yaizu Suisankagaku Industry, Japan). Draw some marks on the camera screen to ensure that the moss can be placed with the same framing as before.
4. Continue recording the growth for at least 6 h and preferably longer to see if growth inhibition is transient. *See* [11] for time-lapse videos.

3.2 Quantitative Toluidine Staining of Chitin Induced Cell Wall Modifications

1. Grow ~16 moss plants on BCD (or BCDAT) plates overlaid with cellophane for 14 days (*see* **Note 2**).
2. Transfer cellophane with the moss onto BCD supplemented with 300 μg/ml chitin for 24 h.
3. Determine fresh weight of five moss plants by removing excess water thoroughly with paper tissue. Put the moss in labelled Eppendorf tubes.
4. Stain the moss for 10 min in 50% EtOH, 0.05% toluidine blue, 50 mM citrate–citric acid buffer, pH 3.5. Perform treatment in triplicates and include an untreated sample.
5. Rinse twice in 1 ml 96% EtOH.
6. Remove cellular content and unbound stain by leaving moss overnight in 96% EtOH at RT. For faster results, incubate for 1 h at 70 °C with one EtOH change after ~30 min.
7. At this point a representative moss from each sample can be put on a glass slide with some water for photography and then put back in the tube for storage.
8. Remove the ethanol and destain for 4 h at 70 °C in acetic acid, ethanol, water (10:95:5).
9. Measure O.D. of supernatant at 600 nm and present data as O.D./μg fresh weight as an average of the three replicates with standard error as error bar.

3.3 Visualization of Chitin Induced Transient Fluorescent Burst

This assay is conducted similarly to the growth inhibition assay but under fluorescent light.

1. Grow moss on BCD media plates overlaid with either cellophane (325P from AA Packaging Ltd., UK) or cellulose acetate discs (pore size 0.45 μm, type WCA from Whatman), for 7–10 days during which protonemal tip growth is rapid and radiates from the center.
2. Transfer the moss on cellophane with forceps to a BCD plate supplemented with 300 μg/ml of chitin and place it under a fluorescent microscope. We use a Leica MZ16F. The fluorescent burst can be observed using both GFP (excitation <488 nm/detection >510 nm) and RFP settings (excitation <563 nm/detection >582 nm) (*see* **Note 3**).
3. Record the fluorescent burst using time lapse photography by taking a picture every second minute. Since the fluorescence only peaks after approximately 4 h of chitin exposure, it can be difficult to guess the optimal light setting for the camera (*see* **Note 4**).

3.4 Quantitative Measure of Chitin Induced Fluorescent Burst

1. Grow the moss on BCD medium overlaid with cellophane for 7–10 days. Use the fresh young protonemal tissue (*see* **Note 5**).
2. Pipette 25 μl water into each well to be used in the experiment in a 96-well plate.
3. Take a small piece of moss with a forceps and place it in the water at the bottom of the wells. The amount of tissue should be just enough to cover the bottom of the well and remain flat. Make sure you try to take the same amount of tissue each time to minimize variation.
4. Add 25 μl of 600 μg/ml chitin to the treated samples and 25 μl of water to the control samples. Run treated samples in at least six replicates (*see* **Note 6**).
5. To each sample well, add a small disc of cellulose acetate punched from a bigger sheet with a paper punch (pore size 0.45 μm, type WCA from Whatman) (*see* **Note 7**).
6. Place the plate in the plate reader. We use a Wallac Victor 2 set to record fluorescence with excitation peak 485 nm and emission 535 nm.
7. Continue recording each fifth minute for 11 h.

4 Notes

1. *Note*: Tip growth is easily disturbed by changes in temperature, lighting, and humidity. Therefore, place chitin containing plate in the same room several hours before transfer to reach

ambient conditions. It is also important to do the transfers as quickly and gently as possible so that the moss encounters as little stress as possible.

2. Make sure the plate is grown without condensation as it is important that the moss are not wet.
3. In our setup the fluorescent light is on continuously throughout the recording but is kept quite dim to reduce possible heat stress or desiccation. We use the Leica light source on 1 and with a ND2 filter. Ideally the fluorescent light is only on when pictures are taken.
4. We use a Sony A7S camera attached directly to the eye piece of the microscope which is much more sensitive than the standard mount Leica DFC420C camera. In our conditions the optimal Sony settings for recording fluorescence of the wild type on cellophane are (1) GFP settings, ISO 10,000 and 6 s exposure or (2) RFP settings on cellulose acetate, ISO 2000 for 2 s. Continue recording for 8–16 h to see the fluorescent burst go down.
5. Negative controls can include cellulose acetate disk without moss tissue.
6. Note the time the treatment is started and add this to the time when the plate reader starts recording. The fluorescence will not start until at least an hour after adding chitin.
7. Make sure each disc lies flat on top of the moss and liquid.

Acknowledgments

This work was supported by grants from the Danish Basic Research Foundation (Center for Comparative Genomics) and the Danish Research Council for Nature and the Universe (Comparative and functional genomics of plant innate immunity, 1323-00267A) to J.M.

References

1. Clarke JT, Warnock RCM, Donoghue PCJ (2011) Establishing a time-scale for plant evolution. New Phytol 192:266–301
2. Schaefer D, Zryd J-P, Knight CD, Cove DJ (1991) Stable transformation of the moss *Physcomitrella patens*. Mol Gen Genet 226: 418–424
3. Mittag J, Šola I, Rusak G, Ludwig-Müller J (2015) *Physcomitrella patens* auxin conjugate synthetase (GH3) double knockout mutants are more resistant to *Pythium* infection than wild type. J Plant Physiol 183:75–83
4. Oliver J, Castro A, Gaggero C, Cascón T, Schmelz E, Castresana C, Ponce de León I (2009) *Pythium* infection activates conserved plant defense responses in mosses. Planta 230:569–579
5. Ponce De León I, Oliver J, Castro A, Gaggero C, Bentancor M, Vidal S (2007) *Erwinia carotovora* elicitors and *Botrytis cinerea* activate defense responses in *Physcomitrella patens*. BMC Plant Biol 7:52
6. Ponce De León I, Schmelz EA, Gaggero C, Castro A, Álvarez A, Montesano M (2012)

Physcomitrella patens activates reinforcement of the cell wall, programmed cell death and accumulation of evolutionary conserved defence signals, such as salicylic acid and 12-oxo-phytodienoic acid, but not jasmonic acid, upon *Botrytis cinerea* infection. Mol Plant Pathol 13:960–974

7. Reboledo G, del Campo R, Alvarez A, Montesano M, Mara H, Ponce de León I (2015) *Physcomitrella patens* activates defense responses against the pathogen *Colletotrichum gloeosporioides*. Int J Mol Sci 16:22280–22298
8. Ponce de León I, Montesano M (2013) Activation of defense mechanisms against pathogens in mosses and flowering plants. Int J Mol Sci 14:3178–3200
9. Lehtonen MT, Akita M, Frank W, Reski R, Valkonen JPT (2012) Involvement of a class III peroxidase and the mitochondrial protein TSPO in oxidative burst upon treatment of moss plants with a fungal elicitor. Mol Plant Microbe Interact 25:363–371
10. Lehtonen MT, Takikawa Y, Rönnholm G, Akita M, Kalkkinen N, Ahola-Iivarinen E, Somervuo P, Varjosalo M, Valkonen JPT (2014) Protein secretome of moss plants (*Physcomitrella patens*) with emphasis on changes induced by a fungal elicitor. J Proteome Res 13(2):447–459
11. Bressendorff S, Azevedo R, Kenchappa CS, Ponce de León I, Rasmussen MW, Erbs G, Newmann M, Petersen M, Mundy J (2016) An innate immunity pathway in the moss *Physcomitrella patens*. Plant Cell 28: 1328–1342
12. Gómez-Gómez L, Felix G, Boller T (1999) A single locus determines sensitivity to bacterial flagellin in *Arabidopsis thaliana*. Plant J 18: 277–284
13. Li J, Zhao-Hui C, Batoux M, Nekrasov V, Roux M, Chinchilla D, Zipfel C, Jones JDG (2009) Specific ER quality control components required for biogenesis of the plant innate immune receptor EFR. Proc Natl Acad Sci U S A 106:15973–15978
14. Beasley CA, Birnbaum EH, Dugger WM, Ting IP (1974) A quantitative procedure for estimating cotton fiber growth. Biotech Histochem 49:85–92
15. Ashton NW, Cove DJ (1977) The isolation and preliminary characterisation of auxotrophic and analogue resistant mutants of the moss *Physcomitrella patens*. Mol Gen Genet 154:87–95

Chapter 28

Methods to Quantify PAMP-Triggered Oxidative Burst, MAP Kinase Phosphorylation, Gene Expression, and Lignification in Brassicas

Simon R. Lloyd, Christopher J. Ridout, and Henk-jan Schoonbeek

Abstract

Quantitative disease resistance (QDR) based on PAMP-triggered immunity (PTI) could be durable and effective against many pathogens (broad spectrum). Development of methods to evaluate PTI responses in crops could therefore accelerate breeding for durable QDR. Most PTI-studies involved model plants such as Arabidopsis and *Nicotiana benthamiana* or cell cultures, and cannot be directly applied to diverse germplasm of crop plants.

We developed methods to measure PTI in *Brassica* crop species (Lloyd et al., Mol Plant Microbe Interact 27:286–295, 2014) which we have elaborated and expanded here to enable their use for screening and evaluating germplasm for potential QDR in breeding programs.

Key words PAMP-triggered immunity, Crop plants, *Brassica napus*, *Brassica oleracea*, Quantitative disease resistance, Reactive oxygen species, Lignin

1 Introduction

The first layer of active defence in plants is based on the perception of pathogen- (or microbe-) associated molecular patterns (PAMPs/MAMPs) leading to PAMP-triggered immunity (PTI). PAMPs are essential molecules, conserved in entire kingdoms of microbes, and are recognized by pattern recognition receptors (PRRs) in plants. Well-characterized bacterial PAMPs are flg22 from flagellin [1, 2] and elf18 from the elongation factor EF-Tu [3, 4]. The fungal cell wall component chitin, or *N*-acetyl-d-glucosamine (GlcNAc), is also a PAMP, recognized by LysM-containing PRRs conferring PTI in plants [5–7]. Additional PAMP-PRR pairs are being identified, reviewed in Macho, Zipfel [8].

Following pathogen recognition defence reactions are initiated, including the production of structural defences such as lignin and callose deposits, and antimicrobial metabolites such as glucosinolates,

Libo Shan and Ping He (eds.), *Plant Pattern Recognition Receptors: Methods and Protocols*, Methods in Molecular Biology, vol. 1578, DOI 10.1007/978-1-4939-6859-6_28, © Springer Science+Business Media LLC 2017

indole alkaloids, and phenylpropanoids. These downstream responses could contribute to quantitative disease resistance (QDR), an important breeding target in crop improvement. Since PAMPs are essential molecules and cannot be lost without affecting pathogen fitness, QDR based on PTI could be durable and effective against many pathogens (broad spectrum). Although measuring PAMP-responses addresses only one aspect of plant–pathogen interactions it measures the ability of a plant to respond to a class of pathogens without the muddling effects of race-specific effectors and damage from pathogen infection. Development of methods to evaluate PTI responses in crops could therefore accelerate breeding for durable QDR. One of the first, and easiest to implement, responses is the transient production of reactive oxygen species (ROS burst). We developed methods to measure PTI in *Brassica* crop species [9] which we have elaborated and expanded here to enable their use for screening and evaluating germplasm for potential QDR in breeding programmes. This includes an adaption of a pretreatment protocol by vacuum-infiltrating leaf discs with water [10] to prepare them for a multitude of assays.

All protocols use $t = 0$ as the moment that PAMPs are added to the plant material. Hence $t = -1$d refers to 1 day before PAMP application, for many assays this is the day that leaf discs are stabbed out in preparation for the actual assay.

2 Materials

2.1 PAMPs

1. flg22 peptide (QRLSTGSRINSAKDDAAGLQIA) (*see* **Note 1**).
2. elf18 peptide (Ac-SKEKFERTKPHVNVGTIG) (*see* **Note 1**).
3. Chitin (*see* **Note 2**).

2.2 ROS Burst

1. Cork borer: size 1 (inside diameter = 4 mm), sharpened.
2. P200 Multichannel Pipette (12 channel).
3. 96-well plates for luminescence (white walled).
4. Sterile water.
5. Luminol stock: 17 mg/ml in DMSO, stored as 1 ml aliquots in dark tubes (*see* **Notes 3** and **4**).
6. Horseradish peroxidase stock: 10 mg/ml in water, stored as 1 ml aliquots (HrPx).
7. Filtertips for handling concentrated PAMP stocks.
8. Tubes or deep-well plates in 96 well lay-out.
9. Luminometer (e.g., Varioskan Flash with SkanIt RE Software) or a high-resolution photon counting camera (e.g., HRPCS218, Photek, St Leonards on Sea, UK).

2.3 Vacuum Infiltration

1. Cork borer 10 mm.
2. 2-ml tubes, and extra pierced lids.
3. Water tight container, higher than 2 ml tubes.
4. Vacuum-pump and desiccator.
5. Liquid nitrogen.

2.4 MAP Kinase Phosphorylation

1. Lacus buffer, make 200 ml and freeze in 10 ml aliquot (−20 °C): 50 mM Tris–HCl pH 7.5, 10 mM $MgCl_2$, 15 mM EGTA, 100 mM NaCl, 2 mM DTT, 1 mM NaF, 1 mM NaMO, 0.5 mM $NaVO_3$ (activated), 30 mM β-glycerophosphate, 0.1% Nonidet P-40.
   Directly before use add these proteinase inhibitors 0.5 mM PMSF, 1% anti-protease (Sigma), 100 nM Calyculin A (LC laboratories http://www.lclabs.com, Woburn, MA, USA).
2. Reagents and spectrophotometer for Bradford protein quantification.
3. SDS loading buffer (Invitrogen).
4. Tank, SDS-PAGE gel and buffer for protein electrophoresis (12% acrylamide NuPage (Invitrogen)).
5. TBS = Tris Buffered Saline = 50 mM Tris, 150 mM NaCl, adjust pH with HCl to pH 7.6.
6. TBST = TBS–Tween (0.1%).
7. Primary antibody: Phospho-p44/42 MAPK (Erk1/2) (Thr202/Tyr204) rabbit monoclonal antibodies (purchased from New English Biolabs (www.neb.uk.com)).
8. Secondary antibody: anti-rabbit-HRP conjugated.
9. HRP western visualization kit.

2.5 Determination of Lignin and Cell Wall-Bound Phenols by TGA-Derivatization

1. Methanol.
2. Thioglycolic acid (TGA; Sigma-Aldrich, catalog number: T3758 or T6750).
3. 32% HCl.
4. Lignin (alkali) (Sigma-Aldrich, catalog number: 471003).
5. 1 M and 0.5 M NaOH.
6. 2 M HCl.
7. 2-ml tubes that stay closed upon heating (screwcap, boilproof or snap cap).
8. Freeze dryer or drying oven.
9. Bead mill, e.g., Tissue Lyser2 (manufactured by Retsch, provided by QIAGEN) or Genogrinder (SPEX; www.spexsampleprep.com) and stainless steel beads (QIAGEN, catalog number: 69989).

10. Shaker/vortex.
11. Heating block/bath up to 100 °C in fumehood.
12. Centrifuge (tabletop for 1.5/2 ml tubes).
13. Spectrophotometer (e.g., BioSpectrometer basic or Varioskan).

3 Methods

3.1 Plant Growth

For assays with leaf material use plants that are between 4 and 7 weeks old, and not yet flowering. Grow plants in normal soil and prevent diseases by avoiding the arrival of inoculum and starting in a clean glasshouse compartment, growth chamber or cabinet. Plants must be grown under normal sunlight or artificial light with a near sunlight spectrum at circa 100 μmol/m^2/s with a photoperiod of maximal 12 h with 18–20 °C/12–15 °C day/night temperatures (*see* **Note 5**). Use leaves of the same age when comparing plant lines, e.g., always leaf 2 or always leaf 3 from at least 4 plants per assay. Do not take leaf discs too close to the leaf edge or big veins.

3.2 ROS Burst

Day before ($t = -1$):

1. Prepare a 96-well white plate with 200 μl water in each well.
2. Stab out leaf discs with the cork borer (4 mm) and float on water, at least eight repetitions, from a total of four leaves.
3. Incubate overnight, keeping covered.

Day 0:

4. Prepare working solution (light sensitive); amounts are for 1 plate.
   (a) One elicitor/concentration per plate: 20 μl Luminol, 20 μl HrPx, 10 ml water.
   Add PAMP to final concentration (*see* **Note 6**).
   Transfer to 12 cell wide pipetting trough.
   (b) Multiple elicitors/concentrations per plate: 25 μl Luminol, 25 μl HrPx, 12.5 ml water.

   Distribute 1 ml each over 12 tubes organized as columns in the 96-well plate.

   Add PAMP to 1 ml in each tube, at your desired concentrations.
5. Remove water from the plate with leaf discs (with multichannel pipette or vacuum aspirator).
6. Prepare the luminometer to start right after the next step (*see* **Note 7**).

7. Using a 12-channel pipette rapidly add the working solution (100 μl per well). Very quickly check if all leaf discs are floating and directly put the plate in the luminometer and start reading, ROS burst can start within 2 min.
8. A standard luminometer program reads every min for an hour. Try to maximize the time read per well to have the highest integrated signal.
9. Export the data to a spread sheet compatible format (Excel or a similar program).
10. Calculate the area under the curve for the transient ROS burst (add up readings at all timepoints). It is important when comparing different varieties to use the same settings and a period that covers the whole curve.

3.3 Leaf-Disc Vacuum Infiltration

To obtain leaf tissue homogenously exposed to elicitors, and allow treating within minutes without the interference of irregular wounding responses, it is important to prepare material in multiple tubes for various simultaneous treatments. It is therefore advisable to do most of the sampling the day before. Once prepared like this, repetitions of different plant lines can be treated nearly simultaneous. Cutting leaf discs followed by vacuum infiltration of water prepares the tissue for a multitude of assays, including lignin quantification, MAPK-activation, RNA extraction, and gene expression.

Day before ($t = -1$):

1. Stab out at least eight leaf discs with the cork borer (10 mm) and insert in 2 ml tubes, (at least three repetitions per treatment/genotype combination).
   (a) For MAPK and RNA, avoid further wounding;
   (b) For lignin, squeeze each leaf disc with tweezers (*see* **Note 10**).
2. Completely fill each tube with water and close with a separate, pierced lid.
3. Submerge under water and vacuum-infiltrate three times 1 min, top up water between rounds if needed, leaf discs should get darker and sink.
4. Incubate overnight, keeping covered.

Day 0:

5. Prepare PAMP solutions, counting 2.5 ml per tube, with water as control.
6. Remove water from the tubes with leaf discs.
7. Take $t = 0$ samples.
8. Add PAMP solution to leaf discs, incubate for required time, *see* below, and drain of all liquid, then flash-freeze in liquid N_2 or add MeOH for extraction of secondary metabolites and preparation for lignin quantification.

Times used:

(a) 5 and 15 min, for MAP Kinase assays;

(b) 30, 60, and 180 min for gene expression;

(c) 2 or 3 days for lignin quantification.

3.4 MAP Kinase Phosphorylation

Adapted from Ségonzac et al. [11].

1. Defrost an aliquot of Lacus buffer and add proteinase inhibitors.
2. Grind leaf discs very finely in liquid N_2 in mortar and pestle or in genogrinder/tissuelyser/bead mill.
3. Thaw in 300–400 μl Lacus buffer on ice.
4. Invert tubes to homogenize.
5. Centrifuge at 16,000 × *g* for 30 min.
6. Filter supernatant through Miracloth into prechilled fresh tube.
7. Quantify using the Bradford assay.
8. Prepare 40 μg to load into 3× SDS loading buffer, boil for 5 min.
9. Run on 12% acrylamide at 120 V.

Western Blotting

10. Transfer to a nitrocellulose membrane (2 h at 75 V in a conventional device or 7 min in an I-blot transfer system (Invitrogen)).
11. Block in 5% (w/v) BSA (Sigma) in TBST for 2 h at room temperature.
12. Wash 3 × 5 min in TBST.
13. Incubate with the primary antibody overnight at 4 °C in 5% BSA-TBST (1/2000); Cell Signalling Phospho-p44/42 MAPK (Erkl/2) (Thr202/Tyr204).
14. Wash 3 × 5 min in TBST.
15. Incubate with the secondary antibody (goat anti-rabbit 1/50,000 in TBST) for 1 h.
16. Wash 3 × 5 min in TBST then 5 min in TBS.
17. Develop using a HrPx visualizing stain, e.g., ECL Plus Western Blotting Detection System (GE healthcare) with a 2-min exposure (*see* **Note 8**).

3.5 RNA Extraction

1. Flash-freeze in liquid nitrogen.
2. Finely grind leaf discs in genogrinder/tissuelyser/bead mill.
3. Extract RNA with TRIzol® or a dedicated plant RNA extraction kit based on silica spin filters (e.g., RNeasy plant; www.Qiagen.com).

4. Do DNAse treatment to degrade genomic DNA (e.g. DNAse turbo, Ambion).
5. Use 1 μg RNA for first strand cDNA synthesis (e.g. Superscript III, Invitrogen).
6. Dilute 20× in water and use 4 μl for qPCR (e.g. SYBR-green mastermix, Sigma) with BnEF1α as a control gene.

3.6 Determination of Lignin and Cell Wall-Bound Phenols by TGA-Derivatization (See Note 9)

Adapted from Bruce, West [12] and Ride [13].

3.6.1 Destaining

1. Break tissue in genogrinder/tissuelyser/bead mill.
2. Add 0.75 ml MeOH, transfer suspension to new tube (use wide bore or cut off tips).
3. Repeat with 0.75 ml MeOH for quantitative transfer.
4. Remove pigments and soluble phenolics through repeated incubation with absolute methanol over 3 days, centrifuge at 14,000 × *g* for 5 min between washes.
5. Dry cleared tissue at 55 °C (or in freeze-dryer) for 1 day.
6. Measure dry weight of the sample.

3.6.2 TGA Extraction

7. Suspend each sample in 0.5 ml 10% TGA in 2 M HCl.
8. Heat samples for 4 h at 100 °C.
9. Centrifuge at 14,000 × g for 15 min.
10. Decant supernatant.
11. Suspend pellet in 0.5 ml H_2O.
12. Centrifuge at 14,000 × *g* for 10 min.
13. Decant supernatant.
14. Add 0.5 ml 1 M NaOH and vortex to solubilize the TGA derivatives.
15. Incubate overnight at 4 °C.
16. Centrifuge at 14,000 × *g* for 10 min.
17. Pipette supernatant into a fresh tube.
18. Add 0.2 ml concentrated HCl.
19. Incubate in ice for 1 h.
20. A reddish brown precipitate should be visible; centrifuge at 14,000 × *g* for 10 min.
21. Decant supernatant.
22. Dissolve pellet in 1 ml 0.5 M NaOH.

23. Measure absorbance at 335 nm; dilute in 0.5 M NaOH if necessary.
24. Use a calibration curve of 0.0005–0.5 g/L lignin to determine the amount of lignin per mg dry weight.

4 Notes

1. Resuspend flg22 or elf18 peptides in sterile H_2O at 10 mM and aliquot, store at −20 °C. Take out single aliquots to dilute to 1 mM and 100 μM stocks and store at −20 °C before use. Peptides in this study were ordered from Peptron (www.peptron.co.kr).
2. Chitin can be obtained from various sources but batches differ in activity, some preparations of partially degraded chitin from crab or shrimp shell work well, others not, always test with a known responsive plant (e.g., young leaves of *Arabidopsis thaliana* Col-0) before using on new plants. When using synthetic oligomers of (GlcNAc) a minimal chain length of 7 or 8 is required [5, 14]. Crab shell chitin (CSC) supplied as NA-COS-Y by Yaizu Suisankagaku Industry CO (Yaizu, Japan) was used in this study. Suspend chitin in sterile H_2O (100 g/L) on a shaker overnight and autoclave before storing aliquots at 100 and 10 g/L at −20 °C.
3. Instead of luminol, 21 nM of the luminol derivative L-012 (8-amino-5-chloro-7-phenylpyrido[3,4-d]pyridazine-1,4(2H,3H)dione) [Wako; 15] can be used. This increases sensitivity but has the risk of elevating background signal.
4. Calyculin A, a phosphatase inhibitor, has been used to sensitize the plant to obtain more intense ROS bursts [10, 16]. The concentration for sensitizing *Brassica* leaf discs is 2 mg/L of working solution. A concentration of 100 mg/L calyculin A triggers an elicitor-independent ROS burst.
5. Day length as such does not massively affect the PAMP response but, when testing cultivars with different flowering times, e.g., spring and winter oil seed rape, long day may trigger flowering in early spring types and alter growth from rosette to flowering stalk, and leaves from a different physiological stage would be compared. We noticed that the spectrum of growth lamps does affect to some extent the response of *Brassica* leaves to PAMPs. The advised spectrum distribution is as close as possible to sunlight.
6. There is no "perfect" concentration of PAMPs, one has to consider that the output of some assays is dose dependent and not just a threshold event. Ideally a concentration is used that is high enough to trigger some response in all lines but not so high that it reaches a plateau in all lines or becomes inhibitory.

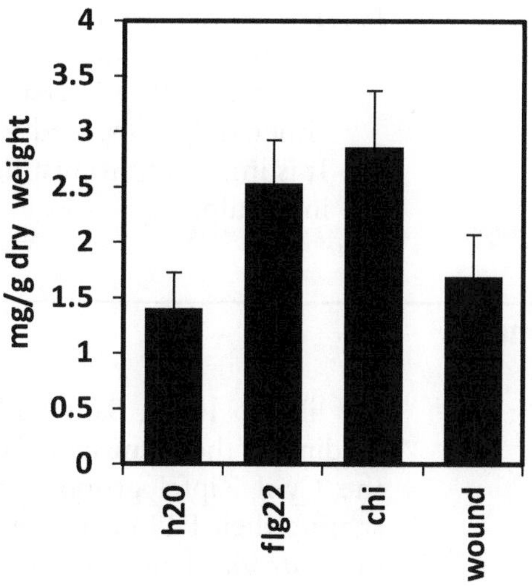

Fig. 1 Production of total cell wall-bound phenols and lignin determined by TGA-derivatization in *B. napus* line Temple. H_20 is no wound and water only, flg22 is wound and 100 nM flg22, chi is wound and 0.5 g/L chitin, wound is wound and water only. Only the combination of PAMP and wound gives a significant increase compared to the control H_20 ($P < 0.05$ using Student's *t*-test)

The following PAMP concentrations are a good starting point for differentiating among *B. napus* and *B. oleracea* lines:

(a) Elf18 100 nM;

(b) Flg22 100 nM;

(c) Chitin: NA-COS-Y brand crab shell chitin; 0.5 g/L appears to be the optimum in brassicas. Exceeding this begins to inhibit the ROS seen.

7. Various equipment to read luminescence in the 96-well plate are available, we used plate reader luminometers (e.g., Varioskan Flash with SkanIt RE Software) or a high-resolution photon counting camera (e.g., HRPCS218, Photek, St Leonards on Sea, UK). Ensure it is set up and programmed to start measuring as soon as your PAMPs and luminol working solution have been added to the leaf discs.
8. It's very important to filter the extract and to let the primary antibody incubate overnight at 4 °C.
9. This method measures all polymerized lignin and most cell wall-bound phenolics. Soluble phenolics extracted by the methanol can be kept for separate quantification of phenols using the Folin-Ciocalteu method [17]. The difference between plant lines might be either in the amount or the speed by which lignin is formed, therefore it may be important to test more than one time point, we did observe lignin formation in *B. napus* after 48 h (Fig. 1).

10. The combination of wounding and PAMP or elicitor gives a much clearer and more defined lignification than each single treatment [9, 13, 18]. To ensure that wounding is even, use either ribbed, beaked forceps or a multi-pinprick device [19]. It is important to just lightly crush the leaf, do not macerate it into pulp.

Acknowledgments

We like to thank Sam Holt and Francesca Stefanato for critical reading of the manuscript, Cecile Ségonzac and Freddy Boutrot in the Cyril Zipfel group (The Sainsbury Lab, Norwich, UK) for sharing their PTI protocols and thinking about adaptations for use on *Brassica* plants, and Yaizu Suisankagaku Industry CO (Yaizu, Japan) for providing CSC. This work was funded by the Biotechnology and Biological Sciences Research Council (BBSRC) grants BB/G042960/1 (as part of the consortium ERA-PG "PRR-CROP"), BB/N005007/1 (as part of the ERA-CAPS consortium "MAQBAT"), the John Innes Institute BIO strategic grant BB/J004553/1 and by a BBSRC doctoral training grant to S.R.L.

References

1. Felix G, Duran JD, Volko S, Boller T (1999) Plants have a sensitive perception system for the most conserved domain of bacterial flagellin. Plant J 18(3):265–276. doi:10.1046/j.1365-313X.1999.00265.x
2. Gomez-Gomez L, Boller T (2000) FLS2: an LRR receptor-like kinase involved in the perception of the bacterial elicitor flagellin in Arabidopsis. Mol Cell 5(6):1003–1011. doi:10.1016/s1097-2765(00)80265-8
3. Kunze G, Zipfel C, Robatzek S, Niehaus K, Boller T, Felix G (2004) The N terminus of bacterial elongation factor Tu elicits innate immunity in Arabidopsis plants. Plant Cell 16(12):3496–3507
4. Zipfel C, Kunze G, Chinchilla D, Caniard A, Jones JDG, Boller T, Felix G (2006) Perception of the bacterial PAMP EF-Tu by the receptor EFR restricts *Agrobacterium*-mediated transformation. Cell 125(4):749–760. doi:10.1016/j.cell.2006.03.037
5. Kaku H, Nishizawa Y, Ishii-Minami N, Akimoto-Tomiyama C, Dohmae N, Takio K, Minami E, Shibuya N (2006) Plant cells recognize chitin fragments for defense signaling through a plasma membrane receptor. Proc Natl Acad Sci U S A 103(29):11086–11091. doi:10.1073/pnas.0508882103
6. Miya A, Albert P, Shinya T, Desaki Y, Ichimura K, Shirasu K, Narusaka Y, Kawakami N, Kaku H, Shibuya N (2007) CERK1, a LysM receptor kinase, is essential for chitin elicitor signaling in Arabidopsis. Proc Natl Acad Sci U S A 104(49):19613–19618. doi:10.1073/pnas.0705147104
7. Wan JR, Tanaka K, Zhang XC, Son GH, Brechenmacher L, Tran HNN, Stacey G (2012) LYK4, a lysin motif receptor-like kinase, is important for chitin signaling and plant innate immunity in Arabidopsis. Plant Physiol 160(1):396–406. doi:10.1104/pp.112.201699
8. Macho AP, Zipfel C (2014) Plant PRRs and the activation of innate immune signaling. Mol Cell 54(2):263–272. doi:10.1016/j.molcel.2014.03.028
9. Lloyd SR, Schoonbeek H-j, Trick M, Zipfel C, Ridout CJ (2014) Methods to study PAMP-triggered immunity in *Brassica* species. Mol Plant Microbe Interact 27(3):286–295. doi:10.1094/MPMI-05-13-0154-FI
10. Schoonbeek H-j, Wang H-H, Stefanato FL, Craze M, Bowden S, Wallington E, Zipfel C, Ridout CJ (2015) Arabidopsis EF-Tu receptor enhances bacterial disease resistance in transgenic wheat. New Phytol 206(2):606–613. doi:10.1111/nph.13356

11. Segonzac C, Feike D, Gimenez-Ibanez S, Hann DR, Zipfel C, Rathjen JP (2011) Hierarchy and roles of pathogen-associated molecular pattern-induced responses in Nicotiana benthamiana. Plant Physiol 156(2):687–699. doi:10.1104/pp.110.171249
12. Bruce RJ, West CA (1989) Elicitation of lignin biosynthesis and isoperoxidase activity by pectic fragments in suspension-cultures of castor bean. Plant Physiol 91(3):889–897. doi:10.1104/pp.91.3.889
13. Ride JP (1975) Lignification in wounded wheat leaves in response to fungi and its possible role in resistance. Physiol Plant Pathol 5(2):125–134. doi:10.1016/0048-4059(75)90016-8
14. Liu TT, Liu ZX, Song CJ, Hu YF, Han ZF, She J, Fan FF, Wang JW, Jin CW, Chang JB, Zhou JM, Chai JJ (2012) Chitin-induced dimerization activates a plant immune receptor. Science 336(6085):1160–1164. doi:10.1126/science.1218867
15. Nishinaka Y, Aramaki Y, Yoshida H, Masuya H, Sugawara T, Ichimori Y (1993) A new sensitive chemiluminescence probe, L-012, for measuring the production of superoxide anion by cells. Biochem Biophys Res Commun 193(2):554–559. doi:10.1006/bbrc.1993.1659
16. Felix G, Regenass M, Spanu P, Boller T (1994) The protein phosphatase inhibitor Calyculin A mimics elicitor action in plant-cells and induces rapid hyperphosphorylation of specific proteins as revealed by pulse labeling with p-33 phosphate. Proc Natl Acad Sci U S A 91(3):952–956. doi:10.1073/pnas.91.3.952
17. Eynck C, Koopmann B, Karlovsky P, von Tiedemann A (2009) Internal resistance in winter oilseed rape inhibits systemic spread of the vascular pathogen Verticillium longisporum. Phytopathology 99(7):802–811. doi:10.1094/phyto-99-7-0802
18. Pearce RB, Ride JP (1982) Chitin and related-compounds as elicitors of the lignification response in wounded wheat leaves. Physiol Plant Pathol 20(1):119–123. doi:10.1016/0048-4059(82)90030-3
19. Chassot C, Buchala AJ, Schoonbeek H, Métraux J-P, Lamotte O (2008) Wounding causes a powerful but transient protection against a major plant pathogen. Plant J 55(4):555–567. doi:10.1111/j.1365-313X.2008.03540.x

Chapter 29

Effectoromics-Based Identification of Cell Surface Receptors in Potato

Emmanouil Domazakis, Xiao Lin, Carolina Aguilera-Galvez, Doret Wouters, Gerard Bijsterbosch, Pieter J. Wolters, and Vivianne G.A.A. Vleeshouwers

Abstract

In modern resistance breeding, effectors have emerged as tools for accelerating and improving the identification of immune receptors. Effector-assisted breeding was pioneered for identifying resistance genes (*R* genes) against *Phytophthora infestans* in potato (*Solanum tuberosum*). Here we show that effectoromics approaches are also well suitable for identifying pathogen recognition receptors (PRRs) that recognize apoplastic effectors. To detect genotypes that recognize apoplastic proteins of *P. infestans*, routine agroinfiltration and potato virus X (PVX) agroinfection methods can be applied. In addition, protein infiltrations are feasible for assessing responses to apoplastic effectors and aid in confirming results obtained from the aforementioned methods. Protocols for the effectoromics pipeline are provided, starting from phenotyping for effector responses, up to genotyping and PRR gene identification.

Key words Pattern recognition receptors (PRRs), Apoplastic effector, Genetic mapping, Agroinfiltration, PVX agroinfection, Protein infiltration, Effectoromics, Yeast protein production, *Solanum*

1 Introduction

Breeding for broad-spectrum and durable disease resistance is one of the most important prerequisites for achieving a stable food production. Potato (*Solanum tuberosum* L.) is the third staple crop world-wide, and suffers from dramatic yield losses due to the devastating late blight pathogen *Phytophthora infestans* (Mont.) de Bary. To understand and improve resistance to late blight in potato, effectors of *P. infestans* have been successfully exploited in breeding [1, 2]. In this 'effectoromics' strategy, effectors are functionally tested for responses in potato germplasm. After detecting responding genotypes, the receptors are being genetically mapped on progeny populations derived from crosses between responding (R)

Libo Shan and Ping He (eds.), *Plant Pattern Recognition Receptors: Methods and Protocols*, Methods in Molecular Biology, vol. 1578, DOI 10.1007/978-1-4939-6859-6_29, © Springer Science+Business Media LLC 2017

and non-responding (NR) genotypes, in which phenotyping is done by high throughput effector assays [3, 4].

Effectoromics was originally focused on detecting resistance (*R*) genes of the NBS-LRR class by assaying cytoplasmic effectors, and protocols were established for a.o. *Solanum* and *Nicotiana* species [5–7]. Recently, the strategy was expanded to identifying pattern recognition receptors (PRRs), or surface receptors, by functionally screening for responses to apoplastic effectors in potato [8].

With the ever increasing amount of pathogen genome data, effectoromics is now being implemented for various other crops and pathosystems. Not only biotrophic pathogens, but also pathogens with other life styles can be the target of effectoromics, and for example toxins can be identified in necrotrophic pathogens [9, 10]. The main requirements for effectoromics, is availability of (1) resistant plant material for sourcing the immune receptors, (2) candidate effectors predicted from pathogen genomes (Box 1), and (3) reliable and efficient functional assays (Box 2/Fig. 1). *Agrobacterium*-based expression systems, such as for agroinfiltration and potato virus X (PVX)-agroinfection, are well established in *Solanaceae* [3, 13, 14] but are less applicable in for example monocots. An advantage of targeting apoplastic effectors is that these are typically more stable than cytoplasmic effectors and can be assayed by protein infiltrations. Protein infiltration may provide a reliable way to assay responses obtained by other systems [4], since it is not dependent on transformation efficiency and cannot be hindered by extreme resistance responses.

Box 1 Prediction of candidate apoplastic effectors

Effector candidates can be identified by applying selection criteria on a database of proteins from the pathogen of interest. Such data might already be available, or could be generated by a genome sequencing and/or RNAseq experiment.

A number of criteria used in various publications [15–18] are generally established for the selection of apoplastic effectors:

1. Presence of a secretion signal. In order for effectors to be secreted in the apoplast, they need to contain a signal peptide for secretion. Signal peptides can be predicted with software such as SignalP [19].
2. Size. Effectors are often small proteins (e.g., <30 kDa).
3. Cysteine content. Apoplastic effectors are typically cysteine-rich proteins; the formation of disulphide bonds is believed to help stabilize the proteins in the apoplastic environment.
4. Expression profile. If such data are available. Functional effectors need to be expressed during the infection process.

Depending on the pathogen, additional criteria may be useful to further prioritize the list of effector candidates, for example: location in the genome (in *Phytophthora* spp., most effectors reside in repeat-rich regions [20] and in *Fusarium* spp. effectors are found close to transposable elements on so-called "pathogenicity" chromosomes [21, 22]), presence or absence of certain domains and motifs (for example including proteins with a predicted LysM domain to find effectors that bind chitin [23], but excluding proteins with motifs like RxLR, which are required for cell entry [24, 25], to filter out cytoplasmic effectors), similarity to known effectors, or information about presence/absence of homologs in virulent or avirulent strains of the pathogen.

Box 2 Effectoromics workflow to identify novel surface receptors

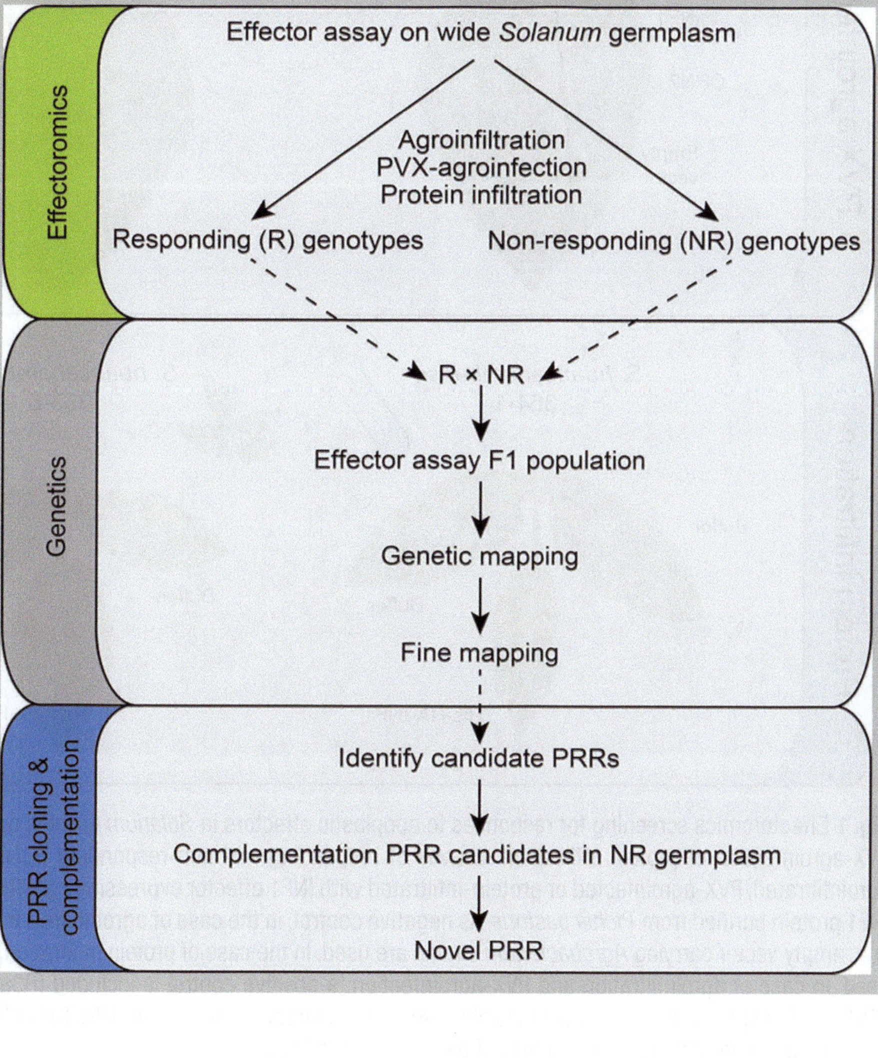

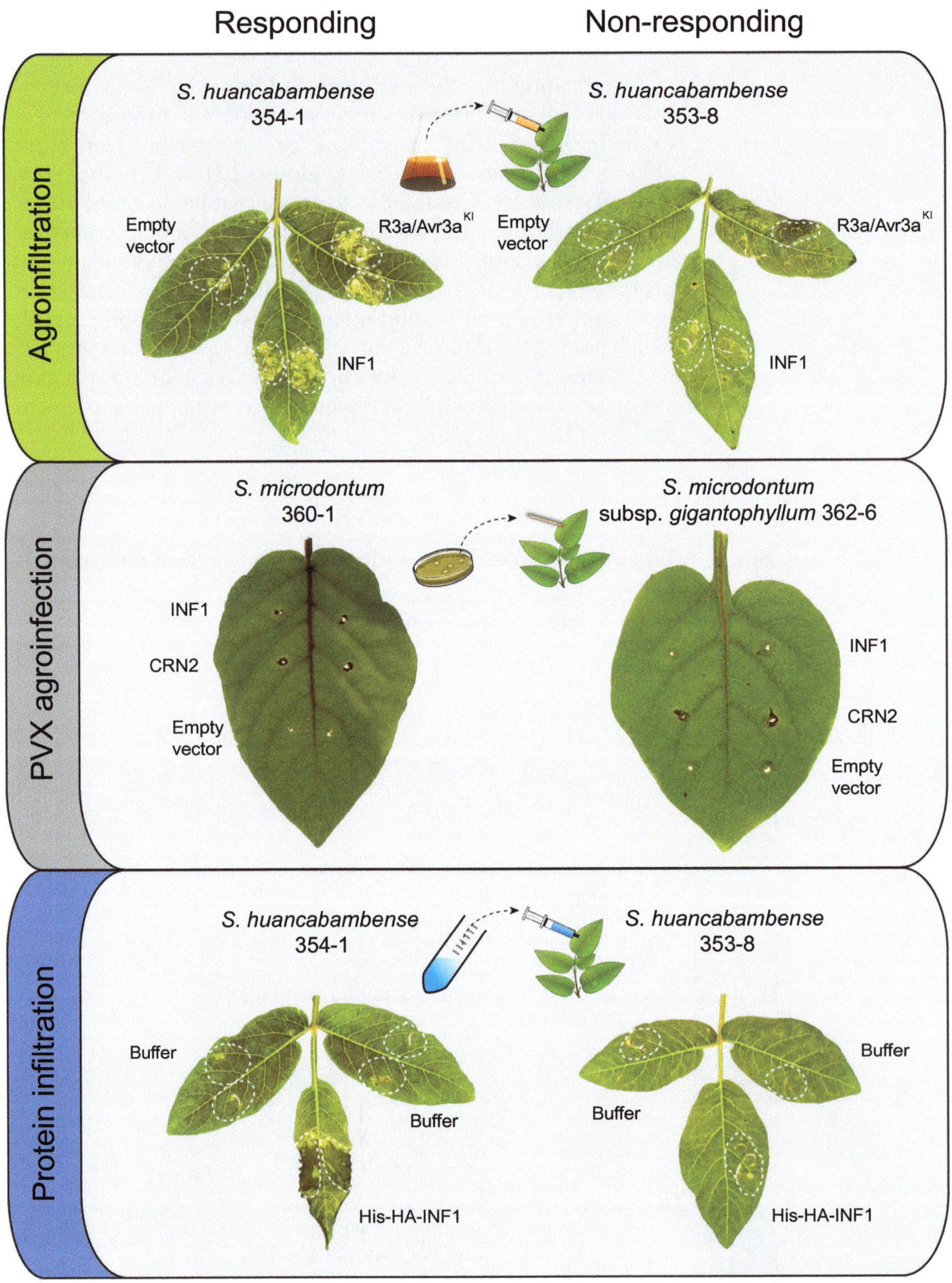

Fig. 1 Effectoromics screening for responses to apoplastic effectors in *Solanum* species using agroinfiltration, PVX-agroinfection or protein infiltration. Leaves of responding and non-responding *Solanum* genotypes are agroinfiltrated, PVX-agroinfected or protein-infiltrated with INF1 effector expressing constructs or with His-HA-INF1 protein purified from *Pichia pastoris*. As negative control, in the case of agroinfiltration and PVX agroinfection, empty vector carrying *Agrobacterium* strains are used. In the case of protein infiltration, buffer infiltration is used. In case of agroinfiltration and PVX-agroinfection, a positive control is included (R3a/Avr3a^{KI} [11, 12] or CRN2 [13], respectively). No cell death is observed for the negative controls and INF1 is only causing cell death in the responding genotypes as confirmed by three methods [4]

Being able to produce high quality apoplastic effector proteins using heterologous systems is of great importance. *Pichia pastoris* has emerged as an appropriate host for secreted protein production [26]. As an eukaryote, it is able to posttranslationally modify target effector proteins and secrete them into the growth medium [27]. This is in a way similar to eukaryotic pathogens secreting effectors in the host apoplast. In addition, compared to *Escherichia coli*-based expression systems, it is able to produce higher quality protein with appropriate folding and does not form aggregates [27].

Recent advances have enabled integration of novel methods, such as proteomics approaches for detecting protein activity and resistance gene enrichment sequencing (RenSeq) for accelerated genetic mapping [28]. In this chapter, we provide protocols for an effectoromics pipeline in potato (Box 2/Fig. 1). These include routine *Agrobacterium*-based methods that are applicable for all types of immune receptors known so far, and with emphasis on apoplastic effector screening methods that are useful for identifying new surface receptors.

2 Materials

2.1 Potato In Vitro Propagation and Maintenance

MS20 medium (per L: 4.4 g Murashige and Skoog basal salt mixture including vitamins, 20 g sucrose in 700 mL of demineralized water (ddH_2O). Adjust pH to 5.8 with NaOH and HCl and bring to a final volume of 1 L. Add 8 g of micro agar and autoclave) [14].

2.2 Agroinfiltration and PVX Agroinfection

1. *Agrobacterium tumefaciens* strains AGL1 [29] or GV3101 [30].
2. LB (lysogeny broth) medium/LBA (LB + agar) (per L: 10 g tryptone, 5 g yeast extract and 10 g NaCl. For LBA add 10 g/L bacteriological agar. Autoclave the media).
3. (Agroinfiltration only) 200 mM acetosyringone (4-Hydroxy-3,5-dimethoxyacetophenone) stock (39.3 mg of acetosyringone in 1 mL of DMSO).
4. (Agroinfiltration only) 1 M MES (2-(N-morpholino)-ethane sulfonic acid) buffer (per 50 mL: 10.66 g of MES monohydrate, adjust pH to 5.6 with NaOH and filter sterilize).
5. (Agroinfiltration only) YEB (yeast extract broth) *Agrobacterium* Growth Medium (per L: 5 g beef extract, 5 g bacteriological peptone, 5 g sucrose, 1 g of yeast extract, 0.492 g of $MgSO_4{\bullet}7H_2O$, autoclave) [31].
6. (Agroinfiltration only) MMA agroinfiltration buffer (per L: 20 g of sucrose, 5 g of MS salts without vitamins, 1.95 g MES, 2 mL 1 M NaOH, 1 mL of 200 mM acetosyringone. Confirm that pH is 5.6. If necessary, adjust with HCL/NaOH).
7. Antibiotics: prepare stock concentration of the appropriate antibiotic. Common antibiotic 1000× stock concentrations:

kanamycin 50 mg/mL, spectinomycin 50 mg/mL, ampicillin/carbenicillin 100 mg/mL (in water) rifampicin 25 mg/mL (in DMSO), chloramphenicol 34 mg/mL, tetracycline 5 mg/mL (in ethanol).

8. (Agroinfiltration only) pK7WG2 vector (Gateway) [32] or other CaMV 35s driven plant expression vector.
9. (PVX-agroinfection only) pGR106 vector [33] or other (CaMV 35s) PVX encoding vector.

2.3 Protein Production in *Pichia pastoris*

1. pPinkα-HC vector (Invitrogen, Carlsbad, USA) [34].
2. Restriction endonucleases and buffers.
3. T4 ligase and buffer.
4. DH5α, TOP10 or similar competent *Escherichia coli* cells.
5. LB medium containing ampicillin or carbenicillin (100 μg/mL).
6. LBA plates containing ampicillin or carbenicillin (100 μg/mL).
7. AOX1 sequencing primers (AOX: 5′-GACTGGTTCCAA TTGACAAGC-3′, AOX: 3′-GCAAATGGCATTCTGACAT CC-5′).
8. Target effector gene primers.
9. Plasmid DNA purification kit.
10. *Pichia pastoris* strain PichiaPink™ 1, 2, 3 or 4 (*ADE2* mutants, strains 2, 3, 4 are also protease deficient strains—*see* manual [34]).
11. Working glycerol stock of PichiaPink™ strains (*see* Invitrogen manual).
12. 20% w/v dextrose solution (per L: 200 g dextrose, autoclave).
13. 1 M sorbitol (per L: 187.17 g sorbitol).
14. 1 M potassium phosphate buffer, pH 6.0 (per L: 132 mL of 1 M K_2HPO_4, 868 mL of 1 M KH_2PO_4 and confirm that the pH is 6.0 ± 0.1, use phosphoric acid or KOH to adjust pH).
15. YNB (per L: 134 g yeast nitrogen base with ammonium sulfate, without amino acids. Filter-sterilize).
16. YPD medium (per L: 1 g yeast extract, 2 g peptone. Autoclave and add 100 mL 20% dextrose) and YPD agar plates (YPD medium containing 8 g/L agar. Autoclave and add 100 mL 20% dextrose).
17. YPDS medium (YPD containing 1 M sorbitol).
18. PAD selection plates (adenine drop-out media) (*see* Invitrogen manual).
19. Sterile Milli-Q water (Merck Millipore, Darmstadt, Germany).
20. Electroporation device with yeast transformation setup (e.g., for Bio-Rad gene pulser and 0.1 cm gap cuvette: 400 Ω resistance, 25 μF capacitance and 1.4 kV voltage).

21. 125 mL and 1 L baffled culture flasks, sterile.
22. 500 mL centrifuge bottles, sterile.
23. 10% glycerol (per L: 100 mL glycerol, autoclaved).
24. 10% methanol (per L: 100 mL absolute methanol, filter sterilized).
25. 0.02% biotin (per 100 mL: 20 mg biotin, filter sterile).
26. BMGY (per L: 1 g yeast extract, 2 g peptone, 700 mL Milli-Q water, autoclave. Then add 100 mL of 1 M potassium phosphate pH 6.0, 100 mL YNB, 2 mL of 0.02% biotin, and 100 mL 10% glycerol).
27. BMMY (per L: dissolve 1 g yeast extract and 2 g peptone in 700 mL Milli-Q water, autoclave. Then add 100 mL of 1 M potassium phosphate pH 6.0, 100 mL YNB, 2 mL 0.02% biotin and 100 mL 10% methanol).
28. Baffled or normal flasks.
29. 96-deep well plates (2 mL capacity).
30. Amicon Ultra 400 mL stirred cell and appropriate filtration membranes (Merck Millipore, Darmstadt, Germany).
31. 4× Laemmli SDS-PAGE loading buffer (per 10 mL: 2 mL 1 M Tris–HCl pH 6.8, 0.8 g SDS, 4 mL 100% glycerol, 0.4 mL 14.7 M β-mercaptoethanol, 1 mL 0.5 M EDTA, 8 mg bromophenol blue).
32. 96-well PCR plates.
33. Antibodies raised against HA (hemagglutinin) or His (histidine) epitopes.

2.4 Genetic Mapping

1. SSR markers [35].
2. 4300 DNA Analyzer (LI-COR, Lincoln, USA).
3. LightScanner® System (Bio Fire, Utah, USA).
4. Phire Green Hot Start II DNA Polymerase (Thermo Fisher Scientific).
5. *S. tuberosum* Group Phureja DM1-3 genome v4.03 (http://solanaceae.plantbiology.msu.edu/pgsc_download.shtml [36]).

3 Methods

3.1 Potato Plant Maintenance and Propagation

Maintain potato plants in vitro in sterile jars containing MS20 medium. Transfer fresh cuttings to new jars and incubate in a climate chamber at 24 °C under long day conditions (16 h light–8 h dark) for 2 weeks [14]. Then transfer them into pots containing sterilized soil in climate regulated greenhouse compartments within the temperature range of 18–22 °C and under 16 h/8 h day/night regime [14].

3.2 Agroinfiltration

1. Inoculate the *Agrobacterium* strains containing the gene of interest in 15 mL of YEB containing 1.5 μL of 200 mM acetosyringone solution, 150 μL of 1 M MES buffer, and the appropriate antibiotics.
2. Incubate on a shaking incubator for 24–48 h at 28 °C at 200 rpm until the culture has grown to an OD_{600} of approximately 1.
3. Harvest the cells by centrifugation at 3000 × *g* for 10 min. Decant supernatant and resuspend the pellet in freshly made MMA buffer to an OD_{600} of 0.4. Gently vortex the cells. For co-infiltration of two strains, mix the cultures in a 1:1 ratio.
4. Incubate the bacterial strains in MMA for 1–6 h at room temperature before infiltrations.
5. Use a 1 mL needleless syringe to infiltrate the lower side of the potato leaf. Use around 4–5-week-old-plants grown from in vitro rooted plantlets (*see* **Notes 1** and **2**). A successful infiltration becomes visible by a change in colour of the infiltrated area from light green to dark green.
6. Use three plants for agroinfiltration with each strain and three leaves per plant. Be sure to include a negative control (e.g., a strain carrying empty vector) and a positive control (e.g., a co-infiltration of matching avirulence and resistance gene) (*see* **Note 3**).
7. Score for cell death after 3 days of infiltration on a scale from 0% (no symptoms) to 100% (confluent cell death) [14] (Fig. 1) (*see* **Notes 3–5**).

3.3 PVX Agroinfection

1. Inoculate the *Agrobacterium* strains containing the gene of interest in 3 mL of YEB. Incubate the cultures at 28 °C, shaking at 200 rpm for 24–48 h.
2. Pipette 100 μL of each *Agrobacterium* strain and spread them on LB agar plates containing the appropriate antibiotics. Incubate at 28 °C for 24–48 h.
3. Use a bacterial spreader to thoroughly collect the *Agrobacterium* culture at the centre of the plate. Dip the tip of a toothpick in the *Agrobacterium* cells.
4. Inoculate each strain on the leaf by piercing the leaf with the toothpick. Use around 2–3-week-old plants from in vitro (*see* **Note 6**). Inoculate at least three plants for each strain and three leaves per plant. Toothpick-inoculate each *Agrobacterium* strain in duplicate, on each side of the mid vein. Be sure to include a positive (e.g., CRN2) and a negative control (empty pGR106 vector) in every treated leaf [13].
5. Score for cell death around the inoculation wound at 14 days post inoculation (Fig. 1) (*see* **Notes 3** and **4**). Record the qualitative/quantitative cell death response (score as 0 for no

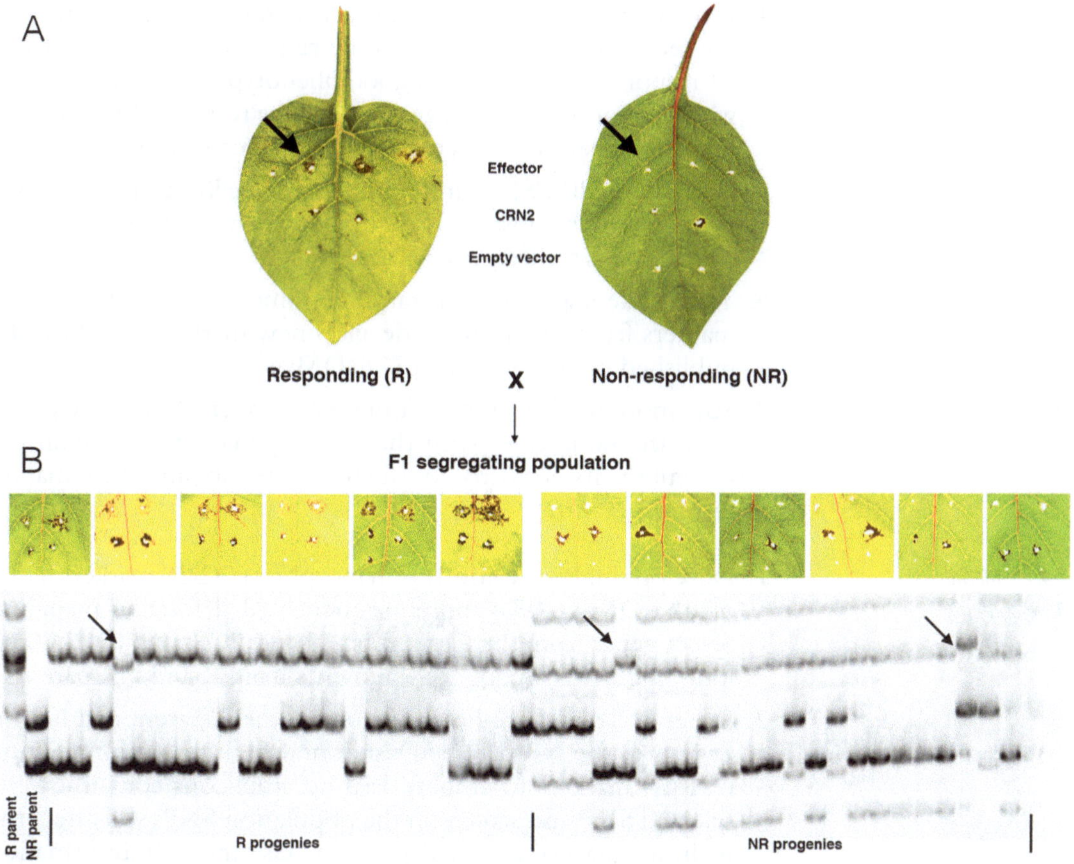

Fig. 2 Map-based cloning of surface immune receptors following the effectoromics screening in diploid *Solanum* species. (**a**) The selected responding (R) *Solanum* genotype is crossed with a non-responding (NR) genotype. The F1 progenies are screened for effector responses as described in this chapter. (**b**) The F1 population shows 1:1 segregation ratio. Eighty SSR markers [35] are used to determine the chromosome where the target gene is located. The depicted SSR marker is linked with the phenotype and three recombinants are indicated (*arrows*)

response, 1 for intermediate and 2 for strong response) for each spot and compare with the negative control [14] (*see* **Notes 3–5**).

3.4 Resistance Gene Mapping in Solanum

1. Identify diploid *Solanum* genotypes that respond to the target effector by agroinfiltration and PVX agroinfection with clear phenotypes (an overview of the map-based cloning strategy is presented in Fig. 2).
2. Cross the genotypes with clear responses (R) with genotypes that do not respond (NR) (Box 2, Fig. 2). Harvest the F1 seeds.
3. Sow 50 F1 seeds in greenhouse and phenotype the response for the target effector. If the F1 population shows a 1:1 segregating ratio, it indicates that the target gene is heterozygous in the responding parents.

4. If the F1 population does not show segregation (indicating that the receptor is homozygous in the response parent), select 2–3 F1 response progenies with good phenotypes for a backcrossing with the non-responding genotypes. The resulting BC1 population should show a segregation for the effector response.
5. Run about 80 SSR markers that are equally distributed over the potato chromosomes [35] to determine on which chromosome the resistance gene is located.
6. Test markers spanning the target chromosome (identify known markers from literature or develop new markers based on the published potato genome [37] (DMv4.03).
7. Sow more seeds to screen a higher number of progeny plants from the same cross with the flanking markers. Maintain the recombinants in vitro for further phenotyping and marker testing.

3.4.1 Marker Development

1. Develop single-nucleotide polymorphism (SNP) markers based on the DM v4.03 genome sequence [37] for fine mapping. Select genes from the target interval and PCR amplify 1200 bp long fragments. Sequence fragments from R and NR parents.
2. Select SNPs that are heterozygous in the R parent and homozygous in the NR parent. Design new high resolution melting (HRM) markers to amplify 150 bp amplicons containing the selected SNP and screen on the population SNPs will affect the melting curve of the PCR product. Use the SNP information to narrow down the target interval.
3. Alternatively, new HRM markers can be developed randomly on exon sequences from within the target interval from DM. 150 bp small amplicons can be used for SNP screening using the LightScanner system.

3.4.2 Fine Mapping

1. Construct a BAC library for the responding genotype (for example, for heterozygous diploid potato genotypes, pick 150,000 BAC clones to obtain 10X coverage of the 900 Mb haploid size ×2).
2. Screen the BAC library using the co-segregating markers identified in the previous steps. Use the markers to select BAC clones that are in coupling phase with the resistance gene to create a minimum tiling path covering the region of interest. Sequence the selected BAC clones.
3. Annotate the genes in the sequenced region and select candidate genes based on their predicted function.
4. Clone the candidate genes in a binary expression vector (e.g., pK7WG2) and co-infiltrate with the matching effector in non-responding potato and *Nicotiana benthamiana* plants to confirm the recognition of the effector from the cloned resistance gene.

5. Generate stable transformants of the candidate gene into an NR potato background (e.g., cultivar Désirée) for further complementation and functional studies.

3.5 Apoplastic Effector Cloning and Transformation of *Pichia pastoris*

We use the amino acid sequence HHHHHHVKLYPYDVPDYAAA (underlined are the 6× His and 1× HA-tag residues with spacer amino acids in between) encoded by the DNA sequence 5′-CATCATCACCATCACCACGTTAAGTTGTACCCATACGACGTTCCAGATTACGCTGCTGCT-3′ in N′ of the mature effector protein to generate N-terminal 6×His-1×HA-tagged effector fusions in pPinkα-HC [34] (*see* **Note** 7).

1. Synthesize codon optimized effector genes for protein production in *Pichia pastoris* (*see* **Notes** 7 and **8**). Six times His and HA tag as well as restriction sites have to be added (StuI in forward primer and KpnI, NaeI, FseI, or SwaI in reverse primer) to facilitate ligation into vector pPinkα-HC (secreted expression, high copy) (*see* **Note 9**).
2. Digest codon optimized effector genes and the pPinkα-HC and vector with the appropriate enzymes and continue cloning as described in the manual [34].
3. Transform competent *E. coli* with pPinkα-HC-effector constructs. Select five colonies for colony PCR confirmation, plasmid isolation and sequencing.
4. Sequence five clones per construct with AOX5′ and AOX3′ primers. Be sure to carefully check for the correct insertion of the effector gene in the vector cassette and the absence of any mutation before and following the cloned gene.
5. Prepare 10 μg of plasmid DNA to be used for transformation by isolating plasmid DNA and digesting with a restriction enzyme that does not cut within your gene (e.g., one of *Mam*I, *Eco*NI, *Spe*I, or *Afl*II).
6. Clean up the digest by ethanol/sodium acetate precipitation and washing as described in the manual [34]. Air-dry, and resuspend in 10 μL sterile, deionized water. Use immediately or store at −20 °C.
7. Prepare electro-competent PichiaPink cells as described in the manual on the day of transformation (do not store cells at −80 °C as transformation efficiency will drop significantly). Use 80 μL of cells and 10 μL of plasmid (10 μg total DNA amount) per electro-transformation.
8. After electroporation, immediately add 1 mL of sterile YPDS medium in the cuvette and incubate at 30 °C for 3–4 h.
9. Plate 500 μL in PAD selection plates and incubate at 30 °C till colonies appear. Usually it takes 3 days for the first transformants to appear.

10. Four to five days after transformation, select 10 of the most well developed, white *P. pastoris* colonies, for each construct (*see* **Notes 10** and **11**).

3.6 High Throughput Protein Production Screening

1. From the original transformation plates, pick 8–10 clones, transfer to fresh PAD selection plates and incubate for 2–3 days at 30 °C.
2. In a 96-deep well plate (2 mL capacity volume), add 200 μL of BMGY per well and inoculate with each clone.
3. Incubate at 28–30 °C on a shaking incubator at 300 rpm for 24 h.
4. Spin down to pellet cells by centrifugation (1500 × *g* for 15 min).
5. Remove supernatants with pipetting and add 300 μL of BMMY for inducing expression.
6. Incubate at 28 °C in a shaking incubator at 300 rpm for 48 h.
7. Spin down to pellet cells and for each culture collect 150 μL supernatant in a clean 96-wells PCR plate.
8. To each supernatant, add 50 μL of 4× Laemmli loading buffer and heat to 95 °C for 10 min.
9. Run SDS-PAGE with all clones per construct and perform western blot with anti-HA or anti-His to detect secreted recombinant proteins.
10. Select the highest producing clone for further protein production. Prepare glycerol stock of this clone by growing it in YPD for 48 h, adding sterile glycerol to 20% v/v and snap freezing in liquid nitrogen.

3.7 Small Scale Recombinant Effector Production

1. Using the highest protein producing *P. pastoris* clone (from PAD selection plates or a −80 °C glycerol stock), inoculate 5 mL of YPD and grow for 1–2 days in order to obtain a starter culture.
2. Inoculate 25 mL of BMGY with 100 μL of the clone pre-culture in a 250 mL baffled flask. Grow at 28 °C in a shaking incubator (250–300 rpm) until culture reaches an OD_{600} of 2–6 (usually 16–20 h).
3. Use this 25 mL culture to inoculate 1 L of BMGY in a 3 or 4 L baffled flask and grow at 28 °C with vigorous shaking (250–300 rpm) until the culture reaches log phase growth (OD_{600} = 2–6) (usually 6–8 h).
4. Harvest the cells by centrifuging in sterile centrifuge bottles at 1500–3000 × *g* for 5 min at room temperature.
5. Decant the supernatant and resuspend the cell pellet in 200 mL of BMMY medium to induce expression.

6. Transfer the cell suspension in a 2 L baffled flask, cover with an air-porous tape such as AirPore™ (Qiagen, Venlo, The Netherlands) or PureLink™ Air Porous Tape (Thermo Fisher Scientific, Waltham, USA). Continue to grow at 28 °C with shaking at 300 rpm.
7. Add 2 mL of 100% Methanol every 24 h to maintain induction.
8. Induce protein expression for 48–72 h.
9. Aliquot the culture in 50 mL conical tubes and centrifuge at 4000 × *g* for 30 min to pellet the cells. Transfer supernatants to new 50 mL conical tubes and repeat centrifugation in order to completely clarify the supernatant.
10. Filter-sterilize the supernatants through a 0.45 μm syringe filter.
11. Transfer sterilized supernatant in the Amicon Ultra 400 mL stirred cell, equipped with a membrane with a pore size (MWCO) that equals half or less the predicted effector molecular weight.
12. Concentrate the supernatant by applying 5.2 bars pressure of nitrogen gas and medium stirring speed till the volume reaches around 10–20 mL. Collect concentrate in a 50 mL tube. Supernatants can be kept for a short term at 4 °C till purification or be stored at −80 °C.
13. Purify recombinant apoplastic effector proteins with a method of choice (*see* **Note 12**).

3.8 Infiltration of Apoplastic Effector Proteins into Leaf Apoplast

1. Dilute purified tagged apoplastic effector proteins in sterile Milli-Q water at the desired molar concentration. Start by testing different molar concentrations, e.g., 10 nM to 10 μM (*see* **Notes 13** and **14**).
2. Use a needleless syringe to infiltrate protein solution in the lower side of a fully expanded leaf. Use Milli-Q water, buffer or another, non-recognized protein (e.g., a cysteine mutant) as a negative control (*see* **Notes 14** and **15**).
3. Score responses at 3, 5, and 7 days post-infiltration, depending on the protein concentration used and the type of response (Fig. 1) (*see* **Notes 4** and **5**).

4 Notes

1. Choose young, healthy, and fully developed leaves for agroinfiltrations.
2. Eye protection should be worn during the agroinfiltration process. Change gloves when infiltrating with different strains to avoid cross contamination.

3. When scoring agroinfiltration, PVX-agroinfection or protein infiltration experiments, always compare your obtained effector responses with the controls. It is good to realize that intensity of responses also depend on transformation efficiency and sensitivity to *Agrobacterium* (and PVX) in the particular genetic background. Therefore, be sure to check for background responses to *Agrobacterium* or PVX.
4. Assay scoring can also be done at different time points, depending on the type of effector and inoculated plant genotypes. In the case of PVX-agroinfection, cell death responses usually start to appear from day 8, and are optimal at day 14. For agroinfiltration, cell death responses can appear from 2 days and reach a maximum at 5 days post infiltration. For protein infiltration, depending on the concentration used, cell death responses can appear from 2 days and reach a maximum at 1 week post infiltration. Keep in mind that responses to effectors are not necessarily associated with cell death [38, 39].
5. Cell death responses triggered by PRR may not be observed at temperatures below 20 °C [40].
6. For high-throughput screening with PVX agroinfection, use 4–5-week-old plants. Those plants have leaves that are sufficiently big for inoculating multiple spots.
7. Tag position may affect protein recognition or function. For apoplastic effectors, a small N-terminal tag is used in most of the cases.
8. Most secreted apoplastic effectors are small. Synthesizing is a convenient option. In addition, when synthesizing, codon optimization is possible which can lead to an increase in protein production yields. Codon optimization services can be requested at most companies that offer gene synthesis services.
9. ATG start codon or the yeast consensus Kozak sequence should not be included when cloning in pPinkα-HC vector as they are on the α-mating factor pre-sequence. You must add a stop codon at the 3′ of the effector gene, as it is not present in vector.
10. When transforming PichiaPink strains, low copy number transformants will appear on the PAD selection plates. Those will have a red-pinkish colour. It is advised not to use those for protein expression.
11. It is not necessary to screen PichiaPink transformants by PCR. All white colonies should be positive, high copy number transformants.
12. Typically, ion exchange chromatography or immobilized metal affinity chromatography (IMAC) are used. IMAC exploits the metal binding affinity of histidine residues fused to the effector

protein. Mainly nickel or cobalt coated beads are utilized for this purpose.

13. Protein concentration is important for an adequate response. Molar concentrations should be used instead of grams.
14. Do not use very high protein concentrations, as they could lead to toxic effects and therefore the experiment will be non-informative. A control for unspecific responses must be performed by including non-responding genotypes in every experiment.
15. Be sure that your protein solutions for infiltration are free of bacterial contamination. Potato plants tend to have a high background response to bacteria.

Acknowledgments

This work was supported by a NWO-VIDI grant 12378 (ED, XL, DW, VGAAV), the China Scholarship Council Program for Graduate Students (XL), Colciendas (CAG), Veenhuizen Tulp Fonds (CAG), J.R. Simplot Company (PJW), and COST FA1208 (XL).

References

1. Vleeshouwers VGAA, Raffaele S, Vossen JH et al (2011) Understanding and exploiting late blight resistance in the age of effectors. Annu Rev Phytopathol 49:507–531. doi:10.1146/annurev-phyto-072910-095326
2. Vleeshouwers VGAA, Rietman H, Krenek P et al (2008) Effector genomics accelerates discovery and functional profiling of potato disease resistance and *Phytophthora infestans* avirulence genes. PLoS One 3(8):e2875. doi:10.1371/journal.pone.0002875
3. Takken FL, Luderer R, Gabriels SH et al (2000) A functional cloning strategy, based on a binary PVX-expression vector, to isolate HR-inducing cDNAs of plant pathogens. Plant J 24(2):275–283. doi:10.1046/j.1365-313x.2000.00866.x
4. Vleeshouwers VGAA, Driesprong JD, Kamphuis LG et al (2006) Agroinfection-based high-throughput screening reveals specific recognition of INF elicitins in *Solanum*. Mol Plant Pathol 7(6):499–510. doi:10.1111/j.1364-3703.2006.00355.x
5. Kapila J, De Rycke R, Van Montagu M et al (1997) An *Agrobacterium*-mediated transient gene expression system for intact leaves. Plant Sci 122(1):101–108. doi:10.1016/s0168-9452(96)04541-4
6. Kanneganti T-D, Huitema E, Kamoun S (2007) *In planta* expression of oomycete and fungal Genes. In: Plant-pathogen interactions, vol 354. Methods Mol Biol, pp 35–43. doi:10.1385/1-59259-966-4:35
7. Du J, Vleeshouwers VGAA (2014) The do's and don'ts of effectoromics. In: Plant-pathogen interactions, vol 1127. Methods Mol Biol, pp 257–268. doi:10.1007/978-1-62703-986-4_19
8. Du J, Verzaux E, Chaparro-Garcia A et al (2015) Elicitin recognition confers enhanced resistance to *Phytophthora infestans* in potato. Nat Plants 1(4):15034. doi:10.1038/nplants.2015.34
9. Oliver RP, Friesen TL, Faris JD et al (2012) *Stagonospora nodorum*: from pathology to genomics and host resistance. Annu Rev Phytopathol 50:23–43. doi:10.1146/annurev-phyto-081211-173019
10. Vleeshouwers VGAA, Oliver RP (2014) Effectors as tools in disease resistance breeding against biotrophic, hemibiotrophic, and necrotrophic plant pathogens. Mol Plant Microbe Interact 27(3):196–206. doi:10.1094/MPMI-10-13-0313-IA
11. Huang S, van der Vossen EA, Kuang H et al (2005) Comparative genomics enabled the isolation of the *R3a* late blight resistance gene in potato. Plant J 42(2):251–261. doi:10.1111/j.1365-313X.2005.02365.x

12. Armstrong MR, Whisson SC, Pritchard L et al (2005) An ancestral oomycete locus contains late blight avirulence gene *Avr3a*, encoding a protein that is recognized in the host cytoplasm. Proc Natl Acad Sci U S A 102(21):7766–7771. doi:10.1073/pnas.0500113102
13. Torto TA, Li S, Styer A et al (2003) EST mining and functional expression assays identify extracellular effector proteins from the plant pathogen *Phytophthora*. Genome Res 13(7): 1675–1685. doi:10.1101/gr.910003
14. Du J, Rietman H, Vleeshouwers VGAA (2014) Agroinfiltration and PVX agroinfection in potato and *Nicotiana benthamiana*. J Vis Exp 83:e50971. doi:10.3791/50971
15. Tan KC, Oliver RP, Solomon PS et al (2010) Proteinaceous necrotrophic effectors in fungal virulence. Funct Plant Biol 37(10):907–912. doi:10.1071/FP10067
16. Liu Z, Zhang Z, Faris JD et al (2012) The cysteine rich necrotrophic effector SnTox1 produced by *Stagonospora nodorum* triggers susceptibility of wheat lines harboring *Snn1*. PLoS Pathog 8(1):e1002467. doi:10.1371/journal.ppat.1002467
17. Saunders DGO, Win J, Cano LM et al. (2012) Using hierarchical clustering of Secreted protein families to classify and rank candidate effectors of rust fungi. PLoS One 7(1):e29847. doi:10.1371/journal.pone.0029847
18. Mirzadi Gohari A, Ware SB, Wittenberg AH et al (2015) Effector discovery in the fungal wheat pathogen *Zymoseptoria tritici*. Mol Plant Pathol 16(9):931–945. doi:10.1111/mpp.12251
19. Petersen TN, Brunak S, von Heijne G et al (2011) SignalP 4.0: discriminating signal peptides from transmembrane regions. Nat Methods 8(10):785–786
20. Haas BJ, Kamoun S, Zody MC et al (2009) Genome sequence and analysis of the Irish potato famine pathogen *Phytophthora infestans*. Nature 461(7262):393–398. doi:10.1038/nature08358
21. Ma LJ, van der Does HC, Borkovich KA et al (2010) Comparative genomics reveals mobile pathogenicity chromosomes in *Fusarium*. Nature 464(7287):367–373. doi:10.1038/nature08850
22. Schmidt SM, Houterman PM, Schreiver I et al (2013) MITEs in the promoters of effector genes allow prediction of novel virulence genes in *Fusarium oxysporum*. BMC Genomics 14(1):119. doi:10.1186/1471-2164-14-119
23. Bolton MD, van Esse HP, Vossen JH et al (2008) The novel *Cladosporium fulvum* lysin motif effector Ecp6 is a virulence factor with orthologues in other fungal species. Mol Microbiol 69(1):119–136. doi:10.1111/j.1365-2958.2008.06270.x
24. Manning VA, Hamilton SM, Karplus PA et al (2008) The Arg-Gly-Asp-containing, solvent-exposed loop of Ptr ToxA is required for internalization. Mol Plant Microbe Interact 21(3):315–325. doi:10.1094/Mpmi-21-3-0315
25. Whisson SC, Boevink PC, Moleleki L et al (2007) A translocation signal for delivery of oomycete effector proteins into host plant cells. Nature 450(7166):115–118. doi:10.1038/nature06203
26. Ahmad M, Hirz M, Pichler H et al (2014) Protein expression in *Pichia pastoris*: recent achievements and perspectives for heterologous protein production. Appl Microbiol Biotechnol 98(12):5301–5317. doi:10.1007/s00253-014-5732-5
27. Brondyk WH, Richard RB, Murray PD (2009) Selecting an appropriate method for expressing a recombinant protein. In: Methods Enzymol, vol 463. Academic Press, New York, pp 131–147. doi:10.1016/S0076-6879(09)63011-1
28. Jupe F, Witek K, Verweij W et al (2013) Resistance gene enrichment sequencing (RenSeq) enables reannotation of the NB-LRR gene family from sequenced plant genomes and rapid mapping of resistance loci in segregating populations. Plant J 76(3):530–544. doi:10.1111/tpj.12307
29. Lazo GR, Stein PA, Ludwig RA (1991) A DNA transformation-competent *Arabidopsis* genomic library in *Agrobacterium*. Biotechnology 9(10):963–967
30. Koncz C, Schell J (1986) The promoter of Tl-DNA gene 5 controls the tissue-specific expression of chimeric genes carried by a novel type of *Agrobacterium* binary vector. Mol Gen Genet 204(3):383–396. doi:10.1007/Bf00331014
31. Van der Hoorn RA, Laurent F, Roth R et al (2000) Agroinfiltration is a versatile tool that facilitates comparative analyses of Avr9/Cf-9-induced and Avr4/Cf-4-induced necrosis. Mol Plant Microbe Interact 13(4):439–446. doi:10.1094/MPMI.2000.13.4.439
32. Karimi M, Inze D, Depicker A (2002) GATEWAY vectors for *Agrobacterium*-mediated plant transformation. Trends Plant Sci 7(5):193–195
33. Lu R, Malcuit I, Moffett P et al (2003) High throughput virus-induced gene silencing

implicates heat shock protein 90 in plant disease resistance. EMBO J 22(21):5690–5699. doi:10.1093/emboj/cdg546

34. PichiaPink™ Expression System manual (2014). Life Technol, MAN0000717
35. Milbourne D, Meyer RC, Collins AJ et al (1998) Isolation, characterisation and mapping of simple sequence repeat loci in potato. Mol Gen Genet 259(3):233–245
36. Hardigan MA, Crisovan E, Hamilton JP et al (2016) Genome reduction uncovers a large dispensable genome and adaptive role for copy number variation in asexually propagated *Solanum tuberosum*. Plant Cell 28(2):388–405. doi:10.1105/tpc.15.00538
37. Potato Genome Sequencing Consortium, Xu X, Pan S et al (2011) Genome sequence and analysis of the tuber crop potato. Nature 475(7355):189–195
38. Jones JD, Dangl JL (2006) The plant immune system. Nature 444(7117):323–329. doi:10.1038/nature05286
39. Katagiri F, Tsuda K (2010) Understanding the plant immune system. Mol Plant Microbe Interact 23(12):1531–1536. doi:10.1094/MPMI-04-10-0099
40. Cheng C, Gao X, Feng B et al (2013) Plant immune response to pathogens differs with changing temperatures. Nat Commun 4:2530. doi:10.1038/ncomms3530

INDEX

Libo Shan and Ping He (eds.), *Plant Pattern Recognition Receptors: Methods and Protocols*, Methods in Molecular Biology, vol. 1578, DOI 10.1007/978-1-4939-6859-6, © Springer Science+Business Media LLC 2017

V

X

Y

Z

MIX
Papier aus verantwortungsvollen Quellen
Paper from responsible sources
FSC® C105338

If you have any concerns about our products,
you can contact us on
ProductSafety@springernature.com

In case Publisher is established outside the EU,
the EU authorized representative is:
Springer Nature Customer Service Center GmbH
Europaplatz 3, 69115 Heidelberg, Germany

Printed by Libri Plureos GmbH
in Hamburg, Germany